MATLAB Guide to Finite Elements

Peter I. Kattan

MATLAB Guide to Finite Elements

An Interactive Approach

Second Edition
With 108 Figures and 25 Tables

 Springer

Peter I. Kattan, PhD
P.O. BOX 1392
Amman 11118
Jordan
pkattan@tedata.net.jo
pkattan@lsu.edu

Additional material to this book can be downloaded from http://extra .springer.com.

ISBN-13   978-3-642-43957-5 Springer Berlin Heidelberg New York

Typesetting: Integra Software Services Pvt. Ltd., Pondicherry, India
Cover design: Erich Kirchner, Heidelberg

Printed on acid-free paper   SPIN: 11301950   42/3100/Integra   5 4 3 2 1 0

Dedicated to My Professor, George Z. Voyiadjis

Preface to the Second Edition

Soon after the first edition of this book was published at the end of 2002, it was realized that a new edition of the book was needed. I received positive feedback from my readers who requested that I provide additional finite elements in other areas like fluid flow and heat transfer. However, I did not want to lengthen the book considerably. Therefore, I decided to add two new chapters thus adding new material while keeping the size of the book reasonable.

The second edition of the book continues with the same successful format that characterized the first edition – which was sold out in less than four years. I continue to emphasize the important features of interactivity of using MATLAB[1] coupled with the simplicity and consistency of presentation of finite elements. One of the most important features also is bypassing the use of numerical integration in favor of exact analytical integration with the use of the MATLAB Symbolic Math Toolbox[2]. The use of this toolbox is emphasized in Chaps. 12, 13, 14, and 16.

In the new edition, two important changes are immediately noted. First, I corrected the handful of typing errors that appeared in the first edition. Second, I added two new chapters. Chap. 16 includes another solid three-dimensional element (the eight-noded brick element) in great detail. The final chapter (Chap. 17) provides a review of the applications of finite elements in other areas like fluid flow, heat transfer, geotechnical engineering, electro-magnetics, structural dynamics, plasticity, etc. In this chapter, I show how the same consistent strategy that was followed in the first sixteen chapters can be used to write MATLAB functions in these areas by providing the MATLAB code for a one-dimensional fluid flow element.

One minor drawback of the first edition as I see it is the absence of a concluding chapter. Therefore, I decided to remedy the situation by adding Chap. 17 as a real concluding chapter to the book. It is clear that this chapter is different from the first sixteen chapters and thus may well provide a well written conclusion to the book.

The second edition still comes with an accompanying CD-ROM that contains the full set of M-files written specifically to be used with this book. These MATLAB functions have been tested with version 7 of MATLAB and should work with any

---

[1] MATLAB is a registered trademark of The MathWorks, Inc.

[2] The MATLAB Symbolic Math Toolbox is a registered trademark of The MathWorks, Inc.

later versions. In addition, the CD-ROM contains a complete solutions manual that includes detailed solutions to all the problems in the book. If the reader does not wish to consult these solutions, then a brief list of answers is provided in printed form at the end of the book.

I would like to thank my family members for their help and continued support without which this book would not have been possible. I would also like to acknowledge the help of the editior at Springer-Verlag (Dr. Thomas Ditzinger) for his assistance in bringing this book out in its present form. Finally, I would like to thank my brother, Nicola, for preparing most of the line drawings in both editions. In this edition, I am providing two email addresses for my readers to contact me (pkattan@tedata.net.jo and pkattan@lsu.edu). The old email address that appeared in the first edition was cancelled in 2004.

December 2006   Peter I. Kattan

Preface to the First Edition

This is a book for people who love finite elements and MATLAB[3]. We will use the popular computer package MATLAB as a matrix calculator for doing finite element analysis. Problems will be solved mainly using MATLAB to carry out the tedious and lengthy matrix calculations in addition to some manual manipulations especially when applying the boundary conditions. In particular the steps of the finite element method are emphasized in this book. The reader will not find ready-made MATLAB programs for use as black boxes. Instead step-by-step solutions of finite element problems are examined in detail using MATLAB. Problems from linear elastic structural mechanics are used throughout the book. The emphasis is not on mass computation or programming, but rather on learning the finite element method computations and understanding of the underlying concepts. In addition to MATLAB, the MATLAB Symbolic Math Toolbox[4] is used in Chaps. 12, 13, and 14.

Many types of finite elements are studied in this book including the spring element, the bar element, two-dimensional and three-dimensional truss elements, plane and space beam and frame elements, two-dimensional elasticity elements for plane stress and plane strain problems, and one three-dimensional solid element. Each chapter deals with only one type of element. Also each chapter starts with a summary of the basic equations for the element followed by a number of examples demonstrating the use of the element using the provided MATLAB functions. Special MATLAB functions for finite elements are provided as M-files on the accompanying CD-ROM to be used in the examples. These functions have been tested successfully with MATLAB versions 5.0, 5.3, and 6.1. They should work with other later versions. Each chapter also ends with a number of problems to be used as practice for students.

This book is written primarily for students studying finite element analysis for the first time. It is intended as a supplementary text to be used with a main textbook for an introductory course on the finite element method. Since the computations of finite elements usually involve matrices and matrix manipulations, it is only natural that students use a matrix-based software package like MATLAB to do the calculations.

---

[3] MATLAB is a registered trademark of The MathWorks, Inc.

[4] The MATLAB Symbolic Math Toolbox is a registered trademark of The MathWorks, Inc.

In fact the word MATLAB stands for MATrix LABoratory. The main features of the book are:

1. The book is divided into fifteen chapters that are well defined ad correlated.
2. The books includes a short tutorial on using MATLAB in Chap. 1.
3. The CD-ROM that accompanies the book includes 75 MATLAB functions (M-files) that are specifically written to be used with this book. These functions comprise what may be called the MATLAB Finite Element Toolbox. It is used mainly for problems in structural mechanics. The provided MATLAB functions are designed to be simple and easy to use.
4. A sequence of six steps is outlined in the first chapter for the finite element method. These six steps are then used systematically in each chapter throughout the book.
5. The book stresses the interactive use of MATLAB. Each example is solved in an interactive session with MATLAB. No ready-made subroutines are provided to be used as black boxes.
6. Answers to the all problems are provided at the end of the book.
7. A solutions manual is also provided on the accompanying CD-ROM. The solutions manual includes detailed solutions to all the problems in the book. It is over 300 pages in length.

The author wishes to thank the editors at Springer-Verlag (especially Dr. Thomas Ditzinger) for their cooperation and assistance during the writing of this book. Special thanks are also given to my family members without whose support and encouragement this book would not have been possible. In particular, I would like to thank Nicola Kattan for preparing most of the figures that appear in the book.

February 2002   Peter I. Kattan

Table of Contents

1 Introduction

This short introductory chapter is divided into two parts. In the first part there is a summary of the steps of the finite element method. The second part includes a short tutorial on MATLAB.

1.1
Steps of the Finite Element Method

There are many excellent textbooks available on finite element analysis like those in [1–18]. Therefore this book will not present any theoretical formulations or derivations of finite element equations. Only the main equations are summarized for each chapter followed by examples. In addition only problems from linear elastic structural mechanics are used throughout the book.

The finite element method is a numerical procedure for solving engineering problems. Linear elastic behavior is assumed throughout this book. The problems in this book are taken from structural engineering but the method can be applied to other fields of engineering as well. In this book six steps are used to solve each problem using finite elements. The six steps of finite element analysis are summarized as follows:

1. Discretizing the domain – this step involves subdividing the domain into elements and nodes. For discrete systems like trusses and frames the system is already discretized and this step is unnecessary. In this case the answers obtained are exact. However, for continuous systems like plates and shells this step becomes very important and the answers obtained are only approximate. In this case, the accuracy of the solution depends on the discretization used. In this book this step will be performed manually (for continuous systems).

2. Writing the element stiffness matrices – the element stiffness equations need to be written for each element in the domain. In this book this step will be performed using MATLAB.

3. Assembling the global stiffness matrix – this will be done using the direct stiffness approach. In this book this step will be performed using MATLAB.

4. Applying the boundary conditions – like supports and applied loads and displacements. In this book this step will be performed manually.
5. Solving the equations – this will be done by partitioning the global stiffness matrix and then solving the resulting equations using Gaussian elimination. In this book the partitioning process will be performed manually while the solution part will be performed using MATLAB with Gaussian elimination.
6. Post-processing – to obtain additional information like the reactions and element forces and stresses. In this book this step will be performed using MATLAB.

It is seen from the above steps that the solution process involves using a combination of MATLAB and some limited manual operations. The manual operations employed are very simple dealing only with discretization (step 1), applying boundary conditions (step 4) and partitioning the global stiffness matrix (part of step 5). It can be seen that all the tedious, lengthy and repetitive calculations will be performed using MATLAB.

1.2
MATLAB Functions for Finite Element Analysis

The CD-ROM accompanying this book includes 84 MATLAB functions (M-files) specifically written by the author to be used for finite element analysis with this book. They comprise what may be called the MATLAB Finite Element Toolbox. These functions have been tested with version 7 of MATLAB and should work with any later versions. The following is a listing of all the functions available on the CD-ROM. The reader can refer to each chapter for specific usage details.

$SpringElementStiffness(k)$
$SpringAssemble(K, k, i, j)$
$SpringElementForces(k, u)$

$LinearBarElementStiffness(E, A, L)$
$LinearBarAssemble(K, k, i, j)$
$LinearBarElementForces(k, u)$
$LinearBarElementStresses(k, u, A)$

$QuadraticBarElementStiffness(E, A, L)$
$QuadraticBarAssemble(K, k, i, j, m)$
$QuadraticBarElementForces(k, u)$
$QuadraticBarElementStresses(k, u, A)$

$PlaneTrussElementLength(x_1, y_1, x_2, y_2)$
$PlaneTrussElementStiffness(E, A, L, theta)$
$PlaneTrussAssemble(K, k, i, j)$
$PlaneTrussElementForce(E, A, L, theta, u)$

PlaneTrussElementStress(E, L, *theta*, u)
PlaneTrussInclinedSupport(T, i, *alpha*)

SpaceTrussElementLength(x_1, y_1, z_1, x_2, y_2, z_2)
SpaceTrussElementStiffness(E, A, L, *thetax*, *thetay*, *thetaz*)
SpaceTrussAssemble(K, k, i, j)
SpaceTrussElementForce(E, A, L, *thetax*, *thetay*, *thetaz*, u)
SpaceTrussElementStress(E, L, *thetax*, *thetay*, *thetaz*, u)

BeamElementStiffness(E, I, L)
BeamAssemble(K, k, i, j)
BeamElementForces(k, u)
BeamElementShearDiagram(f, L)
BeamElementMomentDiagram(f, L)

PlaneFrameElementLength(x_1, y_1, x_2, y_2)
PlaneFrameElementStiffness(E, A, I, L, *theta*)
PlaneFrameAssemble(K, k, i, j)
PlaneFrameElementForces(E, A, I, L, *theta*, u)
PlaneFrameElementAxialDiagram(f, L)
PlaneFrameElementShearDiagram(f, L)
PlaneFrameElementMomentDiagram(f, L)
PlaneFrameInclinedSupport(T, i, *alpha*)

GridElementLength(x_1, y_1, x_2, y_2)
GridElementStiffness(E, G, I, J, L, *theta*)
GridAssemble(K, k, i, j)
GridElementForces(E, G, I, J, L, *theta*, u)

SpaceFrameElementLength(x_1, y_1, z_1, x_2, y_2, z_2)
SpaceFrameElementStiffness(E, G, A, I_y, I_z, J, x_1, y_1, z_1, x_2, y_2, z_2)
SpaceFrameAssemble(K, k, i, j)
SpaceFrameElementForces(E, G, A, I_y, I_z, J, x_1, y_1, z_1, x_2, y_2, z_2, u)
SpaceFrameElementAxialDiagram(f, L)
SpaceFrameElementShearZDiagram(f, L)
SpaceFrameElementShearYDiagram(f, L)
SpaceFrameElementTorsionDiagram(f, L)
SpaceFrameElementMomentZDiagram(f, L)
SpaceFrameElementMomentYDiagram(f, L)

LinearTriangleElementArea(x_i, y_i, x_j, y_j, x_m, y_m)
LinearTriangleElementStiffness(E, NU, t, x_i, y_i, x_j, y_j, x_m, y_m, p)
LinearTriangleAssemble(K, k, i, j, m)
LinearTriangleElementStresses(E, NU, t, x_i, y_i, x_j, y_j, x_m, y_m, p, u)
LinearTriangleElementPStresses(*sigma*)

$QuadTriangleElementArea(x_1, y_1, x_2, y_2, x_3, y_3)$
$QuadTriangleElementStiffness(E, NU, t, x_1, y_1, x_2, y_2, x_3, y_3, p)$
$QuadTriangleAssemble(K, k, i, j, m, p, q, r)$
$QuadTriangleElementStresses(E, NU, t, x_1, y_1, x_2, y_2, x_3, y_3, p, u)$
$QuadTriangleElementPStresses(sigma)$

$BilinearQuadElementArea(x_1, y_1, x_2, y_2, x_3, y_3, x_4, y_4)$
$BilinearQuadElementStiffness(E, NU, t, x_1, y_1, x_2, y_2, x_3, y_3, x_4, y_4, p)$
$BilinearQuadElementStiffness2(E, NU, t, x_1, y_1, x_2, y_2, x_3, y_3, x_4, y_4, p)$
$BilinearQuadAssemble(K, k, i, j, m, n)$
$BilinearQuadElementStresses(E, NU, x_1, y_1, x_2, y_2, x_3, y_3, x_4, y_4, p, u)$
$BilinearQuadElementPStresses(sigma)$

$QuadraticQuadElementArea(x_1, y_1, x_2, y_2, x_3, y_3, x_4, y_4)$
$QuadraticQuadElementStiffness(E, NU, t, x_1, y_1, x_2, y_2, x_3, y_3, x_4, y_4, p)$
$QuadraticQuadAssemble(K, k, i, j, m, p, q, r, s, t)$
$QuadraticQuadElementStresses(E, NU, x_1, y_1, x_2, y_2, x_3, y_3, x_4, y_4, p, u)$
$QuadraticQuadElementPStresses(sigma)$

$TetrahedronElementVolume(x_1, y_1, z_1, x_2, y_2, z_2, x_3, y_3, z_3, x_4, y_4, z_4)$
$TetrahedronElementStiffness(E, NU, x_1, y_1, z_1, x_2, y_2, z_2, x_3, y_3, z_3, x_4, y_4, z_4)$
$TetrahedronAssemble(K, k, i, j, m, n)$
$TetrahedronElementStresses(E, NU, x_1, y_1, z_1, x_2, y_2, z_2, x_3, y_3, z_3, x_4, y_4, z_4, u)$
$TetrahedronElementPStresses(sigma)$

$LinearBrickElementVolume(x_1, y_1, z_1, x_2, y_2, z_2, x_3, y_3, z_3, x_4, y_4, z_4, x_5, y_5, z_5,$
$x_6, y_6, z_6, x_7, y_7, z_7, x_8, y_8, z_8)$
$LinearBrickElementStiffness(E, NU, x_1, y_1, z_1, x_2, y_2, z_2, x_3, y_3, z_3, x_4, y_4, z_4, x_5,$
$y_5, z_5, x_6, y_6, z_6, x_7, y_7, z_7, x_8, y_8, z_8)$
$LinearBrickAssemble(K, k, i, j, m, n, p, q, r, s)$
$LinearBrickElementStresses(E, NU, x_1, y_1, z_1, x_2, y_2, z_2, x_3, y_3, z_3, x_4, y_4, z_4, x_5,$
$y_5, z_5, x_6, y_6, z_6, x_7, y_7, z_7, x_8, y_8, z_8, u)$
$LinearBrickElementPStresses(sigma)$

$FluidFlow1DElementStiffness(K_{xx}, A, L)$
$FluidFlow1DAssemble(K, k, i, j)$
$FluidFlow1DElementVelocities(K_{xx}, L, p)$
$FluidFlow1DElementVFR(K_{xx}, L, p, A)$

1.3
MATLAB Tutorial

In this section a very short MATLAB tutorial is provided. For more details consult the excellent books listed in [19–27] or the numerous freely available tutorials on the internet – see [28–35]. This tutorial is not comprehensive but describes the basic MATLAB commands that are used in this book.

In this tutorial it is assumed that you have started MATLAB on your system successfully and you are ready to type the commands at the MATLAB prompt (which is denoted by double arrows ">>"). Entering scalars and simple operations is easy as is shown in the examples below:

```
» 3*4+5

ans =

    17

» cos (30*pi/180)

ans =

    0.8660

» x=4

x =

    4

» 2/sqrt(3+x)

ans =

    0.7559
```

To suppress the output in MATLAB use a semicolon to end the command line as in the following examples. If the semicolon is not used then the output will be shown by MATLAB:

```
» y=32;
» z=5;
» x=2*y-z;
» w=3*y+4*z

w =

    116
```

MATLAB is case-sensitive, i.e. variables with lowercase letters are different than variables with uppercase letters. Consider the following examples using the variables x and X.

```
» x=1

x =

    1

» X=2

X =

    2

» x

x =

    1
```

Use the help command to obtain help on any particular MATLAB command. The following example demonstrates the use of help to obtain help on the inv command.

```
» help inv

INV     Matrix inverse.
    INV(X) is the inverse of the square matrix X.
    A warning message is printed if X is badly scaled or
    nearly singular.

    See also SLASH, PINV, COND, CONDEST, NNLS, LSCOV.

Overloaded methods
    help sym/inv.m
    help zpk/inv.m
    help tf/inv.m
    help ss/inv.m
    help lti/inv.m
    help frd/inv.m
```

The following examples show how to enter matrices and perform some simple matrix operations:

```
» x=[1 2 3 ; 4 5 6 ; 7 8 9]

x =

    1    2    3
    4    5    6
    7    8    9
```

```
» y=[2 ; 0 ; -3]
```

```
y =
     2
     0
    -3
```

```
» w=x*y
```

```
w =
    -7
   -10
   -13
```

Let us now solve the following system of simultaneous algebraic equations:

$$\begin{bmatrix} 2 & -1 & 3 & 0 \\ 1 & 5 & -2 & 4 \\ 2 & 0 & 3 & -2 \\ 1 & 2 & 3 & 4 \end{bmatrix} \begin{Bmatrix} x_1 \\ x_2 \\ x_3 \\ x_4 \end{Bmatrix} = \begin{Bmatrix} 3 \\ 1 \\ -2 \\ 2 \end{Bmatrix} \tag{1.1}$$

We will use Gaussian elimination to solve the above system of equations. This is performed in MATLAB by using the backslash operator "\" as follows:

```
» A=[2 -1 3 0 ; 1 5 -2 4 ; 2 0 3 -2 ; 1 2 3 4]
```

```
A =

     2    -1     3     0
     1     5    -2     4
     2     0     3    -2
     1     2     3     4
```

```
» b=[3 ; 1 ; -2 ; 2]
```

```
b =

     3
     1
    -2
     2
```

```
» x= A\b

x =

    1.9259
   -1.8148
   -0.8889
    1.5926
```

It is clear that the solution is $x_1 = 1.9259$, $x_2 = -1.8148$, $x_3 = -0.8889$, and $x_4 = 1.5926$. Alternatively, one can use the inverse matrix of A to obtain the same solution directly as follows:

```
» x=inv (A)*b

x =

    1.9259
   -1.8148
   -0.8889
    1.5926
```

It should be noted that using the inverse method usually takes longer that using Gaussian elimination especially for large systems. In this book we will use Gaussian elimination (i.e. the backslash operator "\").

Consider now the following 5×5 matrix D:

```
» D=[1 2 3 4 5 ; 2 4 6 8 9 ; 2 4 6 2 4 ; 1 1 2 3 -2 ; 9 0 2 3 1]

D =

    1    2    3    4    5
    2    4    6    8    9
    2    4    6    2    4
    1    1    2    3   -2
    9    0    2    3    1
```

We can extract the submatrix in rows 2 to 4 and columns 3 to 5 as follows:

```
» E=D (2:4, 3:5)

E =

    6    8    9
    6    2    4
    2    3   -2
```

We can extract the third column of D as follows:

```
» F=D(1:5,3)

F =

    3
    6
    6
    2
    2
```

We can also extract the second row of D as follows:

```
» G=D(2,1:5)

G =

    2    4    6    8    9
```

We can extract the element in row 4 and column 3 as follows:

```
» H=D(4,3)

H =

    2
```

Finally in order to plot a graph of the function $y = f(x)$, we use the MATLAB command plot(x,y) after we have adequately defined both vectors x and y. The following is a simple example.

```
» x=[1 2 3 4 5 6 7 8 9 10]

x =

    1    2    3    4    5    6    7    8    9    10

» y=x.^2

y =

    1    4    9    16    25    36    49    64    81    100

» plot(x,y)
```

Figure 1.1 shows the plot obtained by MATLAB. It is usually shown in a separate graphics window. In this figure no titles are given to the x and y-axes. These titles may be easily added to the figure using the x-label and y-label commands.

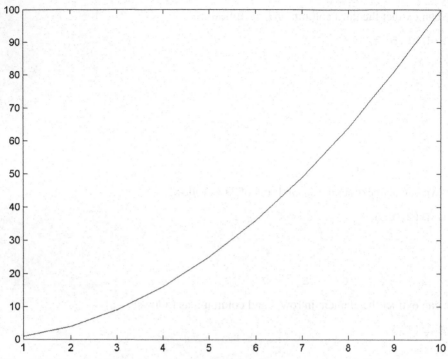

Fig. 1.1. Using the MATLAB `Plot` Command

2 The Spring Element

2.1
Basic Equations

The spring element is a one-dimensional finite element where the local and global coordinates coincide. It should be noted that the spring element is the simplest finite element available. Each spring element has two nodes as shown in Fig. 2.1. Let the stiffness of the spring be denoted by k. In this case the element stiffness matrix is given by (see [1], [8], and [18]).

$$k = \begin{bmatrix} k & -k \\ -k & k \end{bmatrix} \tag{2.1}$$

Fig. 2.1. The Spring Element

Obviously the element stiffness matrix for the spring element is a 2×2 matrix since the spring element has only two degrees of freedom – one at each node. Consequently for a system of spring elements with n nodes, the size of the global stiffness matrix K will be of size $n \times n$ (since we have one degree of freedom at each node). The global stiffness matrix K is obtained by assembling the element stiffness matrices $k_i (i = 1, 2, 3, \ldots, n)$ using the direct stiffness approach. For example the element stiffness matrix k for a spring connecting nodes 4 and 5 in a system will be assembled into the global stiffness matrix K by adding its rows and columns to rows 4 and 5 and columns 4 and 5 of K. A special MATLAB function called *SpringAssemble* is written specifically for this purpose. This process will be illustrated in detail in the examples.

Once the global stiffness matrix K is obtained we have the following system equation:

$$[K]\{U\} = \{F\} \tag{2.2}$$

where U is the global nodal displacement vector and F is the global nodal force vector. At this step the boundary conditions are applied manually to the vectors U and F. Then the matrix (2.2) is solved by partitioning and Gaussian elimination. Finally once the unknown displacements and reactions are found, the element forces are obtained for each element as follows:

$$\{f\} = [k]\{u\} \tag{2.3}$$

where f is the 2×1 element force vector and u is the 2×1 element displacement vector.

2.2
MATLAB Functions Used

The three MATLAB functions used for the spring element are:

"*SpringElementStiffness*(k) – This function calculates the element stiffness matrix for each spring with stiffness k. It returns the 2×2 element stiffness matrix k."

SpringAssemble(K, k, i, j) – This functions assembles the element stiffness matrix k of the spring joining nodes i (at the left end) and j (at the right end) into the global stiffness matrix K. It returns the $n \times n$ global stiffness matrix K every time an element is assembled.

SpringElementForces(k, u) – This function calculates the element force vector using the element stiffness matrix k and the element displacement vector u. It returns the 2×1 element force vector f.

The following is a listing of the MATLAB source code for each function:

```
function y = SpringElementStiffness(k)
%SpringElementStiffness    This function returns the element stiffness
%                          matrix for a spring with stiffness k.
%                          The size of the element stiffness matrix
%                          is 2 x 2.
y = [k -k; -k k];
```

```
function y = SpringAssemble(K,k,i,j)
%SpringAssemble    This function assembles the element stiffness
%                  matrix k of the spring with nodes i and j into the
%                  global stiffness matrix K.
%                  This function returns the global stiffness matrix K
%                  after the element stiffness matrix k is assembled.
K(i,i) = K(i,i) + k(1,1);
K(i,j) = K(i,j) + k(1,2);
```

```
K(j,i) = K(j,i) + k(2,1);
K(j,j) = K(j,j) + k(2,2);
y = K;
```

```
function y = SpringElementForces(k,u)
%SpringElementForces   This function returns the element nodal force
%                      vector given the element stiffness matrix k
%                      and the element nodal displacement vector
u.
y = k * u;
```

Example 2.1:

Consider the two-element spring system shown in Fig. 2.2. Given $k_1 = 100\,\text{kN/m}$, $k_2 = 200\,\text{kN/m}$, and $P = 15\,\text{kN}$, determine:

1. the global stiffness matrix for the system.
2. the displacements at nodes 2 and 3.
3. the reaction at node 1.
4. the force in each spring.

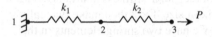

Fig. 2.2. Two-Element Spring System for Example 2.1

Solution:

Use the six steps outlined in Chap. 1 to solve this problem using the spring element.

Step 1 – Discretizing the Domain:

This problem is already discretized. The domain is subdivided into two elements and three nodes. Table 2.1 shows the element connectivity for this example.

Table 2.1. Element Connectivity for Example 2.1

Element Number	Node i	Node j
1	1	2
2	2	3

Step 2 – Writing the Element Stiffness Matrices:

The two element stiffness matrices k_1 and k_2 are obtained by making calls to the MATLAB function *SpringElementStiffness*. Each matrix has size 2×2.

```
» k1=SpringElementStiffness(100)

k1 =

    100   -100
   -100    100
```

```
» k2=SpringElementStiffness(200)

k2 =

    200   -200
   -200    200
```

Step 3 – Assembling the Global Stiffness Matrix:

Since the spring system has three nodes, the size of the global stiffness matrix is 3×3. Therefore to obtain K we first set up a zero matrix of size 3×3 then make two calls to the MATLAB function *SpringAssemble* since we have two spring elements in the system. Each call to the function will assemble one element. The following are the MATLAB commands:

```
» K=zeros(3,3)

K =

    0    0    0
    0    0    0
    0    0    0
```

```
» K=SpringAssemble(K,k1,1,2)

K =

    100   -100    0
   -100    100    0
      0      0    0
```

```
» K=SpringAssemble(K,k2,2,3)

K =

    100   -100      0
   -100    300   -200
      0   -200    200
```

Step 4 – Applying the Boundary Conditions:

The matrix (2.2) for this system is obtained as follows using the global stiffness matrix obtained in the previous step:

$$
\begin{bmatrix}
100 & -100 & 0 \\
-100 & 300 & -200 \\
0 & -200 & 200
\end{bmatrix}
\begin{Bmatrix} U_1 \\ U_2 \\ U_3 \end{Bmatrix}
=
\begin{Bmatrix} F_1 \\ F_2 \\ F_3 \end{Bmatrix}
\tag{2.4}
$$

The boundary conditions for this problem are given as:

$$
U_1 = 0, F_2 = 0, F_3 = 15\,kN
\tag{2.5}
$$

Inserting the above conditions into (2.4) we obtain:

$$
\begin{bmatrix}
100 & -100 & 0 \\
-100 & 300 & -200 \\
0 & -200 & 200
\end{bmatrix}
\begin{Bmatrix} 0 \\ U_2 \\ U_3 \end{Bmatrix}
=
\begin{Bmatrix} F_1 \\ 0 \\ 15 \end{Bmatrix}
\tag{2.6}
$$

Step 5 – Solving the Equations:

Solving the system of equations in (2.6) will be performed by partitioning (manually) and Gaussian elimination (with MATLAB). First we partition (2.6) by extracting the submatrix in rows 2 and 3 and columns 2 and 3. Therefore we obtain:

$$
\begin{bmatrix}
300 & -200 \\
-200 & 200
\end{bmatrix}
\begin{Bmatrix} U_2 \\ U_3 \end{Bmatrix}
=
\begin{Bmatrix} 0 \\ 15 \end{Bmatrix}
\tag{2.7}
$$

The solution of the above system is obtained using MATLAB as follows. Note that the backslash operator "\" is used for Gaussian elimination.

```
» k=K(2:3,2:3)

k =

       300       -200
      -200        200
```

```
» f=[0 ; 15]

f =

       0
      15
```

```
» u=k\f

u =

      0.1500
      0.2250
```

It is now clear that the displacements at nodes 2 and 3 are 0.15 m and 0.225 m, respectively.

Step 6 – Post-processing:

In this step, we obtain the reaction at node 1 and the force in each spring using MATLAB as follows. First we set up the global nodal displacement vector U, then we calculate the global nodal force vector F.

```
» U=[0 ; u]

U =

           0
      0.1500
      0.2250
```

```
» F=K*U

F =

      -15
        0
       15
```

Thus the reaction at node 1 is a force of 15 kN (directed to the left). Finally we set up the element nodal displacement vectors u_1 and u_2, then we calculate the element force vectors f_1 and f_2 by making calls to the MATLAB function *SpringElementForces*.

```
» u1=[0 ; U(2)]

u1 =

          0
     0.1500

» f1=SpringElementForces(k1,u1)

f1 =

       -15
        15

» u2=[U(2) ; U(3)]

u2 =

     0.1500
     0.2250

» f2=SpringElementForces(k2,u2)

f2 =

       -15
        15
```

Thus it is clear that the force in element 1 is 15 kN (tensile) and the force in element 2 is also 15 kN (tensile).

Example 2.2:

Consider the spring system composed of six springs as shown in Fig. 2.3. Given $k = 120\,\text{kN/m}$ and $P = 20\,\text{kN}$, determine:

1. the global stiffness matrix for the system.
2. the displacements at nodes 3, 4, and 5.
3. the reactions at nodes 1 and 2.
4. the force in each spring.

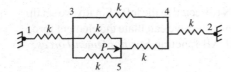

Fig. 2.3. Six-Element Spring System for Example 2.2

Solution:

Use the six steps outlined in Chap. 1 to solve this problem using the spring element.

Step 1 – Discretizing the Domain:

This problem is already discretized. The domain is subdivided into six elements and five nodes. Table 2.2 shows the element connectivity for this example.

Table 2.2. Element Connectivity for Example 2.2

Element Number	Node i	Node j
1	1	3
2	3	4
3	3	5
4	3	5
5	5	4
6	4	2

Step 2 – Writing the Element Stiffness Matrices:

The six element stiffness matrices k_1, k_2, k_3, k_4, k_5, and k_6 are obtained by making calls to the MATLAB function *SpringElementStiffness*. Each matrix has size 2×2.

```
» k1=SpringElementStiffness(120)

k1 =

     120 -120
    -120  120
```

```
» k2=SpringElementStiffness(120)

k2 =


   120    -120
  -120     120

» k3=SpringElementStiffness(120)

k3 =

   120    -120
  -120     120

» k4=SpringElementStiffness(120)

k4 =

   120    -120
  -120     120

» k5=SpringElementStiffness(120)

k5 =

   120    -120
  -120     120

» k6=SpringElementStiffness(120)

k6 =

   120    -120
  -120     120
```

Step 3 – Assembling the Global Stiffness Matrix:

Since the spring system has five nodes, the size of the global stiffness matrix is 5×5. Therefore to obtain K we first set up a zero matrix of size 5×5 then make six calls to the MATLAB function *SpringAssemble* since we have six spring elements in the system. Each call to the function will assemble one element. The following are the MATLAB commands:

```
» K=zeros(5,5)

K =

     0    0    0    0    0
     0    0    0    0    0
     0    0    0    0    0
     0    0    0    0    0
     0    0    0    0    0


» K=SpringAssemble(K,k1,1,3)

K =

    120    0   -120    0    0
      0    0      0    0    0
   -120    0    120    0    0
      0    0      0    0    0
      0    0      0    0    0


» K=SpringAssemble(K,k2,3,4)

K =

    120    0   -120      0    0
      0    0      0      0    0
   -120    0    240   -120    0
      0    0   -120    120    0
      0    0      0      0    0


» K=SpringAssemble(K,k3,3,5)

K =

    120    0   -120      0      0
      0    0      0      0      0
   -120    0    360   -120   -120
      0    0   -120    120      0
      0    0   -120      0    120
```

» K=SpringAssemble(K,k4,3,5)

K =

```
   120    0   -120       0       0
     0    0      0       0       0
  -120    0    480    -120    -240
     0    0   -120     120       0
     0    0   -240       0     240
```

» K=SpringAssemble(K,k5,5,4)

K =

```
   120    0   -120       0       0
     0    0      0       0       0
  -120    0    480    -120    -240
     0    0   -120     240    -120
     0    0   -240    -120     360
```

» K=SpringAssemble(K,k6,4,2)

K =

```
   120      0   -120       0       0
     0    120      0    -120       0
  -120      0    480    -120    -240
     0   -120   -120     360    -120
     0      0   -240    -120     360
```

Step 4 – Applying the Boundary Conditions:

The matrix (2.2) for this system is obtained as follows using the global stiffness matrix obtained in the previous step:

$$
\begin{bmatrix}
120 & 0 & -120 & 0 & 0 \\
0 & 120 & 0 & -120 & 0 \\
-120 & 0 & 480 & -120 & -240 \\
0 & -120 & -120 & 360 & -120 \\
0 & 0 & -240 & -120 & 360
\end{bmatrix}
\begin{Bmatrix}
U_1 \\ U_2 \\ U_3 \\ U_4 \\ U_5
\end{Bmatrix}
=
\begin{Bmatrix}
F_1 \\ F_2 \\ F_3 \\ F_4 \\ F_5
\end{Bmatrix}
\tag{2.8}
$$

The boundary conditions for this problem are given as:

$$U_1 = 0, U_2 = 0, F_3 = 0, F_4 = 0, F_5 = 20\,kN \tag{2.9}$$

Inserting the above conditions into (2.8) we obtain:

$$\begin{bmatrix} 120 & 0 & -120 & 0 & 0 \\ 0 & 120 & 0 & -120 & 0 \\ -120 & 0 & 480 & -120 & -240 \\ 0 & -120 & -120 & 360 & -120 \\ 0 & 0 & -240 & -120 & 360 \end{bmatrix} \begin{Bmatrix} 0 \\ 0 \\ U_3 \\ U_4 \\ U_5 \end{Bmatrix} = \begin{Bmatrix} F_1 \\ F_2 \\ 0 \\ 0 \\ 20 \end{Bmatrix} \tag{2.10}$$

Step 5 – Solving the Equations:

Solving the system of equations in (2.10) will be performed by partitioning (manually) and Gaussian elimination (with MATLAB). First we partition (2.10) by extracting the submatrix in rows 3, 4 and 5 and columns 3, 4 and 5. Therefore we obtain:

$$\begin{bmatrix} 480 & -120 & -240 \\ -120 & 360 & -120 \\ -240 & -120 & 360 \end{bmatrix} \begin{Bmatrix} U_3 \\ U_4 \\ U_5 \end{Bmatrix} = \begin{Bmatrix} 0 \\ 0 \\ 20 \end{Bmatrix} \tag{2.11}$$

The solution of the above system is obtained using MATLAB as follows. Note that the backslash operator "\" is used for Gaussian elimination.

```
» k=K(3:5,3:5)

k =

        480   -120   -240
       -120    360   -120
       -240   -120    360


» f=[0 ; 0 ; 20]

f =

        0
        0
       20
```

```
» u=k\f

u =

     0.0897
     0.0769
     0.1410
```

It is now clear that the displacements at nodes 3, 4 and 5 are 0.0897 m, 0.0769 m, and 0.141 m, respectively.

Step 6 – Post-processing:

In this step, we obtain the reactions at nodes 1 and 2 and the force in each spring using MATLAB as follows. First we set up the global nodal displacement vector U, then we calculate the global nodal force vector F.

```
» U= [0  ;  0  ;  u]

U =

          0
          0
     0.0897
     0.0769
     0.1410


» F=K*U

F =

    -10.7692
     -9.2308
          0
          0
    20.0000
```

Thus the reactions at nodes 1 and 2 are forces of 10.7692 kN (directed to the left) and 9.2308 kN (directed to the left), respectively. Finally we set up the element nodal displacement vectors u_1, u_2, u_3, u_4, u_5, and u_6, then we calculate the element force vectors f_1, f_2, f_3, f_4, f_5, and f_6 by making calls to the MATLAB function *SpringElementForces*.

```
» u1=[0 ; U(3)]

u1 =

          0
     0.0897

» f1=SpringElementForces(k1,u1)

f1 =

    -10.7640
     10.7640

» u2=[U(3) ; U(4)]

u2 =

     0.0897
     0.0769

» f2=SpringElementForces(k2,u2)

f2 =

      1.5360
     -1.5360

» u3=[U(3) ; U(5)]

u3 =
     0.0897
     0.1410

» f3=SpringElementForces(k3,u3)

f3 =

     -6.1560
      6.1560

» u4=[U(3) ; U(5)]

u4 =

     0.0897
     0.1410
```

```
» f4=SpringElementForces(k4,u4)

f4 =

   -6.1560
    6.1560

» u5=[U(5)  ; U(4)]

u5 =

    0.1410
    0.0769

» f5=SpringElementForces(k5,u5)

f5 =

    7.6920
   -7.6920

» u6=[U(4)  ; 0]

u6 =

    0.0769
         0

» f6=SpringElementForces(k6,u6)

f6 =

    9.2280
   -9.2280
```

Thus the forces in the springs are summarized in Table 2.3.

Table 2.3. Summary of Element Forces

Element Number	Force	Type
1	10.764 kN	Tensile
2	1.5360 kN	Compressive
3	6.1560 kN	Tensile
4	6.1560 kN	Tensile
5	7.6920 kN	Compressive
6	9.2280 kN	Compressive

Problems:

Problem 2.1:

Consider the spring system composed of two springs as shown in Fig. 2.4. Given $k_1 = 200\,\text{kN/m}$, $k_2 = 250\,\text{kN/m}$, and $P = 10\,\text{kN}$, determine:

1. the global stiffness matrix for the system.
2. the displacements at node 2.
3. the reactions at nodes 1 and 3.
4. the force in each spring.

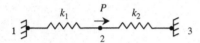

Fig. 2.4. Two-Element Spring System for Problem 2.1

Problem 2.2:

Consider the spring system composed of four springs as shown in Fig. 2.5. Given $k = 170\,\text{kN/m}$ and $P = 25\,\text{kN}$, determine:

1. the global stiffness matrix for the system.
2. the displacements at nodes 2, 3, and 4.
3. the reaction at node 1.
4. the force in each spring.

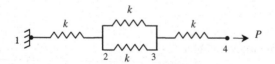

Fig. 2.5. Four-Element Spring System for Problem 2.2

3 The Linear Bar Element

3.1
Basic Equations

The linear bar element is a one-dimensional finite element where the local and global coordinates coincide. It is characterized by linear shape functions and is identical to the spring element except that the stiffness of the bar is not given directly. The linear element has modulus of elasticity E, cross-sectional area A, and length L. Each linear bar element has two nodes as shown in Fig. 3.1. In this case the element stiffness matrix is given by (see [1]).

$$
k = \begin{bmatrix} \dfrac{EA}{L} & -\dfrac{EA}{L} \\ -\dfrac{EA}{L} & \dfrac{EA}{L} \end{bmatrix}
\tag{3.1}
$$

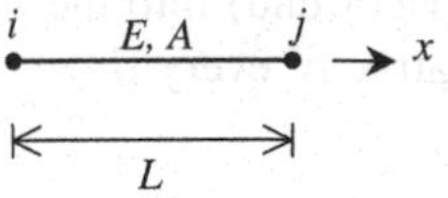

Fig. 3.1. The Linear Bar Element

Obviously the element stiffness matrix for the linear bar element is similar to that of the spring element with the stiffness replaced by EA/L. It is clear that the linear bar element has only two degrees of freedom – one at each node. Consequently for a structure with n nodes, the global stiffness matrix K will be of size $n \times n$ (since we have one degree of freedom at each node). The global stiffness matrix K is assembled by making calls to the MATLAB function *LinearBarAssemble* which is written specifically for this purpose. This process will be illustrated in detail in the examples.

Once the global stiffness matrix K is obtained we have the following structure equation:

$$[K]\{U\} = \{F\} \tag{3.2}$$

where U is the global nodal displacement vector and F is the global nodal force vector. At this step the boundary conditions are applied manually to the vectors U and F. Then the matrix (3.2) is solved by partitioning and Gaussian elimination. Finally once the unknown displacements and reactions are found, the element forces are obtained for each element as follows:

$$\{f\} = [k]\{u\} \tag{3.3}$$

where f is the 2×1 element force vector and u is the 2×1 element displacement vector. The element stresses are obtained by dividing the element forces by the cross-sectional area A.

3.2
MATLAB Functions Used

The four MATLAB functions used for the linear bar element are:

LinearBarElementStiffness(E, A, L) – This function calculates the element stiffness matrix for each linear bar with modulus of elasticity E, cross-sectional area A, and length L. It returns the 2×2 element stiffness matrix k.

LinearBarAssemble(K, k, i, j) – This functions assembles the element stiffness matrix k of the linear bar joining nodes i (at the left end) and j (at the right end) into the global stiffness matrix K. It returns the $n \times n$ global stiffness matrix K every time an element is assembled.

LinearBarElementForces(k, u) – This function calculates the element force vector using the element stiffness matrix k and the element displacement vector u. It returns the 2×1 element force vector f.

LinearBarElementStresses(k, u, A) – This function calculates the element stress vector using the element stiffness matrix k, the element displacement vector u and the cross-sectional area A. It returns the 2×1 element stress vector *sigma* or s.

The following is a listing of the MATLAB source code for each function:

```
function y = LinearBarElementStiffness(E,A,L)
%LinearBarElementStiffness    This function returns the element
%                             stiffness matrix for a linear bar with
%                             modulus of elasticity E, cross-sectional
%                             area A, and length L. The size of the
%                             element stiffness matrix is 2 x 2.
y = [E*A/L -E*A/L ; - -E*A/L -E*A/L];
```

```
function y = LinearBarAssemble(K,k,i,j)
%LinearBarAssemble      This function assembles the element stiffness
%                       matrix k of the linear bar with nodes i and j
%                       into the global stiffness matrix K.
%                       This function returns the global stiffness
%                       matrix K after the element stiffness matrix
%                       k is assembled.
K(i,i) = K(i,i) + k(1,1) ;
K(i,j) = K(i,j) + k(1,2) ;
K(j,i) = K(j,i) + k(2,1) ;
K(j,j) = K(j,j) + k(2,2) ;
y = K;
```

```
function y = LinearBarElementForces(k,u)
%LinearBarElementForces    This function returns the element nodal
%                          force vector given the element stiffness
%                          matrix k and the element nodal displacement
%                          vector u.
y = k * u;
```

```
function y = LinearBarElementStresses(k, u, A)
%LinearBarElementStresses    This function returns the element nodal
%                            stress vector given the element stiffness
%                            matrix k, the element nodal displacement
%                            vector u, and the cross-sectional area A.
y = k * u/A;
```

Example 3.1:

Consider the structure composed of two linear bars as shown in Fig. 3.2. Given $E = 210$ GPa, $A = 0.003\,\text{m}^2$, $P = 10\,\text{kN}$, and node 3 is displaced to the right by $0.002\,\text{m}$, determine:

1. the global stiffness matrix for the structure.
2. the displacement at node 2.
3. the reactions at nodes 1 and 3.
4. the stress in each bar.

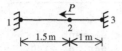

Fig. 3.2. Two-Bar Structure for Problem 3.1

Solution:

Use the six steps outlined in Chap. 1 to solve this problem using the linear bar element.

Step 1 – Discretizing the Domain:

This problem is already discretized. The domain is subdivided into two elements and three nodes. The units used in the MATLAB calculations are kN and meter. Table 3.1 shows the element connectivity for this example.

Table 3.1. Element Connectivity for Example 3.1

Element Number	Node i	Node j
1	1	2
2	2	3

Step 2 – Writing the Element Stiffness Matrices:

The two element stiffness matrices k_1 and k_2 are obtained by making calls to the MATLAB function *LinearBarElementStiffness*. Each matrix has size 2×2.

```
» E=210e6

E =

    210000000

» A=0.003

A =

    0.0030

» L1=1.5

L1 =

    1.5000
```

```
» L2=1

L2 =

    1

» k1=LinearBarElementStiffness(E,A,L1)

k1 =

       420000   -420000
      -420000    420000

» k2=LinearBarElementStiffness(E,A,L2)

k2 =

       630000   -630000
      -630000    630000
```

Step 3 – Assembling the Global Stiffness Matrix:

Since the structure has three nodes, the size of the global stiffness matrix is 3×3. Therefore to obtain K we first set up a zero matrix of size 3×3 then make two calls to the MATLAB function *LinearBarAssemble* since we have two linear bar elements in the structure. Each call to the function will assemble one element. The following are the MATLAB commands:

```
» K=zeros(3,3)

K =

    0    0    0
    0    0    0
    0    0    0

» K=LinearBarAssemble(K,k1,1,2)

K =

       420000   -420000    0
      -420000    420000    0
            0         0    0
```

» K=LinearBarAssemble(K,k2,2,3)

K =

```
     420000     -420000           0
     -20000     1050000     -630000
          0     -630000      630000
```

Step 4 – Applying the Boundary Conditions:

The matrix (3.2) for this structure is obtained as follows using the global stiffness matrix obtained in the previous step:

$$
\begin{bmatrix}
420000 & -420000 & 0 \\
-420000 & 1050000 & -630000 \\
0 & -630000 & 630000
\end{bmatrix}
\begin{Bmatrix} U_1 \\ U_2 \\ U_3 \end{Bmatrix}
=
\begin{Bmatrix} F_1 \\ F_2 \\ F_3 \end{Bmatrix}
\tag{3.4}
$$

The boundary conditions for this problem are given as:

$$
U_1 = 0, F_2 = -10, U_3 = 0.002
\tag{3.5}
$$

Inserting the above conditions into (3.4) we obtain:

$$
\begin{bmatrix}
420000 & -420000 & 0 \\
-420000 & 1050000 & -630000 \\
0 & -630000 & 630000
\end{bmatrix}
\begin{Bmatrix} 0 \\ U_2 \\ 0.002 \end{Bmatrix}
=
\begin{Bmatrix} F_1 \\ -10 \\ F_3 \end{Bmatrix}
\tag{3.6}
$$

Step 5 – Solving the Equations:

Solving the system of equations in (3.6) will be performed by partitioning (manually) and Gaussian elimination (with MATLAB). First we partition (3.6) by extracting the submatrix in row 2 and column 2 which turns out to be a 1×1 matrix. Because of the applied displacement of 0.002 m at node 3, we need to extract the submatrix in row 2 and column 3 which also turns out to be a 1×1 matrix. Therefore we obtain:

$$
[1050000] U_2 + [-630000] (0.002) = \{-10\}
\tag{3.7}
$$

The solution of the above system is obtained using MATLAB as follows. Note that the backslash operator "\" is used for Gaussian elimination.

```
»  k=K(2,2)

k =

    1050000

»  k0=K(2,3)

k0 =

    -630000

»  u0=0.002

u0 =

    0.0020

»  f=[-10]

f =

    -10

»  f0=f-k0*u0

f0 =

    1250

»  u=k\f0

u =

    0.0012
```

It is now clear that the displacement at node 2 is 0.0012 m.

Step 6 – Post-processing:

In this step, we obtain the reactions at nodes 1 and 3, and the stress in each bar using MATLAB as follows. First we set up the global nodal displacement vector U, then we calculate the global nodal force vector F.

```
» U=[0 ; u ; u0]

U =

         0
    0.0012
    0.0020

» F=K*U

F =

    -500.0000
     -10.0000
     510.0000
```

Thus the reactions at nodes 1 and 3 are forces of 500 kN (directed to the left) and 510 kN (directed to the right), respectively. It is clear that force equilibrium is satisfied. Next we set up the element nodal displacement vectors u_1 and u_2, then we calculate the element force vectors f_1 and f_2 by making calls to the MATLAB function *LinearBarElementForces*. Finally, we divide each element force by the cross-sectional area of the element to obtain the element stresses.

```
» u1=[0 ; U(2)]

u1 =

         0
    0.0012

» f1=LinearBarElementForces(k1,u1)

f1 =

    -500.0000
     500.0000

» sigma1=f1/A

sigma1 =

    1.0e+005 *

    -1.6667
     1.6667
```

```
» u2=[U(2) ; U(3)]

u2 =

    0.0012
    0.0020

» f2=LinearBarElementForces(k2,u2)

f2 =

    -510.0000
     510.0000

» sigma2=f2/A

sigma2 =

   1.0e+005 *

    -1.7000
     1.7000
```

Thus it is clear that the stress in element 1 is 1.667×10^5 kN/m^2 (or 166.7 MPa, tensile) and the stress in element 2 is 1.7×10^5 kN/m^2 (or 170 MPa, tensile). Alternatively, we can obtain the element stresses directly by making calls to the MATLAB function *LinearBarElementStresses*. This is performed as follows in which we obtain the same results as above.

```
» s1=LinearBarElementStresses(k1,u1,A)

s1 =

   1.0e+005 *

    -1.6667
     1.6667

» s2=LinearBarElementStresses(k2,u2,A)

s2 =

   1.0e+005 *

    -1.7000
     1.7000
```

Example 3.2:

Consider the tapered bar shown in Fig. 3.3 with $E = 210\,\mathrm{GPa}$ and $P = 18\,\mathrm{kN}$. The cross-sectional areas of the bar at the left and right ends are $0.002\,\mathrm{m}^2$ and $0.012\,\mathrm{m}^2$, respectively. Use linear bar elements to determine the displacement at the free end of the bar.

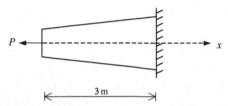

Fig. 3.3. Tapered Bar for Problem 3.2

Solution:

Use the six steps outlined in Chap. 1 to solve this problem using the linear bar element.

Step 1 – Discretizing the Domain:

We will use five linear bar elements in this problem. Therefore, the tapered bar is discretized into five elements and six nodes as shown in Fig. 3.4. For more accurate results more elements can be used (see Problem 3.2). Table 3.2 shows the element connectivity for this example. In this example the length of each element will be 0.6 m. The cross-sectional area for each element will be taken as the cross-sectional area at the middle of the element. This can be calculated using the following formula which gives the cross-sectional area at any distance x from the left end.

$$A(x) = 0.002 + \frac{0.01x}{3} \tag{3.8}$$

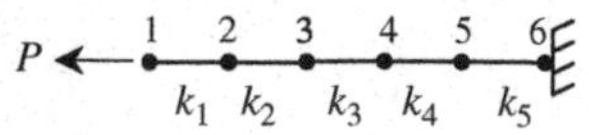

Fig. 3.4. Discretization of the Tapered Bar Example

Table 3.2. Element Connectivity for Example 3.2

Element Number	Node i	Node j
1	1	2
2	2	3
3	3	4
4	4	5
5	5	6

Step 2 – Writing the Element Stiffness Matrices:

The five element stiffness matrices k_1, k_2, k_3, k_4, and k_5 are obtained by making calls to the MATLAB function *LinearBarElementStiffness*. Each matrix has size 2×2.

```
» E=210e6

E =

    210000000

» L=3/5

L =

    0.6000

» A1=0.002+(0.01*0.3/3)

A1 =

    0.0030

» A2=0.002+(0.01*0.9)/3

A2 =

    0.0050

» A3=0.002+(0.01*1.5/3)

A3 =

    0.0070
```

» A4=0.002+(0.01*2.1/3)

A4 =

    0.0090

» A5=0.002+(0.01*2.7/3)

A5 =

    0.0110

» k1=LinearBarElementStiffness(E,A1,L)

k1 =

     1050000   -1050000
    -1050000   1050000

» k2=LinearBarElementStiffness(E,A2,L)

k2 =

  1.0e+006 *

     1.7500   -1.7500
    -1.7500   1.7500

» k3=LinearBarElementStiffness(E,A3,L)

k3 =

     2450000   -2450000
    -2450000   2450000

» k4=LinearBarElementStiffness(E,A4,L)

k4 =

  1.0e+006 *

     3.1500   -3.1500
    -3.1500   3.1500

```
» k5=LinearBarElementStiffness(E,A5,L)

k5 =

        3850000    -3850000
       -3850000     3850000
```

Step 3 – Assembling the Global Stiffness Matrix:

Since the structure has six nodes, the size of the global stiffness matrix is 6×6. Therefore to obtain K we first set up a zero matrix of size 6×6 then make five calls to the MATLAB function *LinearBarAssemble* since we have five linear bar elements in the structure. Each call to the function will assemble one element. The following are the MATLAB commands:

```
» K=zeros(6,6)

K =

     0    0    0    0    0    0
     0    0    0    0    0    0
     0    0    0    0    0    0
     0    0    0    0    0    0
     0    0    0    0    0    0
     0    0    0    0    0    0

» K=LinearBarAssemble(K,k1,1,2)

K =

      1050000   -1050000    0    0    0    0
     -1050000    1050000    0    0    0    0
            0          0    0    0    0    0
            0          0    0    0    0    0
            0          0    0    0    0    0
            0          0    0    0    0    0

» K=LinearBarAssemble(K,k2,2,3)

K =

     1.0e+006  *
```

```
     1.0500   -1.0500          0   0   0   0
    -1.0500    2.8000    -1.7500   0   0   0
          0   -1.7500     1.7500   0   0   0
          0         0          0   0   0   0
          0         0          0   0   0   0
          0         0          0   0   0   0
```

» K=LinearBarAssemble(K,k3,3,4)

K =

   1.0e+006 *

```
     1.0500   -1.0500          0          0   0   0
    -1.0500    2.8000    -1.7500          0   0   0
          0   -1.7500     4.2000    -2.4500   0   0
          0         0    -2.4500     2.4500   0   0
          0         0          0          0   0   0
          0         0          0          0   0   0
```

» K=LinearBarAssemble(K,k4,4,5)

K =

   1.0e+006 *

```
     1.0500   -1.0500          0          0          0   0
    -1.0500    2.8000    -1.7500          0          0   0
          0   -1.7500     4.2000    -2.4500          0   0
          0         0    -2.4500     5.6000    -3.1500   0
          0         0          0    -3.1500     3.1500   0
          0         0          0          0          0   0
```

» K=LinearBarAssemble(K,k5,5,6)

K =

   1.0e+006 *

```
 1.0500   -1.0500        0          0           0          0
-1.0500    2.8000   -1.7500        0           0          0
      0   -1.7500    4.2000   -2.4500         0          0
      0         0   -2.4500    5.6000     -3.1500         0
      0         0         0   -3.1500      7.0000    -3.8500
      0         0         0         0    -x3.8500     3.8500
```

Step 4 – Applying the Boundary Conditions:

The matrix (3.2) for this structure is obtained as follows using the global stiffness matrix obtained in the previous step:

$$
10^6 \begin{bmatrix} 1.05 & -1.05 & 0 & 0 & 0 & 0 \\ -1.05 & 2.80 & -1.75 & 0 & 0 & 0 \\ 0 & -1.75 & 4.2 & -2.45 & 0 & 0 \\ 0 & 0 & -2.45 & 5.60 & -3.15 & 0 \\ 0 & 0 & 0 & -3.15 & 7.00 & -3.85 \\ 0 & 0 & 0 & 0 & -3.85 & 3.85 \end{bmatrix} \begin{Bmatrix} U_1 \\ U_2 \\ U_3 \\ U_4 \\ U_5 \\ U_6 \end{Bmatrix} = \begin{Bmatrix} F_1 \\ F_2 \\ F_3 \\ F_4 \\ F_5 \\ F_6 \end{Bmatrix} \quad (3.9)
$$

The boundary conditions for this problem are given as:

$$
F_1 = -18, \quad F_2 = F_3 = F_4 = F_5 = 0, \quad U_6 = 0 \quad (3.10)
$$

Inserting the above conditions into (3.9) we obtain:

$$
10^6 \begin{bmatrix} 1.05 & -1.05 & 0 & 0 & 0 & 0 \\ -1.05 & 2.80 & -1.75 & 0 & 0 & 0 \\ 0 & -1.75 & 4.20 & -2.45 & 0 & 0 \\ 0 & 0 & -2.45 & 5.60 & -3.15 & 0 \\ 0 & 0 & 0 & -3.15 & 7.00 & -3.85 \\ 0 & 0 & 0 & 0 & -3.85 & 3.85 \end{bmatrix} \begin{Bmatrix} U_1 \\ U_2 \\ U_3 \\ U_4 \\ U_5 \\ 0 \end{Bmatrix} = \begin{Bmatrix} -18 \\ 0 \\ 0 \\ 0 \\ 0 \\ F_6 \end{Bmatrix}
$$

$$(3.11)$$

Step 5 – Solving the Equations:

Solving the system of equations in (3.11) will be performed by partitioning (manually) and Gaussian elimination (with MATLAB). First we partition (3.11) by extracting the submatrix in rows 1 to 5 and columns 1 to 5. Therefore we obtain:

$$10^6 \begin{bmatrix} 1.05 & -1.05 & 0 & 0 & 0 \\ -1.05 & 2.80 & -1.75 & 0 & 0 \\ 0 & -1.75 & 4.20 & -2.45 & 0 \\ 0 & 0 & -2.45 & 5.60 & -3.15 \\ 0 & 0 & 0 & -3.15 & 7.00 \end{bmatrix} \begin{Bmatrix} U_1 \\ U_2 \\ U_3 \\ U_4 \\ U_5 \end{Bmatrix} = \begin{Bmatrix} -18 \\ 0 \\ 0 \\ 0 \\ 0 \end{Bmatrix} \qquad (3.12)$$

The solution of the above system is obtained using MATLAB as follows. Note that the backslash operator "\" is used for Gaussian elimination.

```
» k=K(1:5,1:5)

k =

   1.0e+006 *


    1.0500   -1.0500         0         0         0
   -1.0500    2.8000   -1.7500         0         0
         0   -1.7500    4.2000   -2.4500         0
         0         0   -2.4500    5.6000   -3.1500
         0         0         0   -3.1500    7.0000


» f=[-18 ; 0 ; 0 ; 0 ; 0 ]

f =

   -18
     0
     0
     0
     0

» u= k\f

u =

   1.0e-004 *

   -0.4517
   -0.2802
   -0.1774
   -0.1039
   -0.0468
```

Therefore it is clear that the required displacement at the free end is -0.4517×10^{-4} m or -0.04517 mm (the minus sign indicates that it is directed to the left). No calls are made to the MATLAB functions *LinearBarElementForces* or *LinearBarElementStresses* since the element forces and stresses are not required in this example. Thus Step 6 (Post-processing) will not be performed for this example.

Problems:

Problem 3.1:

Consider the structure composed of three linear bars as shown in Fig. 3.5. Given $E = 70$ GPa, $A = 0.005$ m^2, $P_1 = 10$ kN, and $P_2 = 15$ kN, determine:

1. the global stiffness matrix for the structure.
2. the displacements at nodes 2, 3, and 4.
3. the reaction at node 1.
4. the stress in each bar.

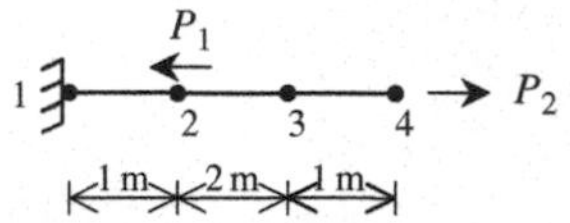

Fig. 3.5. Three-Bar Structure for Problem 3.1

Problem 3.2:

Solve Example 3.2 again using ten linear elements instead of five. Determine the displacement at the free end of the tapered bar and compare your answer with the result obtained in the example.

Problem 3.3:

Consider the structure composed of a spring and a linear bar as shown in Fig. 3.6. Given $E = 200$ GPa, $A = 0.01$ m^2, $k = 1000$ kN/m, and $P = 25$ kN, determine:

1. the global stiffness matrix for the structure.
2. the displacement at node 2.
3. the reactions at nodes 1 and 3.
4. the stress in the bar.
5. the force in the spring.

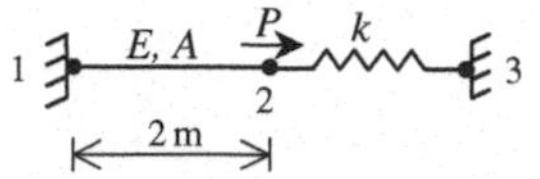

Fig. 3.6. Linear Bar with a Spring for Problem 3.3

4 The Quadratic Bar Element

4.1
Basic Equations

The quadratic bar element is a one-dimensional finite element where the local and global coordinates coincide. It is characterized by quadratic shape functions. The quadratic bar element has modulus of elasticity E, cross-sectional area A, and length L. Each quadratic bar element has three nodes as shown in Fig. 4.1. The third node is at the middle of the element. In this case the element stiffness matrix is given by (see [8]).

$$k = \frac{EA}{3L} \begin{bmatrix} 7 & 1 & -8 \\ 1 & 7 & -8 \\ -8 & -8 & 16 \end{bmatrix} \tag{4.1}$$

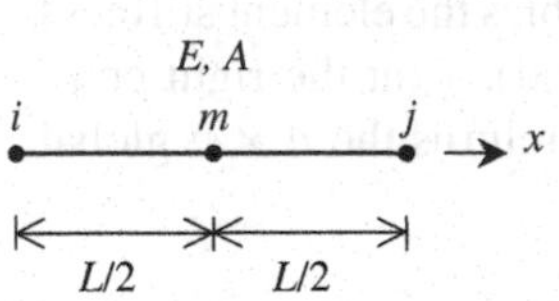

Fig. 4.1. The Quadratic Bar Element

It is clear that the quadratic bar element has three degrees of freedom – one at each node. Consequently for a structure with n nodes, the global stiffness matrix K will be of size $n \times n$ (since we have one degree of freedom at each node). The order of the nodes for this element is very important – the first node is the one at the left end, the second node is the one at the right end, and the third node is the one in the middle of the element. The global stiffness matrix K is assembled by making calls to the MATLAB function *QuadraticBarAssemble* which is written specifically for this purpose. This process will be illustrated in detail in the examples.

Once the global stiffness matrix K is obtained we have the following structure equation:

$$[K]\{U\} = \{F\} \qquad (4.2)$$

where U is the global nodal displacement vector and F is the global nodal force vector. At this step the boundary conditions are applied manually to the vectors U and F. Then the matrix (4.2) is solved by partitioning and Gaussian elimination. Finally once the unknown displacements and reactions are found, the element forces are obtained for each element as follows:

$$\{f\} = [k]\{u\} \qquad (4.3)$$

where f is the 3×1 element force vector and u is the 3×1 element displacement vector. The element stresses are obtained by dividing the element forces by the cross-sectional area A.

4.2
MATLAB Functions Used

The four MATLAB functions used for the quadratic bar element are:

QuadraticBarElementStiffness(E, A, L) – This function calculates the element stiffness matrix for each quadratic bar with modulus of elasticity E, cross-sectional area A, and length L. It returns the 3×3 element stiffness matrix k.

QuadraticBarAssemble(K, k, i, j, m) – This function assembles the element stiffness matrix k of the quadratic bar joining nodes i (at the left end), j (at the right end), and m (in the middle) into the global stiffness matrix K. It returns the $n \times n$ global stiffness matrix K every time an element is assembled.

QuadraticBarElementForces(k, u) – This function calculates the element force vector using the element stiffness matrix k and the element displacement vector u. It returns the 3×1 element force vector f.

QuadraticBarElementStresses(k, u, A) – This function calculates the element stress vector using the element stiffness matrix k, the element displacement vector u and the cross-sectional area A. It returns the 3×1 element stress vector *sigma* or *s*.

The following is a listing of the MATLAB source code for each function:

```
function y = QuadraticBarElementStiffness(E,A,L)
%QuadraticBarElementStiffness    This function returns the element
%                                stiffness matrix for a quadratic bar
%                                with modulus of elasticity E,
%                                cross-sectional area A, and length L.
%                                The size of the element stiffness
%                                matrix is 3 x 3.
y = E*A/(3*L) * [7 1 -8 ; 1 7 -8 ; -8 -8 16];
```

```
function y = QuadraticBarAssemble(K,k,i,j,m)
%QuadraticBarAssemble    This function assembles the element stiffness
%                        matrix k of the quadratic bar with nodes i, j
%                        and m into the global stiffness matrix K.
%                        This function returns the global stiffness
%                        matrix K after the element stiffness matrix
%                        k is assembled.
K(i,i) = K(i,i) + k(1,1);
K(i,j) = K(i,j) + k(1,2);
K(i,m) = K(i,m) + k(1,3);
K(j,i) = K(j,i) + k(2,1);
K(j,j) = K(j,j) + k(2,2);
K(j,m) = K(j,m) + k(2,3);
K(m,i) = K(m,i) + k(3,1);
K(m,j) = K(m,j) + k(3,2);
K(m,m) = K(m,m) + k(3,3);
y = K;
```

```
function y = QuadraticBarElementForces(k,u)
%QuadraticBarElementForces    This function returns the element nodal
%                             force vector given the element stiffness
%                             matrix k and the element nodal
%                             displacement vector u.
y = k * u;
```

```
function y = QuadraticBarElementStresses(k, u, A)
%QuadraticBarElementStresses    This function returns the element
%                               nodal stress vector given the element
%                               stiffness matrix k, the element nodal
%                               displacement vector u, and the
%                               cross-sectional area A.
y = k * u/A;
```

Example 4.1:

Consider the structure composed of two quadratic bars as shown in Fig. 4.2. Given
$E = 210\,\text{GPa}$ and $A = 0.003\,\text{m}^2$, determine:

1. the global stiffness matrix for the structure.
2. the displacements at nodes 2, 3, 4, and 5.

3. the reaction at node 1.
4. the element stresses.

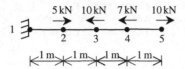

Fig. 4.2. Two Quadratic Bars for Example 4.1

Solution:

Use the six steps outlined in Chap. 1 to solve this problem using the quadratic bar element.

Step 1 – Discretizing the Domain:

This problem is already discretized. The domain is subdivided into two elements and five nodes. The units used in the MATLAB calculations are kN and meter. Table 4.1 shows the element connectivity for this example.

Table 4.1. Element Connectivity for Example 4.1

Element Number	Node i	Node j	Node m
1	1	3	2
2	3	5	4

Step 2 – Writing the Element Stiffness Matrices:

The two element stiffness matrices k_1 and k_2 are obtained by making calls to the MATLAB function *QuadraticBarElementStiffness*. Each matrix has size 3×3.

```
» E=210e6

E =

   210000000

» A=0.003

A =

   0.0030
```

```
» L=2

L =

    2

» k1=QuadraticBarElementStiffness(E,A,L)

k1 =

       735000        105000       -840000
       105000        735000       -840000
      -840000       -840000       1680000

» k2=QuadraticBarElementStiffness(E,A,L)

k2 =

       735000        105000       -840000
       105000        735000       -840000
      -840000       -840000       1680000
```

Step 3 – Assembling the Global Stiffness Matrix:

Since the structure has five nodes, the size of the global stiffness matrix is 5×5. Therefore to obtain K we first set up a zero matrix of size 5×5 then make two calls to the MATLAB function *QuadraticBarAssemble* since we have two quadratic bar elements in the structure. Each call to the function will assemble one element. The following are the MATLAB commands:

```
» K=zeros(5,5)

K =

     0     0     0     0     0
     0     0     0     0     0
     0     0     0     0     0
     0     0     0     0     0
     0     0     0     0     0
```

```
» K=QuadraticBarAssemble(K,k1,1,3,2)
```

K =

```
     735000    -840000     105000     0    0
    -840000    1680000    -840000     0    0
     105000    -840000     735000     0    0
          0          0          0     0    0
          0          0          0     0    0
```

```
» K=QuadraticBarAssemble(K,k2,3,5,4)
```

K =

```
     735000    -840000     105000          0          0
    -840000    1680000    -840000          0          0
     105000    -840000    1470000    -840000     105000
          0          0    -840000    1680000    -840000
          0          0     105000    -840000     735000
```

Step 4 – Applying the Boundary Conditions:

The matrix (4.2) for this structure is obtained as follows using the global stiffness matrix obtained in the previous step:

$$10^3 \begin{bmatrix} 735 & -840 & 105 & 0 & 0 \\ -840 & 1680 & -840 & 0 & 0 \\ 105 & -840 & 1470 & -840 & 105 \\ 0 & 0 & -840 & 1680 & -840 \\ 0 & 0 & 105 & -840 & 735 \end{bmatrix} \begin{Bmatrix} U_1 \\ U_2 \\ U_3 \\ U_4 \\ U_5 \end{Bmatrix} = \begin{Bmatrix} F_1 \\ F_2 \\ F_3 \\ F_4 \\ F_5 \end{Bmatrix} \qquad (4.4)$$

The boundary conditions for this problem are given as:

$$U_1 = 0, \quad F_2 = 5, \quad F_3 = -10, \quad F_4 = -7, \quad F_5 = 10 \qquad (4.5)$$

Inserting the above conditions into (4.4) we obtain:

$$10^3 \begin{bmatrix} 735 & -840 & 105 & 0 & 0 \\ -840 & 1680 & -840 & 0 & 0 \\ 105 & -840 & 1470 & -840 & 105 \\ 0 & 0 & -840 & 1680 & -840 \\ 0 & 0 & 105 & -840 & 735 \end{bmatrix} \begin{Bmatrix} 0 \\ U_2 \\ U_3 \\ U_4 \\ U_5 \end{Bmatrix} = \begin{Bmatrix} F_1 \\ 5 \\ -10 \\ -7 \\ 10 \end{Bmatrix} \qquad (4.6)$$

Step 5 – Solving the Equations:

Solving the system of equations in (4.6) will be performed by partitioning (manually) and Gaussian elimination (with MATLAB). First we partition (4.6) by extracting the submatrix in rows 2 to 5 and columns 2 to 5. Therefore we obtain:

$$10^3 \begin{bmatrix} 1680 & -840 & 0 & 0 \\ -840 & 1470 & -840 & 105 \\ 0 & -840 & 1680 & -840 \\ 0 & 105 & -840 & 735 \end{bmatrix} \begin{Bmatrix} U_2 \\ U_3 \\ U_4 \\ U_5 \end{Bmatrix} = \begin{Bmatrix} 5 \\ -10 \\ -7 \\ 10 \end{Bmatrix} \tag{4.7}$$

The solution of the above system is obtained using MATLAB as follows. Note that the backslash operator "\" is used for Gaussian elimination.

```
» k=K(2:5,2:5)

k =

    1680000     -840000           0           0
    -840000     1470000     -840000      105000
          0     -840000     1680000     -840000
          0      105000     -840000      735000

» f=[5 ; -10 ; -7 ; 10]

f =

     5
   -10
    -7
    10

» u=k\f

u =

   1.0e-004 *

   -0.0417
   -0.1429
   -0.0813
    0.0635
```

It is now clear that the displacements at node 2, 3, 4, and 5 are –0.00417 mm, –0.01429 mm, –0.00813 mm, and 0.00635 mm, respectively.

Step 6 – Post-processing:

In this step, we obtain the reaction at node 1, and the stresses in each bar using MATLAB as follows. First we set up the global nodal displacement vector U, then we calculate the global nodal force vector F.

```
» U=[0 ; u]

U =

  1.0e-004 *

        0
   -0.0417
   -0.1429
   -0.0813
    0.0635

» F=K*U

F =

      2.0000
      5.0000
    -10.0000
     -7.0000
     10.0000
```

Thus the reaction at node 1 is a force of $2\,\text{kN}$ (directed to the right). Obviously force equilibrium is satisfied for this problem. Next we set up the element nodal displacement vectors u_1 and u_2, then we calculate the element stress vectors *sigma1* and *sigma2* by making calls to the MATLAB function *QuadraticBarElement Stresses*.

```
» u1=[0 ; U(3) ; U(2)]

u1 =
  1.0e-004 *

        0
   -0.1429
   -0.0417

» sigma1=QuadraticBarElementStresses(k1,u1,A)
```

```
sigma1 =

   1.0e+003 *

      0.6667
     -2.3333
      1.6667

» u2=[U(3) ; U(5) ; U(4)]

u2 =

   1.0e-004 *

     -0.1429
      0.0635
     -0.0813

» sigma2=QuadraticBarElementStresses(k2,u2,A)

sigma2 =

   1.0e+003 *

     -1.0000
      3.3333
     -2.3333
```

Thus it is clear that the stresses are not constant within each element since we have used quadratic bar elements instead of linear bar elements in this example.

Example 4.2:

Consider the structure composed of two quadratic bars as shown in Fig. 4.3. Given $E = 210\,\text{GPa}$, $A = 0.003\,\text{m}^2$, $P = 10\,\text{kN}$, and node 5 is displaced to the right by $0.002\,\text{m}$, determine:

1. the global stiffness matrix for the structure.
2. the displacements at nodes 2, 3, and 4.
3. the reactions at nodes 1 and 5.
4. the element stresses.

Note that this is exactly the same problem solved in Example 3.1 but now we are using quadratic bar elements instead of linear bar elements.

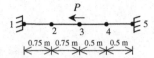

Fig. 4.3. Two Quadratic Bars for Example 4.2

Solution:

Use the six steps outlined in Chap. 1 to solve this problem using the quadratic bar element.

Step 1 – Discretizing the Domain:

This problem is already discretized. The domain is subdivided into two elements and five nodes. The units used in the MATLAB calculations are kN and meter. Table 4.2 shows the element connectivity for this example.

Table 4.2. Element Connectivity for Example 4.2

Element Number	Node i	Node j	Node m
1	1	3	2
2	3	5	4

Step 2 – Writing the Element Stiffness Matrices:

The two element stiffness matrices k_1 and k_2 are obtained by making calls to the MATLAB function *QuadraticBarElementStiffness*. Each matrix has size 3×3.

```
» E=210E6

E =

   210000000

» A=0.003

A =

   0.0030

» L1=1.5
```

```
L1 =

    1.5000

» L2=1

L2 =

    1

» k1=QuadraticBarElementStiffness(E,A,L1)

k1 =

       980000        140000      -1120000
       140000        980000      -1120000
     -1120000      -1120000       2240000

» k2=QuadraticBarElementStiffness(E,A,L2)

k2 =

      1470000        210000      -1680000
       210000       1470000      -1680000
     -1680000      -1680000       3360000
```

Step 3 – Assembling the Global Stiffness Matrix:

Since the structure has five nodes, the size of the global stiffness matrix is 5×5. Therefore to obtain K we first set up a zero matrix of size 5×5 then make two calls to the MATLAB function *QuadraticBarAssemble* since we have two quadratic bar elements in the structure. Each call to the function will assemble one element. The following are the MATLAB commands:

```
» K=zeros(5,5)

K =

     0     0     0     0     0
     0     0     0     0     0
     0     0     0     0     0
     0     0     0     0     0
     0     0     0     0     0
```

```
» K=QuadraticBarAssemble(K,k1,1,3,2)

K =
```

```
   980000   -1120000    140000   0   0
 -1120000    2240000  -1120000   0   0
   140000   -1120000    980000   0   0
        0          0         0   0   0
        0          0         0   0   0
```

```
» K=QuadraticBarAssemble(K,k2,3,5,4)

K =
```

```
   980000   -1120000    140000          0          0
 -1120000    2240000  -1120000          0          0
   140000   -1120000   2450000   -1680000     210000
        0          0  -1680000    3360000   -1680000
        0          0    210000   -1680000    1470000
```

Step 4 – Applying the Boundary Conditions:

The matrix (4.2) for this structure is obtained as follows using the global stiffness matrix obtained in the previous step:

$$
10^3
\begin{bmatrix}
980 & -1120 & 140 & 0 & 0 \\
-1120 & 2240 & -1120 & 0 & 0 \\
140 & -1120 & 2450 & -1680 & 210 \\
0 & 0 & -1680 & 3360 & -1680 \\
0 & 0 & 210 & -1680 & 1470
\end{bmatrix}
\begin{Bmatrix}
U_1 \\ U_2 \\ U_3 \\ U_4 \\ U_5
\end{Bmatrix}
=
\begin{Bmatrix}
F_1 \\ F_2 \\ F_3 \\ F_4 \\ F_5
\end{Bmatrix}
\quad (4.8)
$$

The boundary conditions for this problem are given as:

$$
U_1 = 0, F_2 = 0, F_3 = -10, F_4 = 0, U_5 = 0.002 \quad (4.9)
$$

Inserting the above conditions into (4.8) we obtain:

$$10^3 \begin{bmatrix} 980 & -1120 & 140 & 0 & 0 \\ -1120 & 2240 & -1120 & 0 & 0 \\ 140 & -1120 & 2450 & -1680 & 210 \\ 0 & 0 & -1680 & 3360 & -1680 \\ 0 & 0 & 210 & -1680 & 1470 \end{bmatrix} \begin{Bmatrix} 0 \\ U_2 \\ U_3 \\ U_4 \\ 0.002 \end{Bmatrix} = \begin{Bmatrix} F_1 \\ 0 \\ -10 \\ 0 \\ F_5 \end{Bmatrix}$$

$$(4.10)$$

Step 5 – Solving the Equations:

Solving the system of equations in (4.10) will be performed by partitioning (manually) and Gaussian elimination (with MATLAB). First we partition (4.10) by extracting the submatrix in rows 2 to 4 and columns 2 to 4 which turns out to be a 3×3 matrix. Because of the applied displacement of 0.002 m at node 5, we need also to extract the submatrix in rows 2 to 4 and column 5 which turns out to be a 3×1 vector. Therefore we obtain:

$$10^3 \begin{bmatrix} 2240 & -1120 & 0 \\ -1120 & 2450 & -1680 \\ 0 & -1680 & 3360 \end{bmatrix} \begin{Bmatrix} U_2 \\ U_3 \\ U_4 \end{Bmatrix} + 10^3 \begin{Bmatrix} 0 \\ 210 \\ -1680 \end{Bmatrix} (0.002) = \begin{Bmatrix} 0 \\ -10 \\ 0 \end{Bmatrix}$$

$$(4.11)$$

The solution of the above system is obtained using MATLAB as follows. Note that the backslash operator "\" is used for Gaussian elimination.

```
» k=K(2:4,2:4)

k =

    2240000   -1120000         0
   -1120000    2450000  -1680000
          0   -1680000   3360000

» k0=K(2:4,5)

k0 =

          0
     210000
   -1680000

» u0=0.002
```

```
u0 =

    0.0020

» f=[0 ; -10 ; 0]

f =
     0
   -10
     0

» f0=f-k0*u0

f0 =

     0
  -430
  3360

» u=k\f0

u =

   0.0006
   0.0012
   0.0016
```

It is now clear that the displacements at nodes 2, 3, and 4 are 0.0006 m, 0.0012 m, and 0.0016 m, respectively. Note that we have obtained the same value for the displacement at the middle node of the structure as that obtained in Example 3.1, namely, 0.0012 m.

Step 6 – Post-processing:

In this step, we obtain the reactions at nodes 1 and 5, and the stress in each bar using MATLAB as follows. First we set up the global nodal displacement vector U, then we calculate the global nodal force vector F.

```
» U=[0 ; u ; u0]

U =
        0
   0.0006
   0.0012
   0.0016
   0.0020
```

```
» F=K*U

F =

   -500.0000
          0
    -10.0000
      0.0000
    510.0000
```

Thus the reactions at nodes 1 and 5 are forces of 500 kN (directed to the left) and 510 kN (directed to the right), respectively. It is clear that force equilibrium is satisfied and that these are the same values obtained in Example 3.1. Next we set up the element nodal displacement vectors u_1 and u_2, then we calculate the element stress vectors *sigma1* and *sigma2* by making calls to the MATLAB function *QuadraticBarElementStresses*.

```
» u1=[0 ; U(3) ; U(2)]

u1 =

        0
   0.0012
   0.0006

» sigma1=QuadraticBarElementStresses(k1,u1,A)

sigma1 =

   1.0e+005 *

    -1.6667
     1.6667
          0

» u2=[U(3) ; U(5) ; U(4)]

u2 =

   0.0012
   0.0020
   0.0016

» sigma2=QuadraticBarElementStresses(k2,u2,A)

sigma2 =

   1.0e+005 *
```

```
-1.7000
 1.7000
 0.0000
```

Thus it is clear that the stress in element 1 is $1.667 \times 10^5 \, \text{kN/m}^2$ (or 166.7 MPa, tensile) and the stress in element 2 is $1.7 \times 10^5 \, \text{kN/m}^2$ (or 170 MPa, tensile). Note that these are the same values for the stresses obtained previously in Example 3.1.

Problems:

Problem 4.1:

Consider the tapered bar problem of Example 3.2 (see Fig. 3.3). Solve this problem again using quadratic bars instead of linear bars. Use two quadratic bar elements to model the problem such that the structure has five nodes. Compare your answer with those obtained in Example 3.2 and Problem 3.2.

Problem 4.2:

Consider the structure composed of a spring and a quadratic bar element as shown in Fig. 4.4. Given $k = 2000 \, \text{kN/m}$, $E = 70 \, \text{GPa}$, and $A = 0.001 \, \text{m}^2$, determine:

1. the global stiffness matrix for the structure.
2. the displacements at nodes 2, 3, and 4.
3. the reaction at node 1.
4. the force in the spring.
5. the quadratic bar element stresses.

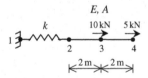

Fig. 4.4. Quadratic Bar and a Spring for Problem 4.2

5 The Plane Truss Element

5.1
Basic Equations

The plane truss element is a two-dimensional finite element with both local and global coordinates. It is characterized by linear shape functions. The plane truss element has modulus of elasticity E, cross-sectional area A, and length L. Each plane truss element has two nodes and is inclined with an angle θ measured counterclockwise from the positive global X axis as shown in Fig. 5.1. Let $C = \cos \theta$ and $S = \sin \theta$. In this case the element stiffness matrix is given by (see [1]).

$$
k = \frac{EA}{L} \begin{bmatrix} C^2 & CS & -C^2 & -CS \\ CS & S^2 & -CS & -S^2 \\ -C^2 & -CS & C^2 & CS \\ -CS & -S^2 & CS & S^2 \end{bmatrix} \tag{5.1}
$$

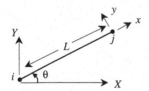

Fig. 5.1. The Plane Truss Element

It is clear that the plane truss element has four degrees of freedom – two at each node. Consequently for a structure with n nodes, the global stiffness matrix K will be of size $2n \times 2n$ (since we have two degrees of freedom at each node). The global stiffness matrix K is assembled by making calls to the MATLAB function *PlaneTrussAssemble* which is written specifically for this purpose. This process will be illustrated in detail in the examples.

Once the global stiffness matrix K is obtained we have the following structure equation:

$$[K]\{U\} = \{F\} \tag{5.2}$$

where U is the global nodal displacement vector and F is the global nodal force vector. At this step the boundary conditions are applied manually to the vectors U and F. Then the matrix (5.2) is solved by partitioning and Gaussian elimination. Finally once the unknown displacements and reactions are found, the force is obtained for each element as follows:

$$f = \frac{EA}{L} \begin{bmatrix} -C & -S & C & S \end{bmatrix} \{u\} \tag{5.3}$$

where f is the force in the element (a scalar) and u is the 4×1 element displacement vector. The element stress is obtained by dividing the element force by the cross-sectional area A.

If there is an inclined support at one of the nodes of the truss then the global stiffness matrix needs to be modified using the following equation:

$$[K]_{new} = [T][K]_{old}[T]^T \tag{5.4}$$

where $[T]$ is a $2n \times 2n$ transformation matrix that is obtained by making a call to the MATLAB function *PlaneTrussInclinedSupport*. The inclined support is assumed to be at node i with an angle of inclination *alpha* as shown in Fig. 5.2.

Fig. 5.2. Inclined Support in a Plane Truss

5.2
MATLAB Functions Used

The six MATLAB functions used for the plane truss element are:

PlaneTrussElementLength(x_1, y_1, x_2, y_2) – This function returns the element length given the coordinates of the first node (x_1, y_1) and the coordinates of the second node (x_2, y_2).

PlaneTrussElementStiffness(*E*, *A*, *L*, *theta*) – This function calculates the element stiffness matrix for each plane truss element with modulus of elasticity *E*, cross-sectional area *A*, length *L*, and angle *theta* (in degrees). It returns the 4×4 element stiffness matrix *k*.

PlaneTrussAssemble(*K*, *k*, *i*, *j*) – This function assembles the element stiffness matrix *k* of the plane truss element joining nodes *i* and *j* into the global stiffness matrix *K*. It returns the $2n \times 2n$ global stiffness matrix *K* every time an element is assembled.

PlaneTrussElementForce(*E*, *A*, *L*, *theta*, *u*) – This function calculates the element force using the modulus of elasticity *E*, the cross-sectional area *A*, the length *L*, the angle *theta* (in degrees), and the element displacement vector *u*. It returns the force in the element as a scalar.

PlaneTrussElementStress(*E*, *L*, *theta*, *u*) – This function calculates the element stress using the modulus of elasticity *E*, the length *L*, the angle *theta* (in degrees), and the element displacement vector *u*. It returns the stress in the element as a scalar.

PlaneTrussInclinedSupport(*T*, *i*, *alpha*) – This function calculates the transformation matrix of the inclined support using the node number *i* of the inclined support and the angle of inclination *alpha* (in degrees). It returns the $2n \times 2n$ transformation matrix.

The following is a listing of the MATLAB source code for each function:

```
function y = PlaneTrussElementLength(x1,y1,x2,y2)
%PlaneTrussElementLength      This function returns the length of the
%                             plane truss element whose first node has
%                             coordinates (x1, y1) and second node has
%                             coordinates (x2, y2).
y = sqrt((x2-x1)*(x2-x1) + (y2-y1)*(y2-y1));
```

```
function y = PlaneTrussElementStiffness(E,A,L, theta)
%PlaneTrussElementStiffness   This function returns the element
%                             stiffness matrix for a plane truss
%                             element with modulus of elasticity E,
%                             cross-sectional area A, length L, and
%                             angle theta (in degrees).
%                             The size of the element stiffness
%                             matrix is 4 x 4.
x = theta*pi/180;
C = cos(x);
S = sin(x);
y = E*A/L*[C*C C*S -C*C -C*S ; C*S S*S -C*S -S*S ;
  -C*C -C*S C*C C*S ; -C*S -S*S C*S S*S] ;
```

```
function y = PlaneTrussAssemble(K,k,i,j)
%PlaneTrussAssemble      This function assembles the element stiffness
%                        matrix k of the plane truss element with nodes
%                        i and j into the global stiffness matrix K.
%                        This function returns the global stiffness
%                        matrix K after the element stiffness matrix
%                        k is assembled.
K(2*i-1,2*i-1) = K(2*i-1,2*i-1) + k(1,1) ;
K(2*i-1,2*i) = K(2*i-1,2*i) + k(1,2) ;
K(2*i-1,2*j-1) = K(2*i-1,2*j-1) + k(1,3) ;
K(2*i-1,2*j) = K(2*i-1,2*j) + k(1,4) ;
K(2*i,2*i-1) = K(2*i,2*i-1) + k(2,1) ;
K(2*i,2*i) = K(2*i,2*i) + k(2,2) ;
K(2*i,2*j-1) = K(2*i,2*j-1) + k(2,3) ;
K(2*i,2*j) = K(2*i,2*j) + k(2,4) ;
K(2*j-1,2*i-1) = K(2*j-1,2*i-1) + k(3,1) ;
K(2*j-1,2*i) = K(2*j-1,2*i) + k(3,2) ;
K(2*j-1,2*j-1) = K(2*j-1,2*j-1) + k(3,3) ;
K(2*j-1,2*j) = K(2*j-1,2*j) + k(3,4) ;
K(2*j,2*i-1) = K(2*j,2*i-1) + k(4,1) ;
K(2*j,2*i) = K(2*j,2*i) + k(4,2) ;
K(2*j,2*j-1) = K(2*j,2*j-1) + k(4,3) ;
K(2*j,2*j) = K(2*j,2*j) + k(4,4) ;
y = K;
```

```
function y = PlaneTrussElementForce(E,A,L,theta,u)
%PlaneTrussElementForce          This function returns the element force
%                                given the modulus of elasticity E, the
%                                cross-sectional area A, the length L,
%                                the angle theta (in degrees), and the
%                                element nodal displacement vector u.
x = theta* pi/180;
C = cos(x);
S = sin(x);
y = E*A/L*[-C -S C S]* u;
```

```
function y = PlaneTrussElementStress(E,L,theta,u)
%PlaneTrussElementStress         This function returns the element stress
%                                given the modulus of elasticity E, the
%                                the length L, the angle theta (in
%                                degrees), and the element nodal
%                                displacement vector u.
x = theta * pi/180;
C = cos(x);
S = sin(x);
y = E/L*[-C -S C S]* u;
```

```
function y = PlaneTrussInclinedSupport(T,i,alpha)
%PlaneTrussInclinedSupport       This function calculates the
%                                tranformation matrix T of the inclined
%                                support at node i with angle of
%                                inclination alpha (in degrees).
x = alpha*pi/180;
T(2*i-1,2*i-1) = cos(x) ;
T(2*i-1,2*i) = sin(x) ;
T(2*i,2*i-1) = -sin(x) ;
T(2*i,2*i) = cos(x) ;
y = T;
```

Example 5.1:

Consider the plane truss shown in Fig. 5.3. Given $E = 210\,\text{GPa}$ and $A = 1 \times 10^{-4}\,\text{m}^2$, determine:

1. the global stiffness matrix for the structure.
2. the horizontal displacement at node 2.
3. the horizontal and vertical displacements at node 3.
4. the reactions at nodes 1 and 2.
5. the stress in each element.

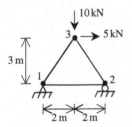

Fig. 5.3. Plane Truss with Three Elements for Example 5.1

Solution:

Use the six steps outlined in Chap. 1 to solve this problem using the plane truss element.

Step 1 – Discretizing the Domain:

This problem is already discretized. The domain is subdivided into three elements and three nodes. The units used in the MATLAB calculations are kN and meter. Table 5.1 shows the element connectivity for this example.

Table 5.1. Element Connectivity for Example 5.1

Element Number	Node i	Node j
1	1	2
2	1	3
3	2	3

Step 2 – Writing the Element Stiffness Matrices:

The three element stiffness matrices k_1, k_2, and k_3 are obtained by making calls to the MATLAB function *PlaneTrussElementStiffness*. Each matrix has size 4×4.

```
» E=210e6

E =

   210000000


» A=1e-4

A =

   1.0000e-004


» L1=4

L1 =

   4

» L2=PlaneTrussElementLength(0,0,2,3)

L2 =

   3.6056
```

```
» L3=PlaneTrussElementLength(0,0,-2,3)

L3 =

    3.6056

» k1=PlaneTrussElementStiffness(E,A,L1,0)

k1 =

    5250      0    -5250      0
       0      0        0      0
   -5250      0     5250      0
       0      0        0      0


» theta2=atan(3/2)*180/pi

theta2 =

   56.3099

» theta3=180-theta2

theta3 =

   123.6901

» k2=PlaneTrussElementStiffness(E,A,L2,theta2)

k2 =

  1.0e+003 *

    1.7921    2.6882   -1.7921   -2.6882
    2.6882    4.0322   -2.6882   -4.0322
   -1.7921   -2.6882    1.7921    2.6882
   -2.6882   -4.0322    2.6882    4.0322


» k3=PlaneTrussElementStiffness(E,A,L3,theta3)

k3 =

  1.0e+003 *
```

```
 1.7921   -2.6882   -1.7921    2.6882
-2.6882    4.0322    2.6882   -4.0322
-1.7921    2.6882    1.7921   -2.6882
 2.6882   -4.0322   -2.6882    4.0322
```

Step 3 – Assembling the Global Stiffness Matrix:

Since the structure has three nodes, the size of the global stiffness matrix is 6×6. Therefore to obtain K we first set up a zero matrix of size 6×6 then make three calls to the MATLAB function *PlaneTrussAssemble* since we have three elements in the structure. Each call to the function will assemble one element. The following are the MATLAB commands:

```
» K=zeros(6,6)

K =

    0    0    0    0    0    0
    0    0    0    0    0    0
    0    0    0    0    0    0
    0    0    0    0    0    0
    0    0    0    0    0    0
    0    0    0    0    0    0

» K=PlaneTrussAssemble(K,k1,1,2)

K =

    5250     0   -5250     0     0     0
       0     0       0     0     0     0
   -5250     0    5250     0     0     0
       0     0       0     0     0     0
       0     0       0     0     0     0
       0     0       0     0     0     0

» K=PlaneTrussAssemble(K,k2,1,3)

K =

  1.0e+003 *

    7.0421    2.6882    -5.2500     0    -1.7921    -2.6882
    2.6882    4.0322          0     0    -2.6882    -4.0322
```

```
-5.2500              0   5.2500   0            0              0
      0              0        0   0            0              0
-1.7921        -2.6882        0   0       1.7921        2.6882
-2.6882        -4.0322        0   0       2.6882        4.0322
```

» K=PlaneTrussAssemble(K,k3,2,3)

K =

   1.0e+003 *

```
   7.0421     2.6882    -5.2500          0   -1.7921    -2.6882
   2.6882     4.0322          0          0   -2.6882    -4.0322
  -5.2500          0     7.0421    -2.6882   -1.7921     2.6882
        0          0    -2.6882     4.0322    2.6882    -4.0322
  -1.7921    -2.6882    -1.7921     2.6882    3.5842     0.0000
  -2.6882    -4.0322     2.6882    -4.0322    0.0000     8.0645
```

Step 4 – Applying the Boundary Conditions:

The matrix (5.2) for this structure is obtained as follows using the global stiffness matrix obtained in the previous step:

$$
10^3 \begin{bmatrix}
7.0421 & 2.6882 & -5.2500 & 0 & -1.7921 & -2.6882 \\
2.6882 & 4.0322 & 0 & 0 & -2.6882 & -4.0322 \\
-5.2500 & 0 & 7.0421 & -2.6882 & -1.7921 & 2.6882 \\
0 & 0 & -2.6882 & 4.0322 & 2.6882 & -4.0322 \\
-1.7921 & -2.6882 & -1.7921 & 2.6882 & 3.5842 & 0 \\
-2.6882 & -4.0322 & 2.6882 & -4.0322 & 0 & 8.0645
\end{bmatrix}
$$

$$
\begin{Bmatrix}
U_{1x} \\
U_{1y} \\
U_{2x} \\
U_{2y} \\
U_{3x} \\
U_{3y}
\end{Bmatrix}
=
\begin{Bmatrix}
F_{1x} \\
F_{1y} \\
F_{2x} \\
F_{2y} \\
F_{3x} \\
F_{3y}
\end{Bmatrix}
\tag{5.5}
$$

The boundary conditions for this problem are given as:

$$
U_{1x} = U_{1y} = U_{2y} = 0, \quad F_{2x} = 0, \quad F_{3x} = 5, \quad F_{3y} = -10 \tag{5.6}
$$

Inserting the above conditions into (5.5) we obtain:

$$
10^3
\begin{bmatrix}
7.041 & 2.6882 & -5.2500 & 0 & -1.7921 & -2.6882 \\
2.6882 & 4.0322 & 0 & 0 & -2.6882 & -4.0322 \\
-5.2500 & 0 & 7.0421 & -2.6882 & -1.7921 & 2.6882 \\
0 & 0 & -2.6882 & 4.0322 & 2.6882 & -4.0322 \\
-1.7921 & -2.6882 & -1.7921 & 2.6882 & 3.5842 & 0 \\
-2.6882 & -4.0322 & 2.6882 & -4.0322 & 0 & 8.0645
\end{bmatrix}
$$

$$
\begin{Bmatrix}
0 \\ 0 \\ U_{2x} \\ 0 \\ U_{3x} \\ U_{3y}
\end{Bmatrix}
=
\begin{Bmatrix}
F_{1x} \\ F_{1y} \\ 0 \\ F_{2y} \\ 5 \\ -10
\end{Bmatrix}
\tag{5.7}
$$

Step 5 – Solving the Equations:

Solving the system of equations in (5.7) will be performed by partitioning (manually) and Gaussian elimination (with MATLAB). First we partition (5.7) by extracting the submatrix in row 3 and column 3, row 3 and columns 5 and 6, rows 5 and 6 and column 3, and rows 5 and 6 and columns 5 and 6. Therefore we obtain:

$$
10^3
\begin{bmatrix}
7.0421 & -1.7921 & 2.6882 \\
-1.7921 & 3.5842 & 0 \\
2.6882 & 0 & 8.0645
\end{bmatrix}
\begin{Bmatrix}
U_{2x} \\ U_{3x} \\ U_{3y}
\end{Bmatrix}
=
\begin{Bmatrix}
0 \\ 5 \\ -10
\end{Bmatrix}
\tag{5.8}
$$

The solution of the above system is obtained using MATLAB as follows. Note that the backslash operator "\" is used for Gaussian elimination.

```
» k=[K(3,3)  K(3,5:6)  ;  K(5:6,3)  K(5:6,5:6)]

k =

  1.0e+003 *

      7.0421     -1.7921      2.6882
     -1.7921      3.5842      0.0000
      2.6882      0.0000      8.0645

» f=[0 ; 5 ; -10]

f =
       0
       5
     -10
```

```
» u=k\f

u =

     0.0011
     0.0020
    -0.0016
```

It is now clear that the horizontal displacement at node 2 is 0.0011 m, and the horizontal and vertical displacements at node 3 are 0.0020 m and −0.0016 m, respectively.

Step 6 – Post-processing:

In this step, we obtain the reactions at nodes 1 and 2, and the stress in each element using MATLAB as follows. First we set up the global nodal displacement vector U, then we calculate the global nodal force vector F.

```
» U=[0 ; 0 ; u(1) ; 0 ; u(2:3)]

U =

          0
          0
     0.0011
          0
     0.0020
    -0.0016

» F=K*U

F =

    -5.0000
     1.2500
    -0.0000
     8.7500
     5.0000
   -10.0000
```

Thus the horizontal and vertical reactions at node 1 are forces of 5 kN (directed to the left) and 1.25 kN (directed upwards). The vertical reaction at node 2 is a force of 8.75 N (directed upwards). Obviously force equilibrium is satisfied for this problem. Next we set up the element nodal displacement vectors u_1, u_2, and u_3 then we calculate the element stresses *sigma1*, *sigma2*, and *sigma3* by making calls to the MATLAB function *PlaneTrussElementStress*.

```
» u1=[U(1) ; U(2) ; U(3) ; U(4)]

u1 =

           0
           0
      0.0011
           0

» u2=[U(1) ; U(2) ; U(5) ; U(6)]

u2 =

           0
           0
      0.0020
     -0.0016

» u3=[U(3) ; U(4) ; U(5) ; U(6)]

u3 =

      0.0011
           0
      0.0020
     -0.0016

» sigma1=PlaneTrussElementStress(E,L1,0,u1)

sigma1 =

      5.8333e+004

» sigma2=PlaneTrussElementStress(E,L2,theta2,u2)

sigma2 =

     -1.5023e+004

» sigma3=PlaneTrussElementStress(E,L3,theta3,u3)

sigma3 =

     -1.0516e+005
```

Thus it is clear that the stress in elements 1, 2, and 3 are 58.3333 MPa (tensile), 15.023 MPa (compressive), and 105.16 MPa (compressive), respectively.

Example 5.2:

Consider the plane truss with an inclined support as shown in Fig. 5.4. Given $E = 70\,\text{GPa}$ and $A = 0.004\,\text{m}^2$, determine:

1. the global stiffness matrix for the structure.
2. the displacements at nodes 2, 3, and 4.
3. the reactions at nodes 1 and 4.
4. the stress in each element.

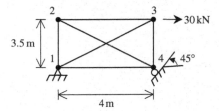

Fig. 5.4. Plane Truss with Inclined Support for Example 5.2

Solution:

Use the six steps outlined in Chap. 1 to solve this problem using the plane truss element.

Step 1 – Discretizing the Domain:

This problem is already discretized. The domain is subdivided into six elements and four nodes. The units used in the MATLAB calculations are kN and meter. Table 5.2 shows the element connectivity for this example.

Table 5.2. Element Connectivity for Example 5.2

Element Number	Node i	Node j
1	1	2
2	1	4
3	1	3
4	2	4
5	2	3
6	3	4

Step 2 – Writing the Element Stiffness Matrices:

The six element stiffness matrices k_1, k_2, k_3, k_4, k_5, and k_6 are obtained by making calls to the MATLAB function *PlaneTrussElementStiffness*. Each matrix has size 4×4.

```
» E=70e6

E =

    70000000

» A=0.004

A =

    0.0040

» L1=3.5

L1 =

    3.5000

» theta1=90

theta1 =

    90

» L2=4

L2 =

    4

» theta2=0

theta2 =

    0

» L3=PlaneTrussElementLength(0,0,4,3.5)
```

```
L3 =

    5.3151

» theta3= atan(3.5/4)*180/pi

theta3 =

    41.1859

» L4=L3
L4 =

    5.3151

» theta4=360-theta3
theta4 =

    318.8141

» L5=4
L5 =

    4

» theta5=0
theta5 =

    0

» L6=3.5
L6 =

    3.5000

» theta6=270
```

```
theta6 =

    270

» k1=PlaneTrussElementStiffness(E,A,L1,theta1)

k1 =

   1.0e+004*

      0.0000      0.0000     -0.0000     -0.0000
      0.0000      8.0000     -0.0000     -8.0000
     -0.0000     -0.0000      0.0000      0.0000
     -0.0000     -8.0000      0.0000      8.0000


» k2=PlaneTrussElementStiffness(E,A,L2,theta2)

k2 =

     70000      0    -70000      0
         0      0         0      0
    -70000      0     70000      0
         0      0         0      0


» k3=PlaneTrussElementStiffness(E,A,L3,theta3)

k3 =

   1.0e+004 *

      2.9837      2.6107     -2.9837     -2.6107
      2.6107      2.2844     -2.6107     -2.2844
     -2.9837     -2.6107      2.9837      2.6107
     -2.6107     -2.2844      2.6107      2.2844


» k4=PlaneTrussElementStiffness(E,A,L4,theta4)

k4 =

   1.0e+004 *
```

```
    2.9837   -2.6107   -2.9837    2.6107
   -2.6107    2.2844    2.6107   -2.2844
   -2.9837    2.6107    2.9837   -2.6107
    2.6107   -2.2844   -2.6107    2.2844
```

» k5=PlaneTrussElementStiffness(E,A,L5,theta5)

k5 =

```
    70000     0    -70000     0
        0     0         0     0
   -70000     0     70000     0
        0     0         0     0
```

» k6=PlaneTrussElementStiffness(E,A,L6,theta6)

k6 =

  1.0e+004 *

```
    0.0000    0.0000   -0.0000   -0.0000
    0.0000    8.0000   -0.0000   -8.0000
   -0.0000   -0.0000    0.0000    0.0000
   -0.0000   -8.0000    0.0000    8.0000
```

Step 3 – Assembling the Global Stiffness Matrix:

Since the structure has four nodes, the size of the global stiffness matrix is 8×8. Therefore to obtain K we first set up a zero matrix of size 8×8 then make six calls to the MATLAB function *PlaneTrussAssemble* since we have six elements in the structure. Each call to the function will assemble one element. The following are the MATLAB commands:

» K=zeros(8,8)

K =

```
    0    0    0    0    0    0    0    0
    0    0    0    0    0    0    0    0
    0    0    0    0    0    0    0    0
    0    0    0    0    0    0    0    0
    0    0    0    0    0    0    0    0
    0    0    0    0    0    0    0    0
    0    0    0    0    0    0    0    0
    0    0    0    0    0    0    0    0
```

```
» K=PlaneTrussAssemble(K,k1,1,2)

K =

   1.0e+004 *

   Columns 1 through 7

     0.0000      0.0000     -0.0000     -0.0000      0      0      0
     0.0000      8.0000     -0.0000     -8.0000      0      0      0
    -0.0000     -0.0000      0.0000      0.0000      0      0      0
    -0.0000     -8.0000      0.0000      8.0000      0      0      0
          0           0           0           0      0      0      0
          0           0           0           0      0      0      0
          0           0           0           0      0      0      0
          0           0           0           0      0      0      0

   Column 8

          0
          0
          0
          0
          0
          0
          0
          0

» K=PlaneTrussAssemble(K,k2,1,4)

K =

   1.0e+004 *

   Columns 1 through 7

     7.0000      0.0000     -0.0000     -0.0000      0      0     -7.0000
     0.0000      8.0000     -0.0000     -8.0000      0      0           0
    -0.0000     -0.0000      0.0000      0.0000      0      0           0
    -0.0000     -8.0000      0.0000      8.0000      0      0           0
          0           0           0           0      0      0           0
          0           0           0           0      0      0           0
    -7.0000           0           0           0      0      0      7.0000
          0           0           0           0      0      0           0
```

```
Column 8

       0
       0
       0
       0
       0
       0
       0
       0

» K=PlaneTrussAssemble(K,k3,1,3)

K =

  1.0e+005 *

Columns 1 through 7

  0.9984    0.2611   -0.0000   -0.0000   -0.2984   -0.2611   -0.7000
  0.2611    1.0284   -0.0000   -0.8000   -0.2611   -0.2284         0
 -0.0000   -0.0000    0.0000    0.0000         0         0         0
 -0.0000   -0.8000    0.0000    0.8000         0         0         0
 -0.2984   -0.2611         0         0    0.2984    0.2611         0
 -0.2611   -0.2284         0         0    0.2611    0.2284         0
 -0.7000         0         0         0         0         0    0.7000
       0         0         0         0         0         0         0


Column 8

       0
       0
       0
       0
       0
       0
       0
       0

» K=PlaneTrussAssemble(K,k4,2,4)

K =

  1.0e+005 *
```

```
Columns 1 through 7

  0.9984   0.2611 -0.0000 -0.0000 -0.2984 -0.2611 -0.7000
  0.2611   1.0284 -0.0000 -0.8000 -0.2611 -0.2284        0
 -0.0000  -0.0000  0.2984 -0.2611       0        0 -0.2984
 -0.0000  -0.8000 -0.2611  1.0284       0        0  0.2611
 -0.2984  -0.2611       0       0  0.2984  0.2611        0
 -0.2611  -0.2284       0       0  0.2611  0.2284        0
 -0.7000        0 -0.2984  0.2611       0        0  0.9984
        0        0  0.2611 -0.2284       0        0 -0.2611

   Column 8

        0
        0
   0.2611
  -0.2284
        0
        0
  -0.2611
   0.2284

» K=PlaneTrussAssemble(K,k5,2,3)

K =

   1.0e+005 *

Columns 1 through 7

  0.9984   0.2611 -0.0000 -0.0000 -0.2984 -0.2611 -0.7000
  0.2611   1.0284 -0.0000 -0.8000 -0.2611 -0.2284        0
 -0.0000  -0.0000  0.9984 -0.2611 -0.7000        0 -0.2984
 -0.0000  -0.8000 -0.2611  1.0284       0        0  0.2611
 -0.2984  -0.2611 -0.7000       0  0.9984  0.2611        0
 -0.2611  -0.2284       0       0  0.2611  0.2284        0
 -0.7000        0 -0.2984  0.2611       0        0  0.9984
        0        0  0.2611 -0.2284       0        0 -0.2611

   Column 8

        0
        0
   0.2611
```

```
        -0.2284
             0
             0
        -0.2611
         0.2284
```

» K=PlaneTrussAssemble(K,k6,3,4)

K =

  1.0e+005 *

Columns 1 through 7

```
  0.9984   0.2611  -0.0000  -0.0000  -0.2984  -0.2611  -0.7000
  0.2611   1.0284  -0.0000  -0.8000  -0.2611  -0.2284        0
 -0.0000  -0.0000   0.9984  -0.2611  -0.7000        0  -0.2984
 -0.0000  -0.8000  -0.2611   1.0284        0        0   0.2611
 -0.2984  -0.2611  -0.7000        0   0.9984   0.2611  -0.0000
 -0.2611  -0.2284        0        0   0.2611   1.0284  -0.0000
 -0.7000        0  -0.2984   0.2611  -0.0000  -0.0000   0.9984
        0        0   0.2611  -0.2284  -0.0000  -0.8000  -0.2611
```

   Column 8

```
             0
             0
         0.2611
        -0.2284
        -0.0000
        -0.8000
        -0.2611
         1.0284
```

Next we need to modify the global stiffness matrix obtained above to take the effect of the inclined support at node 4 by using (5.4). First we set up an identity matrix T of size 8 × 8. Then we make a call to the MATLAB function *PlaneTrussInclinedSupport* as shown below. The new matrix K0 obtained is thus the global stiffness matrix for the structure.

» T=eye (8, 8)

T =

```
    1   0   0   0   0   0   0   0
    0   1   0   0   0   0   0   0
```

```
         0      0      1      0      0      0      0      0
         0      0      0      1      0      0      0      0
         0      0      0      0      1      0      0      0
         0      0      0      0      0      1      0      0
         0      0      0      0      0      0      1      0
         0      0      0      0      0      0      0      1
```

» T=PlaneTrussInclinedSupport (T, 4, 45)

T =

Columns 1 through 7

```
   1.0000        0        0        0        0        0        0
        0   1.0000        0        0        0        0        0
        0        0   1.0000        0        0        0        0
        0        0        0   1.0000        0        0        0
        0        0        0        0   1.0000        0        0
        0        0        0        0        0   1.0000        0
        0        0        0        0        0        0   0.7071
        0        0        0        0        0        0  -0.7071
```

Column 8

```
        0
        0
        0
        0
        0
        0
   0.7071
   0.7071
```

» K0=T*K*T'

K0 =

   1.0e+005 *

Columns 1 through 7

```
   0.9984   0.2611  -0.0000  -0.0000  -0.2984  -0.2611  -0.4950
   0.2611   1.0284  -0.0000  -0.8000  -0.2611  -0.2284        0
  -0.0000  -0.0000   0.9984  -0.2611  -0.7000        0  -0.0264
```

```
-0.0000 -0.8000 -0.2611  1.0284        0        0  0.0231
-0.2984 -0.2611 -0.7000        0  0.9984  0.2611 -0.0000
-0.2611 -0.2284        0        0  0.2611  1.0284 -0.5657
-0.4950        0 -0.0264  0.0231 -0.0000 -0.5657  0.7523
 0.4950        0  0.3956 -0.3461 -0.0000 -0.5657  0.0150

   Column 8

    0.4950
         0
    0.3956
   -0.3461
   -0.0000
   -0.5657
    0.0150
    1.2745
```

Step 4 – Applying the Boundary Conditions:

The matrix (5.2) for this structure is obtained as follows using the global stiffness matrix obtained in the previous step:

$$
10^3
\begin{bmatrix}
99.84 & 26.11 & 0 & 0 & -29.84 & -26.11 & -49.50 & 49.50 \\
26.11 & 102.84 & 0 & -80.00 & -26.11 & -22.84 & 0 & 0 \\
0 & 0 & 99.84 & -26.11 & -70.00 & 0 & -2.64 & 39.56 \\
0 & -80.00 & -26.11 & 102.84 & 0 & 0 & 2.31 & -34.61 \\
-29.84 & -26.11 & -70.00 & 0 & 99.84 & 26.11 & 0 & 0 \\
-26.11 & -22.84 & 0 & 0 & 26.11 & 102.84 & -56.57 & -56.57 \\
-49.50 & 0 & -2.64 & 2.31 & 0 & -56.57 & 75.23 & 1.50 \\
49.50 & 0 & 39.56 & -34.61 & 0 & -56.57 & 1.50 & 127.45
\end{bmatrix}
\begin{Bmatrix}
U_{1x} \\ U_{1y} \\ U_{2x} \\ U_{2y} \\ U_{3x} \\ U_{3y} \\ U'_{4x} \\ U'_{4y}
\end{Bmatrix}
=
\begin{Bmatrix}
F_{1x} \\ F_{1y} \\ F_{2x} \\ F_{2y} \\ F_{3x} \\ F_{3y} \\ F'_{4x} \\ F'_{4y}
\end{Bmatrix}
\qquad (5.9)
$$

The boundary conditions for this problem are given as:

$$
U_{1x} = U_{1y} = U'_{4y} = 0, \quad F_{2x} = F_{2y} = F_{3y} = F'_{4x} = 0, \quad F_{3x} = 30 \quad (5.10)
$$

Inserting the above conditions into equation (5.9) we obtain:

$$
10^3
\begin{bmatrix}
99.84 & 26.11 & 0 & 0 & -29.84 & -26.11 & -49.50 & 49.50 \\
26.11 & 102.84 & 0 & -80.00 & -26.11 & -22.84 & 0 & 0 \\
0 & 0 & 99.84 & -26.11 & -70.00 & 0 & -2.64 & 39.56 \\
0 & -80.00 & -26.11 & 102.84 & 0 & 0 & 2.31 & -34.61 \\
-29.84 & -26.11 & -70.00 & 0 & 99.84 & 26.11 & 0 & 0 \\
-26.11 & -22.84 & 0 & 0 & 26.11 & 102.84 & -56.57 & -56.57 \\
-49.50 & 0 & -2.64 & 2.31 & 0 & -56.57 & 75.23 & 1.50 \\
49.50 & 0 & 39.56 & -34.61 & 0 & -56.57 & 1.50 & 127.45
\end{bmatrix}
$$

$$
\begin{Bmatrix}
0 \\
0 \\
U_{2x} \\
U_{2y} \\
U_{3x} \\
U_{3y} \\
U'_{4x} \\
0
\end{Bmatrix}
=
\begin{Bmatrix}
F_{1x} \\
F_{1y} \\
0 \\
0 \\
30 \\
0 \\
0 \\
F'_{4y}
\end{Bmatrix}
\tag{5.11}
$$

Step 5 – Solving the Equations:

Solving the system of equations in (5.11) will be performed by partitioning (manually) and Gaussian elimination (with MATLAB). First we partition (5.11) by extracting the submatrix in rows 3 to 7 and columns 3 to 7. Therefore we obtain:

$$
10^3
\begin{bmatrix}
99.84 & -26.11 & -70.00 & 0 & -2.64 \\
-26.11 & 102.84 & 0 & 0 & 2.31 \\
-70.00 & 0 & 99.84 & 26.11 & 0 \\
0 & 0 & 26.11 & 102.84 & -56.57 \\
-2.64 & 2.31 & 0 & -56.57 & 75.23
\end{bmatrix}
\begin{Bmatrix}
U_{2x} \\
U_{2y} \\
U_{3x} \\
U_{3y} \\
U'_{4x}
\end{Bmatrix}
=
\begin{Bmatrix}
0 \\
0 \\
30 \\
0 \\
0
\end{Bmatrix}
\tag{5.12}
$$

The solution of the above system is obtained using MATLAB as follows. Note that the backslash operator "\" is used for Gaussian elimination.

```
» k=K0 (3:7, 3:7)

k =

   1.0e+005 *

       0.9984    -0.2611    -0.7000         0    -0.0264
      -0.2611     1.0284          0         0     0.0231
      -0.7000          0     0.9984    0.2611    -0.0000
            0          0     0.2611    1.0284    -0.5657
      -0.0264     0.0231    -0.0000   -0.5657     0.7523
```

```
» f=[0 ; 0; 30 ; 0 ; 0]

f =

    0
    0
   30
    0
    0

» u=k\f

u =

  1.0e-003  *

   0.6053
   0.1590
   0.8129
  -0.3366
  -0.2367
```

It is now clear that the horizontal and vertical displacements at node 2 are 0.6053×10^{-3} m and 0.1590×10^{-3} m, respectively, the horizontal and vertical displacements at node 3 are 0.8129×10^{-3} m and -0.3366×10^{-3} m, respectively, and the inclined displacement at node 4 is -0.2367×10^{-3} m.

Step 6 – Post-processing:

In this step, we obtain the reactions at nodes 1 and 2, and the stress in each element using MATLAB as follows. First we set up the global nodal displacement vector U, then we calculate the global nodal force vector F.

```
» U=[0 ; 0 ; u ; 0]

U =

  1.0e-003  *

        0
        0
   0.6053
   0.1590
   0.8129
```

```
   -0.3366
   -0.2367
         0
```

```
» F=K0*U

F =

    -3.7500
   -26.2500
         0
    -0.0000
    30.0000
     0.0000
         0
    37.1231
```

Thus the horizontal and vertical reactions at node 1 are forces of 3.75 kN (directed to the left) and 26.25 kN (directed downward). The inclined reaction at node 4 is a force of 37.11231 kN. Obviously force equilibrium is satisfied for this problem. Next we set up the element nodal displacement vectors u_1, u_2, u_3, u_4, u_5, and u_6 then we calculate the element stresses *sigma1*, *sigma2*, *sigma3*, *sigma4*, *sigma5*, and *sigma6* by making calls to the MATLAB function *PlaneTrussElementStress*.

```
» u1=[U(1) ; U(2) ; U(3) ; U(4)]

u1 =

  1.0e-003 *

         0
         0
    0.6053
    0.1590
```

```
» sigma1=PlaneTrussElementStress(E,L1,theta1,u1)

sigma1 =

   3.1791e+003
```

```
» u2=[U(1) ; U(2) ; U(7) ; U(8)]

u2 =

  1.0e-003 *

          0
          0
    -0.2367
          0

» sigma2=PlaneTrussElementStress(E,L2,theta2,u2)

sigma2 =

  -4.1425e+003

» u3=[U(1) ; U(2) ; U(5) ; U(6)]

u3 =

  1.0e-003 *

          0
          0
     0.8129
    -0.3366

» sigma3=PlaneTrussElementStress(E,L3,theta3,u3)

sigma3 =

   5.1380e+003

» u4=[U(3) ; U(4) ; U(7) ; U(8)]

u4 =

  1.0e-003 *

     0.6053
     0.1590
    -0.2367
          0
```

```
» sigma4=PlaneTrussElementStress(E,L4,theta4,u4)

sigma4 =

  -6.9666e+003

» u5=[U(3) ; U(4) ; U(5) ; U(6)]

u5 =

  1.0e-003 *

    0.6053
    0.1590
    0.8129
   -0.3366

» sigma5=PlaneTrussElementStress(E,L5,theta5,u5)

sigma5 =

   3.6333e+003

» u6=[U(5) ; U(6) ; U(7) ; U(8)]

u6 =

  1.0e-003 *

    0.8129
   -0.3366
   -0.2367
         0

» sigma6=PlaneTrussElementStress(E,L6,theta6,u6)

sigma6 =

  -6.7311e+003
```

Thus it is clear that the stresses in elements 1, 2, 3, 4, 5 and 6 are 3.1791 MPa (tensile), 4.1425 MPa (compressive), 5.1380 MPa (tensile), 6.9666 MPa (compressive), 3.6333 MPa (tensile), and 6.7311 MPa (compressive), respectively.

Problems:

Problem 5.1:

Consider the plane truss shown in Fig. 5.5. Given $E = 210\,\mathrm{GPa}$ and $A = 0.005\,\mathrm{m}^2$, determine:

1. the global stiffness matrix for the structure.
2. the horizontal and vertical displacements at nodes 2, 3, 4, and 5.
3. the horizontal and vertical reactions at nodes 1 and 6.
4. the stress in each element.

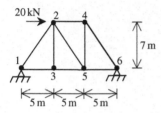

Fig. 5.5. Plane Truss for Problem 5.1

Problem 5.2:

Consider the structure composed of a spring and plane truss as shown in Fig. 5.6. Given $E = 70\,\mathrm{GPa}$, $A = 0.01\,\mathrm{m}^2$, and $k = 3000\,\mathrm{kN/m}$, determine:

1. the global stiffness matrix for the structure.
2. the displacements at nodes 4 and 5.
3. the reactions at nodes 1, 2, and 3.
4. the force in the spring.
5. the stress in each truss element.

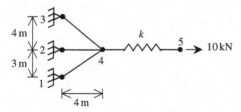

Fig. 5.6. Plane Truss with a Spring for Problem 5.2

6 The Space Truss Element

6.1
Basic Equations

The space truss element is a three-dimensional finite element with both local and global coordinates. It is characterized by linear shape functions. The space truss element has modulus of elasticity E, cross-sectional area A, and length L. Each space truss element has two nodes and is inclined with angles θ_x, θ_y, and θ_z measured from the global $X, Y,$ and Z axes, respectively, to the local x axis as shown in Fig. 6.1. Let $C_x = cos\theta_x$, $C_y = cos\theta_y$, and $C_z = cos\theta_z$. In this case the element stiffness matrix is given by (see [1] and [18]).

$$k = \frac{EA}{L} \begin{bmatrix} C_x^2 & C_xC_y & C_xC_z & -C_x^2 & -C_xC_y & -C_xC_z \\ C_yC_x & C_y^2 & C_yC_z & -C_yC_x & -C_y^2 & -C_yC_z \\ C_zC_x & C_zC_y & C_z^2 & -C_zC_x & -C_zC_y & -C_z^2 \\ -C_x^2 & -C_xC_y & -C_xC_z & C_x^2 & C_xC_y & C_xC_z \\ -C_yC_x & -C_y^2 & -C_yC_z & C_yC_x & C_y^2 & C_yC_z \\ -C_zC_x & -C_zC_y & -C_z^2 & C_zC_x & C_zC_y & C_z^2 \end{bmatrix} \quad (6.1)$$

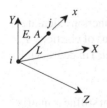

Fig. 6.1. The Space Truss Element

It is clear that the space truss element has six degrees of freedom – three at each node. Consequently for a structure with n nodes, the global stiffness matrix K will be of size $3n \times 3n$ (since we have three degrees of freedom at each node). The global stiffness matrix K is assembled by making calls to the MATLAB function *SpaceTrussAssemble* which is written specifically for this purpose. This process will be illustrated in detail in the examples.

Once the global stiffness matrix K is obtained we have the following structure equation:

$$[K]\{U\} = \{F\} \tag{6.2}$$

where U is the global nodal displacement vector and F is the global nodal force vector. At this step the boundary conditions are applied manually to the vectors U and F. Then the matrix (6.2) is solved by partitioning and Gaussian elimination. Finally once the unknown displacements and reactions are found, the force is obtained for each element as follows:

$$f = \frac{EA}{L} \begin{bmatrix} -C_x & -C_y & -C_z & C_x & C_y & C_z \end{bmatrix} \{u\} \tag{6.3}$$

where f is the force in the element (a scalar) and u is the 6×1 element displacement vector. The element stress is obtained by dividing the element force by the cross-sectional area A.

6.2
MATLAB Functions Used

The five MATLAB functions used for the space truss element are:

SpaceTrussElementLength$(x_1, y_1, z_1, x_2, y_2, z_2)$ – This function returns the element length given the coordinates of the first node (x_1, y_1, z_1) and the coordinates of the second node (x_2, y_2, z_2).

SpaceTrussElementStiffness$(E, A, L, thetax, thetay, thetaz)$ – This function calculates the element stiffness matrix for each space truss element with modulus of elasticity E, cross-sectional area A, length L, and angles *thetax*, *thetay*, and *thetaz* (in degrees). It returns the 6×6 element stiffness matrix k.

SpaceTrussAssemble(K, k, i, j) – This function assembles the element stiffness matrix k of the space truss element joining nodes i and j into the global stiffness matrix K. It returns the $3n \times 3n$ global stiffness matrix K every time an element is assembled.

SpaceTrussElementForce(*E, A, L, thetax, thetay, thetaz, u*) – This function calculates the element force using the modulus of elasticity *E*, the cross-sectional area *A*, the length *L*, the angles *thetax, thetay, thetaz* (in degrees), and the element displacement vector *u*. It returns the force in the element as a scalar.

SpaceTrussElementStress(*E, L, thetax, thetay, thetaz, u*) – This function calculates the element stress using the modulus of elasticity *E*, the length *L*, the angles *thetax, thetay, thetaz* (in degrees), and the element displacement vector *u*. It returns the stress in the element as a scalar.

The following is a listing of the MATLAB source code for each function:

```
function y = SpaceTrussElementLength(x1,y1,z1,x2,y2,z2)
%SpaceTrussElementLength     This function returns the length of the
%                            space truss element whose first node has
%                            coordinates (x1,y1,z1) and second node has
%                            coordinates (x2,y2,z2).
y = sqrt((x2-x1)*(x2-x1) + (y2-y1)*(y2-y1) + (z2-z1)*(z2-z1));
```

```
function y = SpaceTrussElementStiffness(E,A,L,thetax,thetay,thetaz)
%SpaceTrussElementStiffness     This function returns the element
%                               stiffness matrix for a space truss
%                               element with modulus of elasticity E,
%                               cross-sectional area A, length L, and
%                               angles thetax, thetay, thetaz
%                               (in degrees). The size of the element
%                               stiffness matrix is 6 x 6.
x = thetax*pi/180;
u = thetay*pi/180;
v = thetaz*pi/180;
Cx = cos(x);
Cy = cos(u);
Cz = cos(v);
w = [Cx*Cx Cx*Cy Cx*Cz ; Cy*Cx Cy*Cy Cy*Cz ; Cz*Cx Cz*Cy Cz*Cz];
y = E*A/L*[w -w ; -w w];
```

```
function y = SpaceTrussAssemble(K,k,i,j)
%SpaceTrussAssemble     This function assembles the element stiffness
%                       matrix k of the space truss element with nodes
%                       i and j into the global stiffness matrix K.
%                       This function returns the global stiffness
%                       matrix K after the element stiffness matrix
%                       k is assembled.
K(3*i-2,3*i-2) = K(3*i-2,3*i-2) + k(1,1);
K(3*i-2,3*i-1) = K(3*i-2,3*i-1) + k(1,2);
K(3*i-2,3*i) = K(3*i-2,3*i) + k(1,3);
K(3*i-2,3*j-2) = K(3*i-2,3*j-2) + k(1,4);
K(3*i-2,3*j-1) = K(3*i-2,3*j-1) + k(1,5);
```

```
K(3*i-2,3*j) = K(3*i-2,3*j) + k(1,6);
K(3*i-1,3*i-2) = K(3*i-1,3*i-2) + k(2,1);
K(3*i-1,3*i-1) = K(3*i-1,3*i-1) + k(2,2);
K(3*i-1,3*i) = K(3*i-1,3*i) + k(2,3);
K(3*i-1,3*j-2) = K(3*i-1,3*j-2) + k(2,4);
K(3*i-1,3*j-1) = K(3*i-1,3*j-1) + k(2,5);
K(3*i-1,3*j) = K(3*i-1,3*j) + k(2,6);
K(3*i,3*i-2) = K(3*i,3*i-2) + k(3,1);
K(3*i,3*i-1) = K(3*i,3*i-1) + k(3,2);
K(3*i,3*i) = K(3*i,3*i) + k(3,3);
K(3*i,3*j-2) = K(3*i,3*j-2) + k(3,4);
K(3*i,3*j-1) = K(3*i,3*j-1) + k(3,5);
K(3*i,3*j) = K(3*i,3*j) + k(3,6);
K(3*j-2,3*i-2) = K(3*j-2,3*i-2) + k(4,1);
K(3*j-2,3*i-1) = K(3*j-2,3*i-1) + k(4,2);
K(3*j-2,3*i) = K(3*j-2,3*i) + k(4,3);
K(3*j-2,3*j-2) = K(3*j-2,3*j-2) + k(4,4);
K(3*j-2,3*j-1) = K(3*j-2,3*j-1) + k(4,5);
K(3*j-2,3*j) = K(3*j-2,3*j) + k(4,6);
K(3*j-1,3*i-2) = K(3*j-1,3*i-2) + k(5,1);
K(3*j-1,3*i-1) = K(3*j-1,3*i-1) + k(5,2);
K(3*j-1,3*i) = K(3*j-1,3*i) + k(5,3);
K(3*j-1,3*j-2) = K(3*j-1,3*j-2) + k(5,4);
K(3*j-1,3*j-1) = K(3*j-1,3*j-1) + k(5,5);
K(3*j-1,3*j) = K(3*j-1,3*j) + k(5,6);
K(3*j,3*i-2) = K(3*j,3*i-2) + k(6,1);
K(3*j,3*i-1) = K(3*j,3*i-1) + k(6,2);
K(3*j,3*i) = K(3*j,3*i) + k(6,3);
K(3*j,3*j-2) = K(3*j,3*j-2) + k(6,4);
K(3*j,3*j-1) = K(3*j,3*j-1) + k(6,5);
K(3*j,3*j) = K(3*j,3*j) + k(6,6);
y = K;
```

```
function y = SpaceTrussElementForce(E,A,L,thetax,thetay,thetaz,u)
%SpaceTrussElementForce    This function returns the element force
%                          given the modulus of elasticity E, the
%                          cross-sectional area A, the length L,
%                          the angles thetax, thetay, thetaz
%                          (in degrees), and the element nodal
%                          displacement vector u.
x = thetax * pi/180;
w = thetay * pi/180;
v = thetaz * pi/180;
Cx = cos(x);
Cy = cos(w);
Cz = cos(v);
y = E*A/L*[-Cx -Cy -Cz  Cx  Cy  Cz]*u;
```

```
function y = SpaceTrussElementStress(E,L,thetax,thetay,thetaz,u)
%SpaceTrussElementStress      This function returns the element stress
%                             given the modulus of elasticity E, the
%                             length L, the angles thetax, thetay,
%                             thetaz (in degrees), and the element
%                             nodal displacement vector u.
x = thetax * pi/180;
w = thetay * pi/180;
v = thetaz * pi/180;
Cx = cos(x);
Cy = cos(w);
Cz = cos(v);
y = E/L*[-Cx -Cy -Cz  Cx  Cy  Cz]*u;
```

Example 6.1:

Consider the space truss shown in Fig. 6.2. The supports at nodes 1, 2, and 3 are ball-and-socket joints allowing rotation but no translation. Given $E = 200\,\text{GPa}$, $A_{14} = 0.001\,\text{m}^2$, $A_{24} = 0.002\,\text{m}^2$, $A_{34} = 0.001\,\text{m}^2$, and $P = 12\,\text{kN}$, determine:

1. the global stiffness matrix for the structure.
2. the displacements at node 4.
3. the reactions at nodes 1, 2, and 3.
4. the stress in each element.

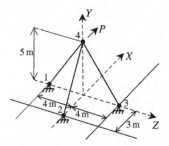

Fig. 6.2. Space Truss with Three Elements for Example 6.1

Solution:

Use the six steps outlined in Chap. 1 to solve this problem using the space truss element.

Step 1 – Discretizing the Domain:

This problem is already discretized. The domain is subdivided into three elements and four nodes. The units used in the MATLAB calculations are kN and meter. Table 6.1 shows the element connectivity for this example.

Table 6.1. Element Connectivity for Example 6.1

Element Number	Node i	Node j
1	1	4
2	2	4
3	3	4

Step 2 – Writing the Element Stiffness Matrices:

The three element stiffness matrices k_1, k_2, and k_3 are obtained by making calls to the MATLAB function *SpaceTrussElementStiffness*. Each matrix has size 6×6.

```
» E=200e6

E =

    200000000

» A1=0.001

A1 =

    0.0010

» A2=0.002

A2 =

    0.0020

» A3=0.001

A3 =

    0.0010

» L1=SpaceTrussElementLength(0,0,-4,0,5,0)

L1 =

    6.4031

» L2=SpaceTrussElementLength(-3,0,0,0,5,0)
```

```
L2 =

    5.8310

» L3=SpaceTrussElementLength(0,0,4,0,5,0)

L3 =

    6.4031

» theta1x=acos(0/L1)*180/pi

theta1x =

    90

» theta1y=acos(5/L1)*180/pi

theta1y =

    38.6598

» theta1z=acos(4/L1)*180/pi

theta1z =

    51.3402

» theta2x=acos(3/L2)*180/pi

theta2x =

    59.0362

» theta2y=acos(5/L2)*180/pi

theta2y =

    30.9638

» theta2z=acos(0/L2)*180/pi

theta2z =

    90
```

```
» theta3x=acos(0/L3)*180/pi

theta3x =

   90

» theta3y=acos(5/L3)*180/pi

theta3y =

   38.6598

» theta3z=acos(-4/L3)*180/pi

theta3z =

   128.6598

» k1=SpaceTrussElementStiffness(E,A1,L1,theta1x,theta1y,theta1z)

k1 =

1.0e+004 *


     0.0000     0.0000     0.0000    -0.0000    -0.0000    -0.0000
     0.0000     1.9046     1.5236    -0.0000    -1.9046    -1.5236
     0.0000     1.5236     1.2189    -0.0000    -1.5236    -1.2189
    -0.0000    -0.0000    -0.0000     0.0000     0.0000     0.0000
    -0.0000    -1.9046    -1.5236     0.0000     1.9046     1.5236
    -0.0000    -1.5236    -1.2189     0.0000     1.5236     1.2189


» k2=SpaceTrussElementStiffness(E,A2,L2,theta2x,theta2y,theta2z)

k2 =

  1.0e+004 *


     1.8159     3.0264     0.0000    -1.8159    -3.0264    -0.0000
     3.0264     5.0441     0.0000    -3.0264    -5.0441    -0.0000
     0.0000     0.0000     0.0000    -0.0000    -0.0000    -0.0000
    -1.8159    -3.0264    -0.0000     1.8159     3.0264     0.0000
    -3.0264    -5.0441    -0.0000     3.0264     5.0441     0.0000
    -0.0000    -0.0000    -0.0000     0.0000     0.0000     0.0000
```

```
» k3=SpaceTrussElementStiffness(E,A3,L3,theta3x,theta3y,theta3z)

k3 =

   1.0e+004  *

    0.0000    0.0000   -0.0000   -0.0000   -0.0000    0.0000
    0.0000    1.9046   -1.5236   -0.0000   -1.9046    1.5236
   -0.0000   -1.5236    1.2189    0.0000    1.5236   -1.2189
   -0.0000   -0.0000    0.0000    0.0000    0.0000   -0.0000
   -0.0000   -1.9046    1.5236    0.0000    1.9046   -1.5236
    0.0000    1.5236   -1.2189   -0.0000   -1.5236    1.2189
```

Step 3 – Assembling the Global Stiffness Matrix:

Since the structure has four nodes, the size of the global stiffness matrix is 12×12. Therefore to obtain K we first set up a zero matrix of size 12×12 then make three calls to the MATLAB function *SpaceTrussAssemble* since we have three elements in the structure. Each call to the function will assemble one element. The following are the MATLAB commands:

```
» K=zeros(12,12)

K =

     0     0     0     0     0     0     0     0     0     0     0     0
     0     0     0     0     0     0     0     0     0     0     0     0
     0     0     0     0     0     0     0     0     0     0     0     0
     0     0     0     0     0     0     0     0     0     0     0     0
     0     0     0     0     0     0     0     0     0     0     0     0
     0     0     0     0     0     0     0     0     0     0     0     0
     0     0     0     0     0     0     0     0     0     0     0     0
     0     0     0     0     0     0     0     0     0     0     0     0
     0     0     0     0     0     0     0     0     0     0     0     0
     0     0     0     0     0     0     0     0     0     0     0     0
     0     0     0     0     0     0     0     0     0     0     0     0
     0     0     0     0     0     0     0     0     0     0     0     0
```

```
» K=SpaceTrussAssemble(K,k1,1,4)

K =

   1.0e+004  *
```

Columns 1 through 7

```
 0.0000        0.0000        0.0000        0     0     0     0
 0.0000        1.9046        1.5236        0     0     0     0
 0.0000        1.5236        1.2189        0     0     0     0
      0             0             0        0     0     0     0
      0             0             0        0     0     0     0
      0             0             0        0     0     0     0
      0             0             0        0     0     0     0
      0             0             0        0     0     0     0
      0             0             0        0     0     0     0
-0.0000       -0.0000       -0.0000        0     0     0     0
-0.0000       -1.9046       -1.5236        0     0     0     0
-0.0000       -1.5236       -1.2189        0     0     0     0
```

Columns 8 through 12

```
0     0    -0.0000       -0.0000       -0.0000
0     0    -0.0000       -1.9046       -1.5236
0     0    -0.0000       -1.5236       -1.2189
0     0         0             0             0
0     0         0             0             0
0     0         0             0             0
0     0         0             0             0
0     0         0             0             0
0     0         0             0             0
0     0     0.0000        0.0000        0.0000
0     0     0.0000        1.9046        1.5236
0     0     0.0000        1.5236        1.2189
```

» K=SpaceTrussAssemble(K,k2,2,4)

K =

  1.0e+004 *

Columns 1 through 7

```
 0.0000   0.0000   0.0000        0        0        0     0
 0.0000   1.9046   1.5236        0        0        0     0
 0.0000   1.5236   1.2189        0        0        0     0
      0        0        0   1.8159   3.0264   0.0000     0
      0        0        0   3.0264   5.0441   0.0000     0
      0        0        0   0.0000   0.0000   0.0000     0
```

```
     0          0          0          0          0          0   0
     0          0          0          0          0          0   0
     0          0          0          0          0          0   0
-0.0000    -0.0000    -0.0000    -1.8159    -3.0264    -0.0000   0
-0.0000    -1.9046    -1.5236    -3.0264    -5.0441    -0.0000   0
-0.0000    -1.5236    -1.2189    -0.0000    -0.0000    -0.0000   0
```

Columns 8 through 12

```
     0     0    -0.0000    -0.0000    -0.0000
     0     0    -0.0000    -1.9046    -1.5236
     0     0    -0.0000    -1.5236    -1.2189
     0     0    -1.8159    -3.0264    -0.0000
     0     0    -3.0264    -5.0441    -0.0000
     0     0    -0.0000    -0.0000    -0.0000
     0     0          0          0          0
     0     0          0          0          0
     0     0          0          0          0
     0     0     1.8159     3.0264     0.0000
     0     0     3.0264     6.9486     1.5236
     0     0     0.0000     1.5236     1.2189
```

» K=SpaceTrussAssemble(K,k3,3,4)

K =

  1.0e+004 *

Columns 1 through 7

```
 0.0000   0.0000   0.0000         0          0         0          0
 0.0000   1.9046   1.5236         0          0         0          0
 0.0000   1.5236   1.2189         0          0         0          0
      0        0        0    1.8159     3.0264    0.0000          0
      0        0        0    3.0264     5.0441    0.0000          0
      0        0        0    0.0000     0.0000    0.0000          0
      0        0        0         0          0         0     0.0000
      0        0        0         0          0         0     0.0000
      0        0        0         0          0         0    -0.0000
-0.0000  -0.0000  -0.0000   -1.8159    -3.0264   -0.0000    -0.0000
-0.0000  -1.9046  -1.5236   -3.0264    -5.0441   -0.0000    -0.0000
-0.0000  -1.5236  -1.2189   -0.0000    -0.0000   -0.0000     0.0000
```

```
Columns 8 through 12

        0            0        -0.0000     -0.0000     -0.0000
        0            0        -0.0000     -1.9046     -1.5236
        0            0        -0.0000     -1.5236     -1.2189
        0            0        -1.8159     -3.0264     -0.0000
        0            0        -3.0264     -5.0441     -0.0000
        0            0        -0.0000     -0.0000     -0.0000
   0.0000      -0.0000       -0.0000     -0.0000      0.0000
   1.9046      -1.5236       -0.0000     -1.9046      1.5236
  -1.5236       1.2189        0.0000      1.5236     -1.2189
  -0.0000       0.0000        1.8159      3.0264      0.0000
  -1.9046       1.5236        3.0264      8.8532           0
   1.5236      -1.2189        0.0000           0      2.4378
```

Step 4 – Applying the Boundary Conditions:

The matrix (6.2) for this structure is obtained as follows using the global stiffness matrix obtained in the previous step (the numbers are written below using one decimal place only although the MATLAB calculations are performed with at least four decimal places):

$$
10^4
\begin{bmatrix}
0.0 & 0.0 & 0.0 & 0 & 0 & 0 & 0 & 0 & 0 & -0.0 & -0.0 & -0.0 \\
0.0 & 1.9 & 1.5 & 0 & 0 & 0 & 0 & 0 & 0 & -0.0 & -1.9 & -1.5 \\
0.0 & 1.5 & 1.2 & 0 & 0 & 0 & 0 & 0 & 0 & -0.0 & -1.5 & -1.2 \\
0 & 0 & 0 & 1.8 & 3.0 & 0.0 & 0 & 0 & 0 & -1.8 & -3.0 & -0.0 \\
0 & 0 & 0 & 3.0 & 5.0 & 0.0 & 0 & 0 & 0 & -3.0 & -5.0 & -0.0 \\
0 & 0 & 0 & 0.0 & 0.0 & 0.0 & 0 & 0 & 0 & -0.0 & -0.0 & -0.0 \\
0 & 0 & 0 & 0 & 0 & 0 & 0.0 & 0.0 & -0.0 & -0.0 & -0.0 & -0.0 \\
0 & 0 & 0 & 0 & 0 & 0 & 0.0 & 1.9 & -1.5 & -0.0 & -1.9 & 1.5 \\
0 & 0 & 0 & 0 & 0 & 0 & -0.0 & -1.5 & 1.2 & 0.0 & 1.5 & -1.2 \\
-0.0 & -0.0 & -0.0 & -1.8 & -3.0 & -0.0 & -0.0 & -0.0 & 0.0 & 1.8 & 3.0 & 0.0 \\
-0.0 & -1.9 & -1.5 & -3.0 & -5.0 & -0.0 & -0.0 & -1.9 & 1.5 & 3.0 & 8.9 & 0 \\
-0.0 & -1.5 & -1.2 & -0.0 & -0.0 & -0.0 & 0.0 & 1.5 & -1.2 & 0.0 & 0 & 2.4
\end{bmatrix}
$$

$$
\begin{Bmatrix}
U_{1x} \\ U_{1y} \\ U_{1z} \\ U_{2x} \\ U_{2y} \\ U_{2z} \\ U_{3x} \\ U_{3y} \\ U_{3z} \\ U_{4x} \\ U_{4y} \\ U_{4z}
\end{Bmatrix}
=
\begin{Bmatrix}
F_{1x} \\ F_{1y} \\ F_{1z} \\ F_{2x} \\ F_{2y} \\ F_{2z} \\ F_{3x} \\ F_{3y} \\ F_{3z} \\ F_{4x} \\ F_{4y} \\ F_{4z}
\end{Bmatrix}
\tag{6.4}
$$

The boundary conditions for this problem are given as:

$$U_{1x} = U_{1y} = U_{1z} = U_{2x} = U_{2y} = U_{2z} = U_{3x} = U_{3y} = U_{3z} = 0$$
$$F_{4x} = 12, \quad F_{4y} = F_{4z} = 0 \tag{6.5}$$

Inserting the above conditions into (6.4) we obtain:

$$10^4 \begin{bmatrix}
0.0 & 0.0 & 0.0 & 0 & 0 & 0 & 0 & 0 & 0 & -0.0 & -0.0 & -0.0 \\
0.0 & 1.9 & 1.5 & 0 & 0 & 0 & 0 & 0 & 0 & -0.0 & -1.9 & -1.5 \\
0.0 & 1.5 & 1.2 & 0 & 0 & 0 & 0 & 0 & 0 & -0.0 & -1.5 & -1.2 \\
0 & 0 & 0 & 1.8 & 3.0 & 0.0 & 0 & 0 & 0 & -1.8 & -3.0 & -0.0 \\
0 & 0 & 0 & 3.0 & 5.0 & 0.0 & 0 & 0 & 0 & -3.0 & -5.0 & -0.0 \\
0 & 0 & 0 & 0.0 & 0.0 & 0.0 & 0 & 0 & 0 & -0.0 & -0.0 & -0.0 \\
0 & 0 & 0 & 0 & 0 & 0 & 0.0 & 0.0 & -0.0 & -0.0 & -0.0 & -0.0 \\
0 & 0 & 0 & 0 & 0 & 0 & 0.0 & 1.9 & -1.5 & -0.0 & -1.9 & 1.5 \\
0 & 0 & 0 & 0 & 0 & 0 & -0.0 & -1.5 & 1.2 & 0.0 & 1.5 & -1.2 \\
-0.0 & -0.0 & -0.0 & -1.8 & -3.0 & -0.0 & -0.0 & -0.0 & 0.0 & 1.8 & 3.0 & 0.0 \\
-0.0 & -1.9 & -1.5 & -3.0 & -5.0 & -0.0 & -0.0 & -1.9 & 1.5 & 3.0 & 8.9 & 0 \\
-0.0 & -1.5 & -1.2 & -0.0 & -0.0 & -0.0 & 0.0 & 1.5 & -1.2 & 0.0 & 0 & 2.4
\end{bmatrix}$$

$$\begin{Bmatrix}
F_{1x} \\ F_{1y} \\ F_{1z} \\ F_{2x} \\ F_{2y} \\ F_{2z} \\ F_{3x} \\ F_{3y} \\ F_{3z} \\ U_{4x} \\ U_{4y} \\ U_{4z}
\end{Bmatrix} = \begin{Bmatrix}
0 \\ 0 \\ 0 \\ 0 \\ 0 \\ 0 \\ 0 \\ 0 \\ 0 \\ 12 \\ 0 \\ 0
\end{Bmatrix} \tag{6.6}$$

Step 5 – Solving the Equations:

Solving the system of equations in (6.6) will be performed by partitioning (manually) and Gaussian elimination (with MATLAB). First we partition (6.6) by extracting the submatrix in rows 10 to 12 and columns 10 to 12. Therefore we obtain:

$$10^4 \begin{bmatrix} 1.8 & 3.0 & 0.0 \\ 3.0 & 8.9 & 0 \\ 0.0 & 0 & 2.4 \end{bmatrix} \begin{Bmatrix} U_{4x} \\ U_{4y} \\ U_{4z} \end{Bmatrix} = \begin{Bmatrix} 12 \\ 0 \\ 0 \end{Bmatrix} \tag{6.7}$$

The solution of the above system is obtained using MATLAB as follows. Note that the backslash operator "\" is used for Gaussian elimination.

```
» k=K(10:12,10:12)

k =

   1.0e+004 *

           1.8159      3.0264      0.0000
           3.0264      8.8532           0
           0.0000           0      2.4378


» f=[12 ; 0 ; 0]

f =

      12
       0
       0

» u=k\f

u =

      0.0015
     -0.0005
     -0.0000
```

It is now clear that the horizontal displacements at node 4 in the X and Z directions are 0.0015 m and 0 m (due to symmetry), respectively, while the vertical displacement in the Y direction at node 4 is –0.0005 m.

Step 6 – Post-processing:

In this step, we obtain the reactions at nodes 1, 2, and 3, and the stress in each element using MATLAB as follows. First we set up the global nodal displacement vector U, then we calculate the global nodal force vector F.

```
» U=[0 ; 0 ; 0 ; 0 ; 0 ; 0 ; 0 ; 0 ; u]

U =

            0
            0
            0
            0
```

```
         0
         0
         0
         0
         0
    0.0015
   -0.0005
   -0.0000
```

» F=K*U

F =

```
    0.0000
   10.0000
    8.0000
  -12.0000
  -20.0000
   -0.0000
    0.0000
   10.0000
   -8.0000
   12.0000
   -0.0000
   -0.0000
```

Thus the reactions at node 1 are forces of 0, 10, and 8 kN along the X, Y, and Z directions, respectively. The reactions at node 2 are forces of $-12, -20$, and 0 kN along the X, Y, and Z directions, respectively. The reactions at node 3 are forces of 0, 10, and -8 kN along the X, Y, and Z directions, respectively. Obviously force equilibrium is satisfied for this problem. Next we set up the element nodal displacement vectors u_1, u_2, and u_3 then we calculate the element stresses *sigma1*, *sigma2*, and *sigma3* by making calls to the MATLAB function *SpaceTrussElementStress*.

» u1=[U(1) ; U(2) ; U(3) ; U(10) ; U(11) ; U(12)]

u1 =

```
         0
         0
         0
    0.0015
   -0.0005
   -0.0000
```

```
» u2=[U(4) ; U(5) ; U(6) ; U(10) ; U(11) ; U(12)]

u2 =

         0
         0
         0
    0.0015
   -0.0005
   -0.0000

» u3=[U(7) ; U(8) ; U(9) ; U(10) ; U(11) ; U(12)]

u3 =

         0
         0
         0
    0.0015
   -0.0005
   -0.0000

» sigma1=SpaceTrussElementStress(E,L1,theta1x,theta1y,theta1z,u1)

sigma1 =

  -1.2806e+004

» sigma2=SpaceTrussElementStress(E,L2,theta2x,theta2y,theta2z,u2)

sigma2 =

   1.1662e+004

» sigma3=SpaceTrussElementStress(E,L3,theta3x,theta3y,theta3z,u3)

sigma3 =

  -1.2806e+004
```

Thus it is clear that the stresses in elements 1, 2, and 3 are 12.806 MPa (compressive), 11.662 MPa (tensile), and 12.806 MPa (compressive), respectively. The symmetry in the results regarding elements 1 and 3 is obvious.

Problems:

Problem 6.1:

Consider the space truss shown in Fig. 6.3. The supports at nodes 1, 2, 3, and 4 are ball-and-socket joints allowing rotation but no translation. Given $E = 200\,\text{GPa}$, $A = 0.003\,\text{m}^2$, $P_1 = 15\,\text{kN}$, and $P_2 = 20\,\text{kN}$, determine:

1. the global stiffness matrix for the structure.
2. the displacements at node 5.
3. the reactions at nodes 1, 2, 3, and 4.
4. the stress in each element.

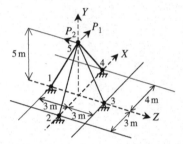

Fig. 6.3. Space Truss with Four Elements for Problem 6.1

7 The Beam Element

7.1
Basic Equations

The beam element is a two-dimensional finite element where the local and global coordinates coincide. It is characterized by linear shape functions. The beam element has modulus of elasticity E, moment of inertia I, and length L. Each beam element has two nodes and is assumed to be horizontal as shown in Fig. 7.1. In this case the element stiffness matrix is given by the following matrix, assuming axial deformation is neglected (see [1]).

$$
k = \frac{EI}{L^3} \begin{bmatrix} 12 & 6L & -12 & 6L \\ 6L & 4L^2 & -6L & 2L^2 \\ -12 & -6L & 12 & -6L \\ 6L & 2L^2 & -6L & 4L^2 \end{bmatrix}
\tag{7.1}
$$

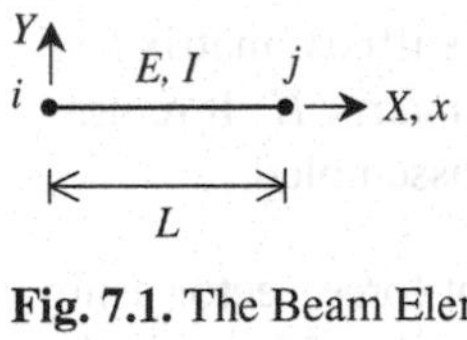

Fig. 7.1. The Beam Element

It is clear that the beam element has four degrees of freedom – two at each node (a transverse displacement and a rotation). The sign convention used is that the displacement is positive if it points upwards and the rotation is positive if it is counterclockwise. Consequently for a structure with n nodes, the global stiffness matrix K will be of size $2n \times 2n$ (since we have two degrees of freedom at each node). The global stiffness matrix K is assembled by making calls to the MATLAB function *BeamAssemble* which is written specifically for this purpose. This process will be illustrated in detail in the examples.

Once the global stiffness matrix K is obtained we have the following structure equation:

$$[K]\{U\} = \{F\} \tag{7.2}$$

where U is the global nodal displacement vector and F is the global nodal force vector. At this step the boundary conditions are applied manually to the vectors U and F. Then the matrix (7.2) is solved by partitioning and Gaussian elimination. Finally once the unknown displacements and reactions are found, the nodal force vector is obtained for each element as follows:

$$\{f\} = [k]\{u\} \tag{7.3}$$

where $\{f\}$ is the 4×1 nodal force vector in the element and u is the 4×1 element displacement vector. The first and second elements in each vector $\{u\}$ are the transverse displacement and rotation, respectively, at the first node, while the third and fourth elements in each vector $\{u\}$ are the transverse displacement and rotation, respectively, at the second node.

7.2
MATLAB Functions Used

The five MATLAB functions used for the beam element are:

BeamElementStiffness(E, I, L) – This function calculates the element stiffness matrix for each beam element with modulus of elasticity E, moment of inertia I, and length L. It returns the 4×4 element stiffness matrix k.

BeamAssemble(K, k, i, j) – This function assembles the element stiffness matrix k of the beam element joining nodes i and j into the global stiffness matrix K. It returns the $2n \times 2n$ global stiffness matrix K every time an element is assembled.

BeamElementForces(k, u) – This function calculates the element force vector using the element stiffness matrix k and the element displacement vector u. It returns the 4×1 element force vector f.

BeamElementShearDiagram(f, L) – This function plots the shear force diagram for the element with nodal force vector f and length L.

BeamElementMomentDiagram(f, L) – This function plots the bending moment diagram for the element with nodal force vector f and length L.

The following is a listing of the MATLAB source code for each function:

```
function y = BeamElementStiffness(E,I,L)
%BeamElementStiffness          This function returns the element
%                              stiffness matrix for a beam
%                              element with modulus of elasticity E,
%                              moment of inertia I, and length L.
%                              The size of the element stiffness
%                              matrix is 4 x 4.
y = E*I/(L*L*L) * [12  6*L -12  6*L ; 6*L 4*L*L -6*L 2*L*L ;
    -12 -6*L 12 -6*L ; 6*L 2*L*L -6*L 4*L*L];
```

```
function y = BeamAssemble(K,k,i,j)
%BeamAssemble                  This function assembles the element stiffness
%                              matrix k of the beam element with nodes
%                              i and j into the global stiffness matrix K.
%                              This function returns the global stiffness
%                              matrix K after the element stiffness matrix
%                              k is assembled.
K(2*i-1,2*i-1) = K(2*i-1,2*i-1) + K(1,1);
K(2*i-1,2*i) = K(2*i-1,2*i) + K(1,2);
K(2*i-1,2*j-1) = K(2*i-1,2*j-1) + K(1,3);
K(2*i-1,2*j) = K(2*i-1,2*j) + K(1,4);
K(2*i,2*i-1) = K(2*i,2*i-1) + K(2,1);
K(2*i,2*i) = K(2*i,2*i) + K(2,2);
K(2*i,2*j-1) = K(2*i,2*j-1) + K(2,3);
K(2*i,2*j) = K(2*i,2*j) + K(2,4);
K(2*j-1,2*i-1) = K(2*j-1,2*i-1) + K(3,1);
K(2*j-1,2*i) = K(2*j-1,2*i) + K(3,2);
K(2*j-1,2*j-1) = K(2*j-1,2*j-1) + K(3,3);
K(2*j-1,2*j) = K(2*j-1,2*j) + K(3,4);
K(2*j,2*i-1) = K(2*j,2*i-1) + K(4,1);
K(2*j,2*i) = K(2*j,2*i) + K(4,2);
K(2*j,2*j-1) = K(2*j,2*j-1) + K(4,3);
K(2*j,2*j) = K(2*j,2*j) + K(4,4);
y = K;
```

```
function y = BeamElementForces(k,u)
%BeamElementForces             This function returns the element nodal force
%                              vector given the element stiffness matrix k
%                              and the element nodal displacement vector u.
y = k * u;
```

```
function y = BeamElementShearDiagram(f, L)
%BeamElementShearDiagram       This function plots the shear force
%                              diagram for the beam element with nodal
%                              force vector f and length L.
x = [0 ; L];
z = [f(1) ; -f(3)];
hold on;
title('Shear Force Diagram');
plot(x,z);
y1 = [0 ; 0];
plot(x,y1,'k')
```

```
function y = BeamElementMomentDiagram(f, L)
%BeamElementMomentDiagram            This function plots the bending moment
%                                    diagram for the beam element with
nodal
%                                    force vector f and length L.
x = [0 ; L];
z = [-f(2) ; f(4)];
hold on;
title('Bending Moment Diagram');
plot(x,z);
y1 = [0 ; 0];
plot(x,y1,'k')
```

Example 7.1:

Consider the beam shown in Fig. 7.2. Given $E = 210\,\text{GPa}, I = 60 \times 10^{-6}\,\text{m}^4$, $P = 20\,\text{kN}$, and $L = 2\,\text{m}$, determine:

1. the global stiffness matrix for the structure.
2. the vertical displacement at node 2.
3. the rotations at nodes 2 and 3.
4. the reactions at nodes 1 and 3.
5. the forces (shears and moments) in each element.
6. the shear force diagram for each element.
7. the bending moment diagram for each element.

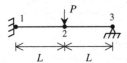

Fig. 7.2. Beam With Two Elements for Example 7.1

Solution:

Use the six steps outlined in Chap. 1 to solve this problem using the beam element.

Step 1 – Discretizing the Domain:

We will put a node (node 2) at the location of the concentrated force so that we may determine the required quantities (displacement, rotation, shear, moment) at that point. Alternatively, we can consider the structure composed of one beam element and two nodes only and use equivalent nodal forces to account for the concentrated load. However, it is required to find the vertical displacement under the concentrated load. Furthermore, the method of equivalent nodal forces will be illustrated for a different problem in Example 7.2. Therefore, the domain is subdivided into two elements and three nodes. The units used in the MATLAB calculations are kN and meter. Table 7.1 shows the element connectivity for this example.

Table 7.1. Element Connectivity for Example 7.1

Element Number	Node i	Node j
1	1	2
2	2	3

Step 2 – Writing the Element Stiffness Matrices:

The two element stiffness matrices k_1 and k_2 are obtained by making calls to the MATLAB function *BeamElementStiffness*. Each matrix has size 4×4.

```
» E=210e6

E =

   210000000

» I=60e-6

I =

   6.0000e-005

» L=2

L =

     2

» k1=BeamElementStiffness(E,I,L)

k1 =

      18900      18900     -18900      18900
      18900      25200     -18900      12600
     -18900     -18900      18900     -18900
      18900      12600     -18900      25200
```

```
» k2=BeamElementStiffness(E,I,L)

k2 =

        18900        18900       -18900        18900
        18900        25200       -18900        12600
       -18900       -18900        18900       -18900
        18900        12600       -18900        25200
```

Step 3 – Assembling the Global Stiffness Matrix:

Since the structure has three nodes, the size of the global stiffness matrix is 6×6. Therefore to obtain K we first set up a zero matrix of size 6×6 then make two calls to the MATLAB function *BeamAssemble* since we have two beam elements in the structure. Each call to the function will assemble one element. The following are the MATLAB commands:

```
» K=zeros(6,6)

K =

        0    0    0    0    0    0
        0    0    0    0    0    0
        0    0    0    0    0    0
        0    0    0    0    0    0
        0    0    0    0    0    0
        0    0    0    0    0    0
```

```
» K=BeamAssemble(K,k1,1,2)

K =

        18900        18900       -18900        18900        0    0
        18900        25200       -18900        12600        0    0
       -18900       -18900        18900       -18900        0    0
        18900        12600       -18900        25200        0    0
            0            0            0            0        0    0
            0            0            0            0        0    0
```

» K=BeamAssemble(K,k2,2,3)

K =

18900	18900	-18900	18900	0	0
18900	25200	-18900	12600	0	0
-18900	-18900	37800	0	-18900	18900
18900	12600	0	50400	-18900	12600
0	0	-18900	-18900	18900	-18900
0	0	18900	12600	-18900	25200

Step 4 – Applying the Boundary Conditions:

The matrix (7.2) for this structure is obtained as follows using the global stiffness matrix obtained in the previous step:

$$10^3 \begin{bmatrix} 18.9 & 18.9 & -18.9 & 18.9 & 0 & 0 \\ 18.9 & 25.2 & -18.9 & 12.6 & 0 & 0 \\ -18.9 & -18.9 & 37.8 & 0 & -18.9 & 18.9 \\ 18.9 & 12.6 & 0 & 50.4 & -18.9 & 12.6 \\ 0 & 0 & -18.9 & -18.9 & 18.9 & -18.9 \\ 0 & 0 & 18.9 & 12.6 & -18.9 & 25.2 \end{bmatrix} \begin{Bmatrix} U_{1y} \\ \phi_1 \\ U_{2y} \\ \phi_2 \\ U_{3y} \\ \phi_3 \end{Bmatrix} = \begin{Bmatrix} F_{1y} \\ M_1 \\ F_{2y} \\ M_2 \\ F_{3y} \\ M_3 \end{Bmatrix} \quad (7.4)$$

The boundary conditions for this problem are given as:

$$U_{1y} = \phi_1 = 0, \quad F_{2y} = -20, \quad M_2 = 0, U_{3y} - 0, \quad M_3 = 0 \quad (7.5)$$

Inserting the above conditions into (7.4) we obtain:

$$10^3 \begin{bmatrix} 18.9 & 18.9 & -18.9 & 18.9 & 0 & 0 \\ 18.9 & 25.2 & -18.9 & 12.6 & 0 & 0 \\ -18.9 & -18.9 & 37.8 & 0 & -18.9 & 18.9 \\ 18.9 & 12.6 & 0 & 50.4 & -18.9 & 12.6 \\ 0 & 0 & -18.9 & -18.9 & 18.9 & -18.9 \\ 0 & 0 & 18.9 & 12.6 & -18.9 & 25.2 \end{bmatrix} \begin{Bmatrix} 0 \\ 0 \\ U_{2y} \\ \phi_2 \\ 0 \\ \phi_3 \end{Bmatrix} = \begin{Bmatrix} F_{1y} \\ M_1 \\ -20 \\ 0 \\ F_{3y} \\ 0 \end{Bmatrix} \quad (7.6)$$

Step 5 – Solving the Equations:

Solving the system of equations in (7.6) will be performed by partitioning (manually) and Gaussian elimination (with MATLAB). First we partition (7.6) by extracting the submatrices in rows 3 to 4 and columns 3 to 4, rows 3 to 4 and column 6, row 6 and columns 3 to 4, and row 6 and column 6. Therefore we obtain:

$$10^3 \begin{bmatrix} 37.8 & 0 & 18.9 \\ 0 & 50.4 & 12.6 \\ 18.9 & 12.6 & 25.2 \end{bmatrix} \begin{Bmatrix} U_{2y} \\ \phi_2 \\ \phi_3 \end{Bmatrix} = \begin{Bmatrix} -20 \\ 0 \\ 0 \end{Bmatrix} \tag{7.7}$$

The solution of the above system is obtained using MATLAB as follows. Note that the backslash operator "\" is used for Gaussian elimination.

```
» k=[K(3:4,3:4) K(3:4,6) ; K(6,3:4) K(6,6)]

k =


      37800           0       18900
          0       50400       12600
      18900       12600       25200


» f=[-20 ; 0 ; 0]

f =

    -20
      0
      0

» u=k\f

u =

   1.0e-003 *

   -0.9259
   -0.1984
    0.7937
```

It is now clear that the vertical displacement at node 2 is 0.9259 m (downward) while the rotations at nodes 2 and 3 are 0.1984 rad (clockwise) and 0.7937 rad (counterclockwise), respectively.

Step 6 – Post-processing:

In this step, we obtain the reactions at nodes 1 and 3, and the forces (shears and moments) in each beam element using MATLAB as follows. First we set up the global nodal displacement vector U, then we calculate the global nodal force vector F.

```
» U=[0 ; 0 ; u(1) ; u(2) ; 0 ; u(3)]

U =

   1.0e-003 *

           0
           0
     -0.9259
     -0.1984
           0
      0.7937

» F=K*U

F =

     13.7500
     15.0000
    -20.0000
           0
      6.2500
     -0.0000
```

Thus the reactions at node 1 are a vertical force of 13.75 kN (upward) and a moment of 15 kN.m (counterclockwise) while the reaction at node 3 is a vertical force of 6.25 kN (upward). It is clear that force equilibrium is satisfied. Next we set up the element nodal displacement vectors u_1 and u_2, then we calculate the element force vectors f_1 and f_2 by making calls to the MATLAB function *BeamElementForces*.

```
» u1=[U(1) ; U(2) ; U(3) ; U(4)]

u1 =

   1.0e-003 *

           0
           0
     -0.9259
     -0.1984

» u2=[U(3) ; U(4) ; U(5) ; U(6)]

u2 =

   1.0e-003 *
```

```
    -0.9259
    -0.1984
         0
     0.7937
```

» f1=BeamElementForces(k1,u1)

f1 =

```
    13.7500
    15.0000
   -13.7500
    12.5000
```

» f2=BeamElementForces(k2,u2)

f2 =

```
    -6.2500
   -12.5000
     6.2500
    -0.0000
```

Thus the shear force and bending moment at each end (node) of each element are given above. Element 1 has a shear force of 13.75 kN and a bending moment of 15 kN.m at its left end while it has a shear force of – 13.75 kN and a bending moment of 12.5 kN.m at its right end. Element 2 has a shear force of – 6.25 kN and a bending moment of – 12.5 kN.m at its left end while it has a shear force of 6.25 kN and a bending moment of 0 kN.m at its right end. Obviously the hinge at the right end has zero moment.

Finally we call the MATLAB functions *BeamElementShearDiagram* and *BeamElementMomentDiagram* to draw the shear force diagram and bending moment diagram, respectively, for each element. This process is illustrated below.

Example 7.2:

Consider the beam shown in Fig. 7.7. Given $E = 210\,\text{GPa}$, $I = 5 \times 10^{-6}\,\text{m}^4$, and $w = 7\,\text{kN/m}$, determine:

1.   the global stiffness matrix for the structure.
2.   the rotations at nodes 1, 2 and 3.
3.   the reactions at nodes 1, 2, 3 and 4.
4.   the forces (shears and moments) in each element.
5.   the shear force diagram for each element.
6.   the bending moment diagram for each element.

» BeamElementShearDiagram(f1,L)

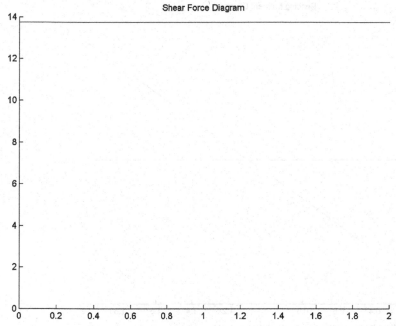

Fig. 7.3. Shear Force Diagram for Element 1

» BeamElementShearDiagram(f2,L)

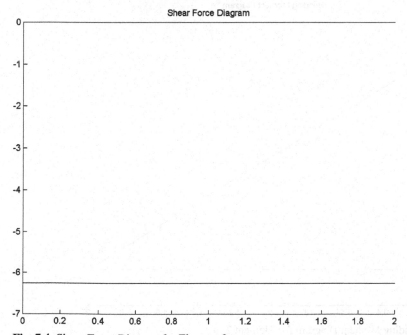

Fig. 7.4. Shear Force Diagram for Element 2

» `BeamElementMomentDiagram(f1,L)`

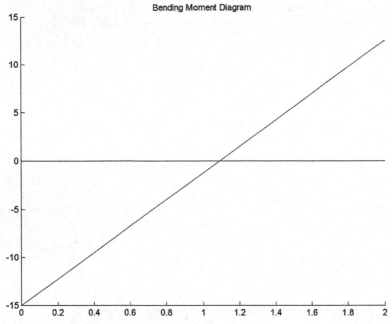

Fig. 7.5. Bending Moment Diagram for Element 1

» `BeamElementMomentDiagram(f2,L)`

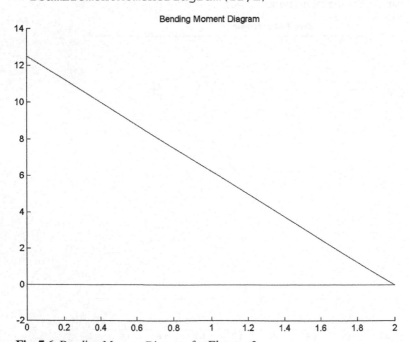

Fig. 7.6. Bending Moment Diagram for Element 2

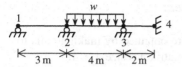

Fig. 7.7. Beam with Distributed Load for Example 7.2

Solution:

Use the six steps outlined in Chap. 1 to solve this problem using the beam element. We need first to replace the distributed loading on element 2 by equivalent nodal loads. This is performed as follows for element 2 with a uniformly distributed load. The resulting beam with equivalent nodal loads is shown in Fig. 7.8. Table 7.2 shows the element connectivity for this example.

$$\{p_2\} = \begin{Bmatrix} -\dfrac{wL}{2} \\ -\dfrac{wL^2}{12} \\ -\dfrac{wL}{2} \\ \dfrac{wL^2}{12} \end{Bmatrix} = \begin{Bmatrix} -14\,kN \\ -9.333\,kN.m \\ -14\,kN \\ 9.333\,kN.m \end{Bmatrix} \tag{7.8}$$

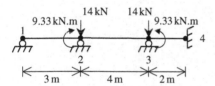

Fig. 7.8. Equivalent nodal loads for example 7.2

Table 7.2. Element Connectivity for Example 7.2

Element Number	Node i	Node j
1	1	2
2	2	3
3	3	4

Step 1 – Discretizing the Domain:

The domain is already subdivided into three elements and four nodes. The units used in the MATLAB calculations are kN and meter.

Step 2 – Writing the Element Stiffness Matrices:

The three element stiffness matrices k_1, k_2 and k_3 are obtained by making calls to the MATLAB function *BeamElementStiffness*. Each matrix has size 4×4.

```
» E=210e6

E =

    210000000

» I=5e-6

I =

    5.0000e-006

» L1=3

L1 =

    3

» L2=4

L2 =

    4

» L3=2

L3 =

    2

» k1=BeamElementStiffness(E,I,L1)

k1 =

   1.0e+003 *

      0.4667      0.7000     -0.4667      0.7000
      0.7000      1.4000     -0.7000      0.7000
     -0.4667     -0.7000      0.4667     -0.7000
      0.7000      0.7000     -0.7000      1.4000
```

» k2=BeamElementStiffness(E,I,L2)

k2 =

  1.0e+003 *

0.1969	0.3937	-0.1969	0.3937
0.3937	1.0500	-0.3937	0.5250
-0.1969	-0.3937	0.1969	-0.3937
0.3937	0.5250	-0.3937	1.0500

» k3=BeamElementStiffness(E,I,L3)

k3 =

1575	1575	-1575	1575
1575	2100	-1575	1050
-1575	-1575	1575	-1575
1575	1050	-1575	2100

Step 3 – Assembling the Global Stiffness Matrix:

Since the structure has four nodes, the size of the global stiffness matrix is 8×8. Therefore to obtain K we first set up a zero matrix of size 8×8 then make three calls to the MATLAB function *BeamAssemble* since we have three beam elements in the structure. Each call to the function will assemble one element. The following are the MATLAB commands:

» K=zeros(8,8)

K =

0	0	0	0	0	0	0	0
0	0	0	0	0	0	0	0
0	0	0	0	0	0	0	0
0	0	0	0	0	0	0	0
0	0	0	0	0	0	0	0
0	0	0	0	0	0	0	0
0	0	0	0	0	0	0	0
0	0	0	0	0	0	0	0

```
» K=BeamAssemble(K,k1,1,2)

K =

   1.0e+003 *

   Columns 1 through 7

      0.4667      0.7000     -0.4667      0.7000      0      0      0
      0.7000      1.4000     -0.7000      0.7000      0      0      0
     -0.4667     -0.7000      0.4667     -0.7000      0      0      0
      0.7000      0.7000     -0.7000      1.4000      0      0      0
           0           0           0           0      0      0      0
           0           0           0           0      0      0      0
           0           0           0           0      0      0      0
           0           0           0           0      0      0      0

   Column 8

           0
           0
           0
           0
           0
           0
           0
           0

» K=BeamAssemble(K,k2,2,3)

K =

   1.0e+003 *

   Columns 1 through 7

      0.4667    0.7000   -0.4667    0.7000           0           0    0
      0.7000    1.4000   -0.7000    0.7000           0           0    0
     -0.4667   -0.7000    0.6635   -0.3063     -0.1969      0.3937    0
      0.7000    0.7000   -0.3063    2.4500     -0.3937      0.5250    0
           0         0   -0.1969   -0.3937      0.1969     -0.3937    0
           0         0    0.3937    0.5250     -0.3937      1.0500    0
           0         0         0         0           0           0    0
           0         0         0         0           0           0    0
```

```
Column 8

         0
         0
         0
         0
         0
         0
         0
         0
```

» K=BeamAssemble(K,k3,3,4)

K =

   1.0e+003 *

   Columns 1 through 7

```
  0.4667   0.7000  -0.4667   0.7000        0        0        0
  0.7000   1.4000  -0.7000   0.7000        0        0        0
 -0.4667  -0.7000   0.6635  -0.3063  -0.1969   0.3937        0
  0.7000   0.7000  -0.3063   2.4500  -0.3937   0.5250        0
       0        0  -0.1969  -0.3937   1.7719   1.1812  -1.5750
       0        0   0.3937   0.5250   1.1812   3.1500  -1.5750
       0        0        0        0  -1.5750  -1.5750   1.5750
       0        0        0        0   1.5750   1.0500  -1.5750
```

   Column 8

```
         0
         0
         0
         0
    1.5750
    1.0500
   -1.5750
    2.1000
```

Step 4 – Applying the Boundary Conditions:

The matrix (7.2) for this structure is obtained as follows using the global stiffness matrix obtained in the previous step. Note that we only show the numbers to two decimal places although the MATLAB calculations are performed using at least four decimal places.

$$
10^3
\begin{bmatrix}
0.47 & 0.70 & -0.47 & 0.70 & 0 & 0 & 0 & 0 \\
0.70 & 1.40 & -0.70 & 0.70 & 0 & 0 & 0 & 0 \\
-0.47 & -0.70 & 0.66 & -0.31 & -0.20 & 0.39 & 0 & 0 \\
0.70 & 0.70 & -0.31 & 2.45 & -0.39 & 0.53 & 0 & 0 \\
0 & 0 & -0.20 & -0.39 & 1.77 & 1.18 & -1.58 & 1.58 \\
0 & 0 & 0.39 & 0.53 & 1.18 & 3.15 & -1.58 & 1.05 \\
0 & 0 & 0 & 0 & -1.58 & -1.58 & 1.58 & -1.58 \\
0 & 0 & 0 & 0 & 1.58 & 1.05 & -1.58 & 2.10
\end{bmatrix}
$$

$$
\begin{Bmatrix}
U_{1y} \\ \phi_1 \\ U_{2y} \\ \phi_2 \\ U_{3y} \\ \phi_3 \\ U_{4y} \\ \phi_4
\end{Bmatrix}
=
\begin{Bmatrix}
F_{1y} \\ M_1 \\ F_{2y} \\ M_2 \\ F_{3y} \\ M_3 \\ F_{4y} \\ M_4
\end{Bmatrix}
\tag{7.9}
$$

The boundary conditions for this problem are given as:

$$
U_{1y} = M_1 = U_{2y} = U_{3y} = U_{4y} = \phi_4 = 0, \quad M_2 = -9.333, \quad M_3 = 9.333 \tag{7.10}
$$

Inserting the above conditions into (7.9) we obtain:

$$
10^3
\begin{bmatrix}
0.47 & 0.70 & -0.47 & 0.70 & 0 & 0 & 0 & 0 \\
0.70 & 1.40 & -0.70 & 0.70 & 0 & 0 & 0 & 0 \\
-0.47 & -0.70 & 0.66 & -0.31 & -0.20 & 0.39 & 0 & 0 \\
0.70 & 0.70 & -0.31 & 2.45 & -0.39 & 0.53 & 0 & 0 \\
0 & 0 & -0.20 & -0.39 & 1.77 & 1.18 & -1.58 & 1.58 \\
0 & 0 & 0.39 & 0.53 & 1.18 & 3.15 & -1.58 & 1.05 \\
0 & 0 & 0 & 0 & -1.58 & -1.58 & 1.58 & -1.58 \\
0 & 0 & 0 & 0 & 1.58 & 1.05 & -1.58 & 2.10
\end{bmatrix}
$$

$$
\begin{Bmatrix}
0 \\ \phi_1 \\ 0 \\ \phi_2 \\ 0 \\ \phi_3 \\ 0 \\ 0
\end{Bmatrix}
=
\begin{Bmatrix}
F_{1y} \\ 0 \\ F_{2y} \\ -9.333 \\ F_{3y} \\ 9.333 \\ F_{4y} \\ M_4
\end{Bmatrix}
\tag{7.11}
$$

Step 5 – Solving the Equations:

Solving the system of equations in (7.11) will be performed by partitioning (manually) and Gaussian elimination (with MATLAB). First we partition (7.11) by extracting the submatrices (scalars in this case) in rows 2, 4, and 6, and columns 2, 4, and 6. Therefore we obtain:

$$
10^3 \begin{bmatrix} 1.40 & 0.70 & 0 \\ 0.70 & 2.45 & 0.53 \\ 0 & 0.53 & 3.15 \end{bmatrix} \begin{Bmatrix} \phi_1 \\ \phi_2 \\ \phi_3 \end{Bmatrix} = \begin{Bmatrix} 0 \\ -9.333 \\ 9.333 \end{Bmatrix} \tag{7.12}
$$

The solution of the above system is obtained using MATLAB as follows. Note that the backslash operator "\" is used for Gaussian elimination.

```
» k=[K(2,2) K(2,4) K(2,6) ; K(4,2) K(4,4) K(4,6) ;
      K(6,2) K(6,4) K(6,6)]

k =

      1400      700        0
       700     2450      525
         0      525     3150


» f=[0 ; -9.333 ; 9.333]

f =

         0
   -9.3330
    9.3330

» u=k\f

u =

    0.0027
   -0.0054
    0.0039
```

It is now clear that the rotations at nodes 1, 2, and 3 are 0.0027 rad (counterclockwise), 0.0054 rad (clockwise), and 0.0039 rad (counterclockwise), respectively.

Step 6 – Post-processing:

In this step, we obtain the reactions at nodes 1, 2, 3, and 4, and the forces (shears and moments) in each beam element using MATLAB as follows. First we set up the global nodal displacement vector U, then we calculate the global nodal force vector F.

```
» U = [0 ; u(1) ; 0 ; u(2) ; 0 ; u(3) ; 0 ; 0 ]

U =

          0
     0.0027
          0
    -0.0054
          0
     0.0039
          0
          0
```

```
» F= K*U

F =

    -1.8937
    -0.0000
     1.2850
    -9.3330
     6.6954
     9.3330
    -6.0867
     4.0578
```

Thus the vertical reactions at nodes 1, 2, 3, and 4 are vertical forces of 1.8937 kN (downward), 1.2850 kN (upward), 6.6954 kN (upward), and 6.0867 kN (downward), respectively. The moment at the fixed support (node 4) is 4.0578 kN.m (counterclockwise). It is clear that force equilibrium is satisfied. Next we set up the element nodal displacement vectors u_1, u_2, and u_3, then we calculate the element force vectors f_1, f_2, and f_3 by making calls to the MATLAB function *BeamElementForces*.

```
» u1=[U(1) ; U(2) ; U(3) ; U(4)]

u1 =

        0
   0.0027
        0
  -0.0054

» u2=[U(3) ; U(4) ; U(5) ; U(6)]

u2 =

        0
  -0.0054
        0
   0.0039

» u3= [U(5) ; U(6) ; U(7) ; U(8)]

u3 =

        0
   0.0039
        0
        0

» f1=BeamElementForces(k1,u1)

f1 =

  -1.8937
  -0.0000
   1.8937
  -5.6810

» f2=BeamElementForces(k2,u2)

f2 =

  -0.6087
  -3.6520
   0.6087
   1.2173
```

```
» f3=BeamElementForces(k3,u3)

f3 =

    6.0867
    8.1157
   -6.0867
    4.0578
```

Note that the forces for element 2 need to be modified because of the distributed load. In order to obtain the correct forces for element 2 we need to subtract from f2 the vector of equivalent nodal loads given in (7.8). This is performed using MATLAB as follows:

```
» f2=f2-[-14 ; -9.333 ; -14 ; 9.333]

f2 =

   13.3913
    5.6810
   14.6087
   -8.1157
```

Therefore the forces for each element are given above. Element 1 has a shear force of -1.8937 kN and a bending moment of 0 kN.m at its left end while it has a shear force of 1.8937 kN and a bending moment of -5.6810 kN.m at its right end. Element 2 has a shear force of 13.3913 kN and a bending moment of 5.6810 kN.m at its left end while it has a shear force of 14.6087 kN and a bending moment of -8.1157 kN.m at its right end. Element 3 has a shear force of 6.0867 kN and a bending moment of 8.1157 kN.m at its left end while it has a shear force of -6.0867 kN and a bending moment of 4.0578 kN.m at its right end. Obviously the roller at the left end has zero moment.

Finally we call the MATLAB functions *BeamElementShearDiagram* and *Beam ElementMomentDiagram* to draw the shear force diagram and bending moment diagram, respectively, for each element. This process is illustrated below.

Problems:

Problem 7.1:

Consider the beam shown in Fig. 7.15. Given $E = 200$ GPa, $I = 70 \times 10^{-5}$ m^4, and $M = 15$ kN.m, determine:

1. the global stiffness matrix for the structure.
2. the rotations at nodes 1, 2 and 3.
3. the reactions at nodes 1, 2, and 3.
4. the forces (shears and moments) in each element.
5. the shear force diagram for each element.
6. the bending moment diagram for each element.

» BeamElementShearDiagram(f1,L1)

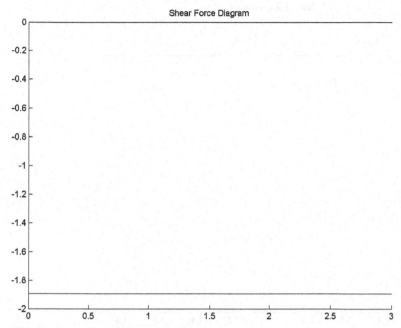

Fig. 7.9. Shear Force Diagram for Element 1

» BeamElementShearDiagram(f2,L2)

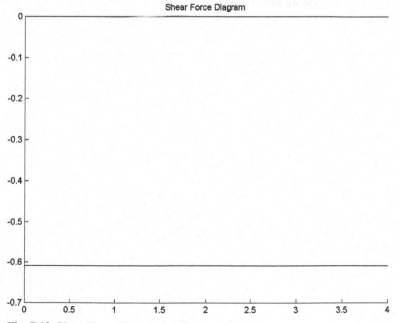

Fig. 7.10. Shear Force Diagram for Element 2

» BeamElementShearDiagram(f3,L3)

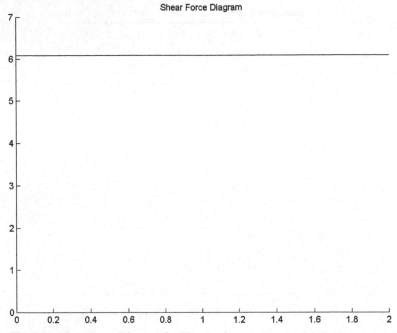

Fig. 7.11. Shear Force Diagram for Element 3

» BeamElementMomentDiagram(f1,L1)

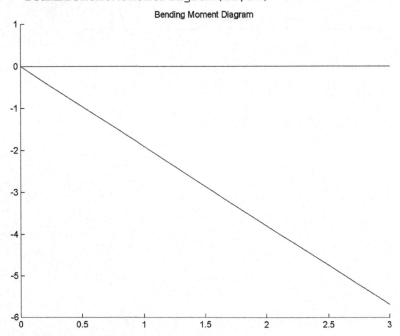

Fig. 7.12. Bending Moment Diagram for Element 1

» `BeamElementMomentDiagram(f2,L2)`

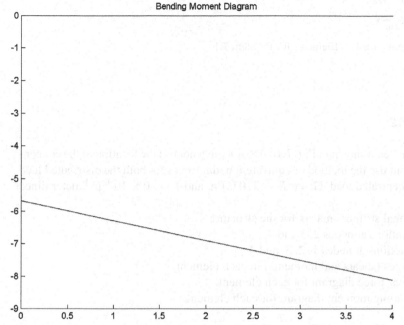

Fig. 7.13. Bending Moment Diagram for Element 2

» `BeamElementMomentDiagram(f3,L3)`

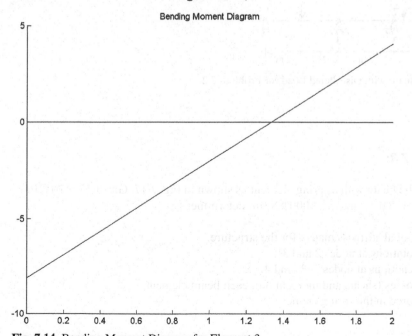

Fig. 7.14. Bending Moment Diagram for Element 3

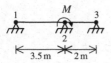

Fig. 7.15. Beam with Two Elements for Problem 7.1

Problem 7.2:

Consider the beam shown in Fig. 7.16. Do not put a node at the location of the concentrated load but use the method of equivalent nodal forces for both the distributed load and the concentrated load. Given $E = 210\,\text{GPa}$, and $I = 50 \times 10^{-6}\,\text{m}^4$, determine:

1. the global stiffness matrix for the structure.
2. the rotations at nodes 2, 3 and 4.
3. the reactions at nodes 1, 2, 3, and 4.
4. the forces (shears and moments) in each element.
5. the shear force diagram for each element.
6. the bending moment diagram for each element.

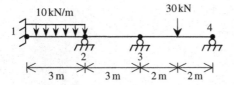

Fig. 7.16. Beam with Distributed Load for Problem 7.2

Problem 7.3:

Consider the beam with a spring element as shown in Fig. 7.17. Given $E = 70\,\text{GPa}$, $I = 40 \times 10^{-6}\,\text{m}^4$, and $k = 5000\,\text{kN/m}$, determine:

1. the global stiffness matrix for the structure.
2. the rotations at nodes 2 and 3.
3. the reactions at nodes 1, 3, and 4.
4. the forces (shears and moments) in each beam element.
5. the force in the spring element.

6. the shear force diagram for each beam element.
7. the bending moment diagram for each beam element.

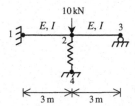

Fig. 7.17. Beam with a Spring for Problem 7.3

8 The Plane Frame Element

8.1
Basic Equations

The plane frame element is a two-dimensional finite element with both local and global coordinates. The plane frame element has modulus of elasticity E, moment of inertia I, cross-sectional area A, and length L. Each plane frame element has two nodes and is inclined with an angle θ measured counterclockwise from the positive global X axis as shown in Fig. 8.1. Let $C = cos\theta$ and $S = sin\theta$. In this case the element stiffness matrix is given by the following matrix including axial deformation (see [1] and [18]).

$$
k = \frac{E}{L} \times
$$

$$
\begin{bmatrix}
AC^2 + \frac{12I}{L^2}S^2 & \left(A - \frac{12I}{L^2}\right)CS & -\frac{6I}{L}S & -\left(AC^2 + \frac{12I}{L^2}S^2\right) & -\left(A - \frac{12I}{L^2}\right)CS & -\frac{6I}{L}S \\
\left(A - \frac{12I}{L^2}\right)CS & AS^2 + \frac{12I}{L^2}C^2 & \frac{6I}{L}C & -\left(A - \frac{12I}{L^2}\right)CS & -\left(AS^2 + \frac{12I}{L^2}C^2\right) & \frac{6I}{L}C \\
-\frac{6I}{L}S & \frac{6I}{L}C & 4I & \frac{6I}{L}S & -\frac{6I}{L}C & 2I \\
-\left(AC^2 + \frac{12I}{L^2}S^2\right) & -\left(A - \frac{12I}{L^2}\right)CS & \frac{6I}{L}S & AC^2 + \frac{12I}{L^2}S^2 & \left(A - \frac{12I}{L^2}\right)CS & \frac{6I}{L}S \\
-\left(A - \frac{12I}{L^2}\right)CS & -\left(AS^2 + \frac{12I}{L^2}C^2\right) & -\frac{6I}{L}C & \left(A - \frac{12I}{L^2}\right)CS & AS^2 + \frac{12I}{L^2}C^2 & -\frac{6I}{L}C \\
-\frac{6I}{L}S & \frac{6I}{L}C & 2I & \frac{6I}{L}S & -\frac{6I}{L}C & 4I
\end{bmatrix}
$$

$$(8.1)$$

It is clear that the plane frame element has six degrees of freedom – three at each node (two displacements and a rotation). The sign convention used is that displacements are positive if they point upwards and rotations are positive if they are counterclockwise. Consequently for a structure with n nodes, the global stiffness matrix K will be of size $3n \times 3n$ (since we have three degrees of freedom at each node).

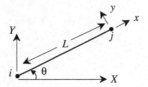

Fig. 8.1. The Plane Frame Element

The global stiffness matrix K is assembled by making calls to the MATLAB function *PlaneFrameAssemble* which is written specifically for this purpose. This process will be illustrated in detail in the examples.

Once the global stiffness matrix K is obtained we have the following structure equation:

$$[K]\{U\} = \{F\} \tag{8.2}$$

where U is the global nodal displacement vector and F is the global nodal force vector. At this step the boundary conditions are applied manually to the vectors U and F. Then the matrix (8.2) is solved by partitioning and Gaussian elimination. Finally once the unknown displacements and reactions are found, the nodal force vector is obtained for each element as follows:

$$\{f\} = [k'][R]\{u\} \tag{8.3}$$

where $\{f\}$ is the 6×1 nodal force vector in the element and u is the 6×1 element displacement vector. The matrices $[k']$ and $[R]$ are given by the following:

$$[k'] = \begin{bmatrix} \frac{EA}{L} & 0 & 0 & -\frac{EA}{L} & 0 & 0 \\ 0 & \frac{12EI}{L^3} & \frac{6EI}{L^2} & 0 & -\frac{12EI}{L^3} & \frac{6EI}{L^2} \\ 0 & \frac{6EI}{L^2} & \frac{4EI}{L} & 0 & -\frac{6EI}{L^2} & \frac{2EI}{L} \\ -\frac{EA}{L} & 0 & 0 & \frac{EA}{L} & 0 & 0 \\ 0 & -\frac{12EI}{L^3} & -\frac{6EI}{L^2} & 0 & \frac{12EI}{L^3} & -\frac{6EI}{L^2} \\ 0 & \frac{6EI}{L^2} & \frac{2EI}{L} & 0 & -\frac{6EI}{L^2} & \frac{4EI}{L} \end{bmatrix} \tag{8.4}$$

$$[R] = \begin{bmatrix} C & S & 0 & 0 & 0 & 0 \\ -S & C & 0 & 0 & 0 & 0 \\ 0 & 0 & 1 & 0 & 0 & 0 \\ 0 & 0 & 0 & C & S & 0 \\ 0 & 0 & 0 & -S & C & 0 \\ 0 & 0 & 0 & 0 & 0 & 1 \end{bmatrix} \tag{8.5}$$

The first and second elements in each vector $\{u\}$ are the two displacements while the third element is the rotation, respectively, at the first node, while the fourth and fifth elements in each vector are the two displacements while the sixth element is the rotation, respectively, at the second node.

If there is an inclined support at one of the nodes of the plane frame then the global stiffness matrix needs to be modified using the following equation:

$$[K]_{new} = [T][K]_{old}[T]^T \tag{8.6}$$

where $[T]$ is a $3n \times 3n$ transformation matrix that is obtained by making a call to the MATLAB function *PlaneFrameInclinedSupport*. The inclined support in a plane frame is handled in the same way it is handled in a plane truss except that the size of the transformation matrix is now $3n \times 3n$ instead of $2n \times 2n$. The inclined support is assumed to be at node i with an angle of inclination *alpha* as shown in Fig. 8.2.

Fig. 8.2. Inclined Support in a Plane Frame

8.2
MATLAB Functions Used

The eight MATLAB functions used for the plane frame element are:

PlaneFrameElementLength(x_1, y_1, x_2, y_2) – This function returns the element length given the coordinates of the first node (x_1, y_1) and the coordinates of the second node (x_2, y_2).

PlaneFrameElementStiffness$(E, A, I, L, theta)$ – This function calculates the element stiffness matrix for each plane frame element with modulus of elasticity E, cross-sectional area A, moment of inertia I, and length L. It returns the 6×6 element stiffness matrix k.

PlaneFrameAssemble(K, k, i, j) – This function assembles the element stiffness matrix k of the plane frame element joining nodes i and j into the global stiffness matrix K. It returns the $3n \times 3n$ global stiffness matrix K every time an element is assembled.

PlaneFrameElementForces($E, A, I, L, theta, u$) – This function calculates the element force vector using the modulus of elasticity E, cross-sectional area A, moment of inertia I, length L, and the element displacement vector u. It returns the 6×1 element force vector f.

PlaneFrameElementAxialDiagram(f, L) – This function plots the axial force diagram for the element with nodal force vector f and length L.

PlaneFrameElementShearDiagram(f, L) – This function plots the shear force diagram for the element with nodal force vector f and length L.

PlaneFrameElementMomentDiagram(f, L) – This function plots the bending moment diagram for the element with nodal force vector f and length L.

PlaneFrameInclinedSupport($T, i, alpha$) – This function calculates the transformation matrix of the inclined support using the node number i of the inclined support and the angle of inclination *alpha* (in degrees). It returns the $3n \times 3n$ transformation matrix.

The following is a listing of the MATLAB source code for each function:

```
function y = PlaneFrameElementLength(x1,y1,x2,y2)
%PlaneFrameElementLength     This function returns the length of the
%                            plane frame element whose first node has
%                            coordinates (x1,y1) and second node has
%                            coordinates (x2,y2).
y = sqrt((x2-x1)*(x2-x1) + (y2-y1)*(y2-y1));
```

```
function y = PlaneFrameElementStiffness(E,A,I,L,theta)
%PlaneFrameElementStiffness  This function returns the element
%                            stiffness matrix for a plane frame
%                            element with modulus of elasticity E,
%                            cross-sectional area A, moment of
%                            inertia I, length L, and angle
%                            theta (in degrees).
%                            The size of the element stiffness
%                            matrix is 6 x 6.
x = theta*pi/180;
C = cos(x);
S = sin(x);
w1 = A*C*C + 12*I*S*S/(L*L);
w2 = A*S*S + 12*I*C*C/(L*L);
w3 = (A-12*I/(L*L))*C*S;
w4 = 6*I*S/L;
w5 = 6*I*C/L;
y = E/L*[w1 w3 -w4 -w1 -w3 -w4 ; w3 w2 w5 -w3 -w2 w5;
   -w4 w5 4*I w4 -w5 2*I ; -w1 -w3 w4 w1 w3 w4;
   -w3 -w2 -w5 w3 w2 -w5 ; -w4 w5 2*I w4 -w5 4*I];
```

```
function y = PlaneFrameAssemble(K,k,i,j)
%PlaneFrameAssemble      This function assembles the element stiffness
%                        matrix k of the plane frame element with nodes
%                        i and j into the global stiffness matrix K.
%                        This function returns the global stiffness
%                        matrix K after the element stiffness matrix
%                        k is assembled.
K(3*i-2,3*i-2) = K(3*i-2,3*i-2) + k(1,1);
K(3*i-2,3*i-1) = K(3*i-2,3*i-1) + k(1,2);
K(3*i-2,3*i) = K(3*i-2,3*i) + k(1,3);
K(3*i-2,3*j-2) = K(3*i-2,3*j-2) + k(1,4);
K(3*i-2,3*j-1) = K(3*i-2,3*j-1) + k(1,5);
K(3*i-2,3*j) = K(3*i-2,3*j) + k(1,6);
K(3*i-1,3*i-2) = K(3*i-1,3*i-2) + k(2,1);
K(3*i-1,3*i-1) = K(3*i-1,3*i-1) + k(2,2);
K(3*i-1,3*i) = K(3*i-1,3*i) + k(2,3);
K(3*i-1,3*j-2) = K(3*i-1,3*j-2) + k(2,4);
K(3*i-1,3*j-1) = K(3*i-1,3*j-1) + k(2,5);
K(3*i-1,3*j) = K(3*i-1,3*j) + k(2,6);
K(3*i,3*i-2) = K(3*i,3*i-2) + k(3,1);
K(3*i,3*i-1) = K(3*i,3*i-1) + k(3,2);
K(3*i,3*i) = K(3*i,3*i) + k(3,3);
K(3*i,3*j-2) = K(3*i,3*j-2) + k(3,4);
K(3*i,3*j-1) = K(3*i,3*j-1) + k(3,5);
K(3*i,3*j) = K(3*i,3*j) + k(3,6);
K(3*j-2,3*i-2) = K(3*j-2,3*i-2) + k(4,1);
K(3*j-2,3*i-1) = K(3*j-2,3*i-1) + k(4,2);
K(3*j-2,3*i) = K(3*j-2,3*i) + k(4,3);
K(3*j-2,3*j-2) = K(3*j-2,3*j-2) + k(4,4);
K(3*j-2,3*j-1) = K(3*j-2,3*j-1) + k(4,5);
K(3*j-2,3*j) = K(3*j-2,3*j) + k(4,6);
K(3*j-1,3*i-2) = K(3*j-1,3*i-2) + k(5,1);
K(3*j-1,3*i-1) = K(3*j-1,3*i-1) + k(5,2);
K(3*j-1,3*i) = K(3*j-1,3*i) + k(5,3);
K(3*j-1,3*j-2) = K(3*j-1,3*j-2) + k(5,4);
K(3*j-1,3*j-1) = K(3*j-1,3*j-1) + k(5,5);
K(3*j-1,3*j) = K(3*j-1,3*j) + k(5,6);
K(3*j,3*i-2) = K(3*j,3*i-2) + k(6,1);
K(3*j,3*i-1) = K(3*j,3*i-1) + k(6,2);
K(3*j,3*i) = K(3*j,3*i) + k(6,3);
K(3*j,3*j-2) = K(3*j,3*j-2) + k(6,4);
K(3*j,3*j-1) = K(3*j,3*j-1) + k(6,5);
K(3*j,3*j) = K(3*j,3*j) + k(6,6);
y = K;
```

```
function y = PlaneFrameElementForces(E,A,I,L,theta,u)
%PlaneFrameElementForces This function returns the element force
%                        vector given the modulus of elasticity E,
%                        the cross-sectional area A, the moment of
%                        inertia I, the length L, the angle theta
%                        (in degrees), and the element nodal
%                        displacement vector u.
```

```
x = theta * pi/180;
C = cos(x);
S = sin(x);
w1 = E*A/L;
w2 = 12*E*I/(L*L*L);
w3 = 6*E*I/(L*L);
w4 = 4*E*I/L;
w5 = 2*E*I/L;
kprime = [w1 0 0 -w1 0 0 ; 0 w2 w3 0 -w2 w3 ;
    0 w3 w4 0 -w3 w5 ; -w1 0 0 w1 0 0 ;
    0 -w2 -w3 0 w2 -w3 ; 0 w3 w5 0 -w3 w4];
T = [C S 0 0 0 0 ; -S C 0 0 0 0 ; 0 0 1 0 0 0 ;
    0 0 0 C S 0 ; 0 0 0 -S C 0 ; 0 0 0 0 0 1];
y = kprime*T* u;
```

```
function y = PlaneFrameElementAxialDiagram(f, L)
%PlaneFrameElementAxialDiagram      This function plots the axial force
%                                   diagram for the plane frame element
%                                   with nodal force vector f and length
%                                   L.
x = [0 ; L];
z = [-f(1) ; f(4)];
hold on;
title('Axial Force Diagram');
plot(x,z);
y1 = [0 ; 0];
plot(x,y1,'k')
```

```
function y = PlaneFrameElementShearDiagram(f, L)
%PlaneFrameElementShearDiagram      This function plots the shear force
%                                   diagram for the plane frame element
%                                   with nodal force vector f and length
%                                   L.
x = [0 ; L];
z = [f(2) ; -f(5)];
hold on;
title('Shear Force Diagram');
plot(x,z);
y1 = [0 ; 0];
plot(x,y1,'k')
```

```
function y = PlaneFrameElementMomentDiagram(f, L)
%PlaneFrameElementMomentDiagram     This function plots the bending
%                                   moment diagram for the plane frame
%                                   element with nodal force vector f
%                                   and length L.
x = [0 ; L];
z = [-f(3) ; f(6)];
```

```
hold on;
title('Bending Moment Diagram');
plot(x,z);
y1 = [0 ; 0];
plot(x,y1,'k')
```

```
function y = PlaneFrameInclinedSupport(T,i,alpha)
%PlaneFrameInclinedSupport      This function calculates the
%                               tranformation matrix T of the inclined
%                               support at node i with angle of
%                               inclination alpha (in degrees).
x = alpha*pi/180;
T(3*i-2,3*i-2) = cos(x);
T(3*i-2,3*i-1) = sin(x);
T(3*i-2,3*i) = 0;
T(3*i-1,3*i-2) = -sin(x);
T(3*i-1,3*i-1) = cos(x);
T(3*i-1,3*i) = 0;
T(3*i,3*i-2) = 0;
T(3*i,3*i-1) = 0;
T(3*i,3*i) = 1;
y = T;
```

Example 8.1:

Consider the plane frame shown in Fig. 8.3. Given $E = 210\,\text{GPa}$, $A = 2 \times 10^{-2}\,\text{m}^2$, and $I = 5 \times 10^{-5}\,\text{m}^4$, determine:

1. the global stiffness matrix for the structure.
2. the displacements and rotations at nodes 2 and 3.
3. the reactions at nodes 1 and 4.
4. the axial force, shear force, and bending moment in each element.
5. the axial force diagram for each element.
6. the shear force diagram for each element.
7. the bending moment diagram for each element.

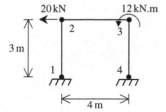

Fig. 8.3. Plane Frame with Three Elements for Example 8.1

Solution:

Use the six steps outlined in Chap. 1 to solve this problem using the plane frame element.

Step 1 – Discretizing the Domain:

This problem is already discretized. The domain is subdivided into three elements and four nodes. The units used in the MATLAB calculations are kN and meter. Table 8.1 shows the element connectivity for this example.

Table 8.1. Element Connectivity for Example 8.1

Element Number	Node i	Node j
1	1	2
2	2	3
3	3	4

Step 2 – Writing the Element Stiffness Matrices:

The three element stiffness matrices k_1, k_2, and k_3 are obtained by making calls to the MATLAB function *PlaneFrameElementStiffness*. Each matrix has size 6×6.

```
» E=210e6

E =

   210000000

» A=2e-2

A =

   0.0200

» I=5e-5

I =

   5.0000e-005
```

```
» L1=3

L1 =

     3

» L2=4

L2 =

     4

» L3=3

L3 =

     3

» k1=PlaneFrameElementStiffness(E,A,I,L1,90)

k1 =

  1.0e+006 *

    0.0047    0.0000   -0.0070   -0.0047   -0.0000   -0.0070
    0.0000    1.4000    0.0000   -0.0000   -1.4000    0.0000
   -0.0070    0.0000    0.0140    0.0070   -0.0000    0.0070
   -0.0047   -0.0000    0.0070    0.0047    0.0000    0.0070
   -0.0000   -1.4000   -0.0000    0.0000    1.4000   -0.0000
   -0.0070    0.0000    0.0070    0.0070   -0.0000    0.0140

» k2=PlaneFrameElementStiffness(E,A,I,L2,0)

k2 =

  1.0e+006 *

    1.0500         0         0   -1.0500         0         0
         0    0.0020    0.0039         0   -0.0020    0.0039
         0    0.0039    0.0105         0   -0.0039    0.0053
   -1.0500         0         0    1.0500         0         0
         0   -0.0020   -0.0039         0    0.0020   -0.0039
         0    0.0039    0.0053         0   -0.0039    0.0105
```

» k3=PlaneFrameElementStiffness(E,A,I,L3,270)

k3 =

    1.0e+006 *

       0.0047   0.0000   0.0070   -0.0047   -0.0000   0.0070
       0.0000   1.4000   -0.0000   -0.0000   -1.4000   -0.0000
       0.0070   -0.0000   0.0140   -0.0070   0.0000   0.0070
      -0.0047   -0.0000   -0.0070   0.0047   0.0000   -0.0070
      -0.0000   -1.4000   0.0000   0.0000   1.4000   0.0000
       0.0070   -0.0000   0.0070   -0.0070   0.0000   0.0140

Step 3 – Assembling the Global Stiffness Matrix:

Since the structure has four nodes, the size of the global stiffness matrix is 12×12.
Therefore to obtain K we first set up a zero matrix of size 12×12 then make three
calls to the MATLAB function *PlaneFrameAssemble* since we have three plane frame
elements in the structure. Each call to the function will assemble one element. The
following are the MATLAB commands:

» K=zeros(12,12)

K =

     0   0   0   0   0   0   0   0   0   0   0   0
     0   0   0   0   0   0   0   0   0   0   0   0
     0   0   0   0   0   0   0   0   0   0   0   0
     0   0   0   0   0   0   0   0   0   0   0   0
     0   0   0   0   0   0   0   0   0   0   0   0
     0   0   0   0   0   0   0   0   0   0   0   0
     0   0   0   0   0   0   0   0   0   0   0   0
     0   0   0   0   0   0   0   0   0   0   0   0
     0   0   0   0   0   0   0   0   0   0   0   0
     0   0   0   0   0   0   0   0   0   0   0   0
     0   0   0   0   0   0   0   0   0   0   0   0
     0   0   0   0   0   0   0   0   0   0   0   0

» K=PlaneFrameAssemble(K,k1,1,2)

K =

    1.0e+006 *

    Columns 1 through 7

```
  0.0047    0.0000   -0.0070   -0.0047   -0.0000   -0.0070   0
  0.0000    1.4000    0.0000   -0.0000   -1.4000    0.0000   0
 -0.0070    0.0000    0.0140    0.0070   -0.0000    0.0070   0
 -0.0047   -0.0000    0.0070    0.0047    0.0000    0.0070   0
 -0.0000   -1.4000   -0.0000    0.0000    1.4000   -0.0000   0
 -0.0070    0.0000    0.0070    0.0070   -0.0000    0.0140   0
       0         0         0         0         0         0   0
       0         0         0         0         0         0   0
       0         0         0         0         0         0   0
       0         0         0         0         0         0   0
       0         0         0         0         0         0   0
       0         0         0         0         0         0   0
```

Columns 8 through 12

```
       0         0         0         0         0
       0         0         0         0         0
       0         0         0         0         0
       0         0         0         0         0
       0         0         0         0         0
       0         0         0         0         0
       0         0         0         0         0
       0         0         0         0         0
       0         0         0         0         0
       0         0         0         0         0
       0         0         0         0         0
       0         0         0         0         0
```

» K=PlaneFrameAssemble(K,k2,2,3)

K =

1.0e+006 *

Columns 1 through 7

```
  0.0047    0.0000   -0.0070   -0.0047   -0.0000   -0.0070          0
  0.0000    1.4000    0.0000   -0.0000   -1.4000    0.0000          0
 -0.0070    0.0000    0.0140    0.0070   -0.0000    0.0070          0
 -0.0047   -0.0000    0.0070    1.0547    0.0000    0.0070   -1.0500
 -0.0000   -1.4000   -0.0000    0.0000    1.4020    0.0039          0
```

```
-0.0070   0.0000   0.0070    0.0070    0.0039    0.0245         0
       0        0        0   -1.0500         0         0    1.0500
       0        0        0         0   -0.0020   -0.0039         0
       0        0        0         0    0.0039    0.0053         0
       0        0        0         0         0         0         0
       0        0        0         0         0         0         0
       0        0        0         0         0         0         0
```

Columns 8 through 12

```
        0              0         0         0         0
        0              0         0         0         0
        0              0         0         0         0
        0              0         0         0         0
  -0.0020         0.0039         0         0         0
  -0.0039         0.0053         0         0         0
        0              0         0         0         0
   0.0020        -0.0039         0         0         0
  -0.0039         0.0105         0         0         0
        0              0         0         0         0
        0              0         0         0         0
        0              0         0         0         0
```

» K=PlaneFrameAssemble(K,k3,3,4)

K =

1.0e+006 *

Columns 1 through 7

```
  0.0047   0.0000  -0.0070  -0.0047  -0.0000  -0.0070         0
  0.0000   1.4000   0.0000  -0.0000  -1.4000   0.0000         0
 -0.0070   0.0000   0.0140   0.0070  -0.0000   0.0070         0
 -0.0047  -0.0000   0.0070   1.0547   0.0000   0.0070   -1.0500
 -0.0000  -1.4000  -0.0000   0.0000   1.4020   0.0039         0
 -0.0070   0.0000   0.0070   0.0070   0.0039   0.0245         0
       0        0        0  -1.0500         0         0    1.0547
       0        0        0        0   -0.0020   -0.0039    0.0000
```

0	0	0	0	0.0039	0.0053	0.0070
0	0	0	0	0	0	-0.0047
0	0	0	0	0	0	-0.0000
0	0	0	0	0	0	0.0070

Columns 8 through 12

0	0	0	0	0
0	0	0	0	0
0	0	0	0	0
0	0	0	0	0
-0.0020	0.0039	0	0	0
-0.0039	0.0053	0	0	0
0.0000	0.0070	-0.0047	-0.0000	0.0070
1.4020	-0.0039	-0.0000	-1.4000	-0.0000
-0.0039	0.0245	-0.0070	0.0000	0.0070
-0.0000	-0.0070	0.0047	0.0000	-0.0070
-1.4000	0.0000	0.0000	1.4000	0.0000
-0.0000	0.0070	-0.0070	0.0000	0.0140

Step 4 – Applying the Boundary Conditions:

The matrix (8.2) for this structure is obtained as follows using the global stiffness matrix obtained in the previous step:

$$
10^3
\begin{bmatrix}
4.7 & 0.0 & -7.0 & -4.7 & -0.0 & -7.0 & 0 & 0 & 0 & 0 & 0 & 0 \\
0.0 & 1400.0 & 0.0 & -0.0 & -1400.0 & 0.0 & 0 & 0 & 0 & 0 & 0 & 0 \\
-7.0 & 0.0 & 14.0 & 7.0 & -0.0 & 7.0 & 0 & 0 & 0 & 0 & 0 & 0 \\
-4.7 & -0.0 & 7.0 & 1054.7 & 0.0 & 7.0 & -1050.0 & 0 & 0 & 0 & 0 & 0 \\
-0.0 & -1400.0 & -0.0 & 0.0 & 1402.0 & 3.9 & 0 & -2.0 & 3.9 & 0 & 0 & 0 \\
-7.0 & 0.0 & 7.0 & 7.0 & 3.9 & 24.5 & 0 & -3.9 & 5.3 & 0 & 0 & 0 \\
0 & 0 & 0 & -1050.0 & 0 & 0 & 1054.7 & 0.0 & 7.0 & -4.7 & -0.0 & 7.0 \\
0 & 0 & 0 & 0 & -2.0 & -3.9 & 0.0 & 1402.0 & -3.9 & -0.0 & -1400.0 & -0.0 \\
0 & 0 & 0 & 0 & 3.9 & 5.3 & 7.0 & -3.9 & 24.5 & -7.0 & 0.0 & 7.0 \\
0 & 0 & 0 & 0 & 0 & 0 & -4.7 & -0.0 & -7.0 & 4.7 & 0.0 & -7.0 \\
0 & 0 & 0 & 0 & 0 & 0 & -0.0 & -1400.0 & 0.0 & 0.0 & 1400.0 & 0.0 \\
0 & 0 & 0 & 0 & 0 & 0 & 7.0 & -0.0 & 7.0 & -7.0 & 0.0 & 14.0
\end{bmatrix}
$$

$$
\begin{Bmatrix}
U_{1x} \\ U_{1y} \\ \phi_1 \\ U_{2x} \\ U_{2y} \\ \phi_2 \\ U_{3x} \\ U_{3y} \\ \phi_3 \\ U_{4x} \\ U_{4y} \\ \phi_4
\end{Bmatrix}
=
\begin{Bmatrix}
F_{1x} \\ F_{1y} \\ M_1 \\ F_{2x} \\ F_{2y} \\ M_2 \\ F_{3x} \\ F_{3y} \\ M_3 \\ F_{4x} \\ F_{4y} \\ M_4
\end{Bmatrix}
\tag{8.7}
$$

The boundary conditions for this problem are given as:

$$U_{1x} = U_{1y} = \phi_1 = U_{4x} = U_{4y} = \phi_4 = 0$$
$$F_{2x} = -20, \quad F_{2y} = M_2 = F_{3x} = F_{3y} = 0, \quad M_3 = 12 \tag{8.8}$$

Inserting the above conditions into (8.7) we obtain:

$$10^3 \begin{bmatrix}
4.7 & 0.0 & -7.0 & -4.7 & -0.0 & -7.0 & 0 & 0 & 0 & 0 & 0 & 0 \\
0.0 & 1400.0 & 0.0 & -0.0 & -1400.0 & 0.0 & 0 & 0 & 0 & 0 & 0 & 0 \\
-7.0 & 0.0 & 14.0 & 7.0 & -0.0 & 7.0 & 0 & 0 & 0 & 0 & 0 & 0 \\
-4.7 & -0.0 & 7.0 & 1054.7 & 0.0 & 7.0 & -1050.0 & 0 & 0 & 0 & 0 & 0 \\
-0.0 & -1400.0 & -0.0 & 0.0 & 1402.0 & 3.9 & 0 & -2.0 & 3.9 & 0 & 0 & 0 \\
-7.0 & 0.0 & 7.0 & 7.0 & 3.9 & 24.5 & 0 & -3.9 & 5.3 & 0 & 0 & 0 \\
0 & 0 & 0 & -1050.0 & 0 & 0 & 1054.7 & 0.0 & 7.0 & -4.7 & -0.0 & 7.0 \\
0 & 0 & 0 & 0 & -2.0 & -3.9 & 0.0 & 1402.0 & -3.9 & -0.0 & -1400.0 & -0.0 \\
0 & 0 & 0 & 0 & 3.9 & 5.3 & 7.0 & -3.9 & 24.5 & -7.0 & 0.0 & 7.0 \\
0 & 0 & 0 & 0 & 0 & 0 & -4.7 & -0.0 & -7.0 & 4.7 & 0.0 & -7.0 \\
0 & 0 & 0 & 0 & 0 & 0 & -0.0 & -1400.0 & 0.0 & 0.0 & 1400.0 & 0.0 \\
0 & 0 & 0 & 0 & 0 & 0 & 7.0 & -0.0 & 7.0 & -7.0 & 0.0 & 14.0
\end{bmatrix}$$

$$\begin{Bmatrix}
0 \\ 0 \\ 0 \\ U_{2x} \\ U_{2y} \\ \phi_2 \\ U_{3x} \\ U_{3y} \\ \phi_3 \\ 0 \\ 0 \\ 0
\end{Bmatrix} = \begin{Bmatrix}
F_{1x} \\ F_{1y} \\ M_1 \\ -20 \\ 0 \\ 0 \\ 0 \\ 0 \\ 12 \\ F_{4x} \\ F_{4y} \\ M_4
\end{Bmatrix} \tag{8.9}$$

Step 5 – Solving the Equations:

Solving the system of equations in (8.9) will be performed by partitioning (manually) and Gaussian elimination (with MATLAB). First we partition (8.9) by extracting the submatrix in rows 4 to 9 and columns 4 to 9. Therefore we obtain:

$$10^3 \begin{bmatrix}
1054.7 & 0.0 & 7.0 & -1050.0 & 0 & 0 \\
0.0 & 1402.0 & 3.9 & 0 & -2.0 & 3.9 \\
7.0 & 3.9 & 24.5 & 0 & -3.9 & 5.3 \\
-1050.0 & 0 & 0 & 1054.7 & 0.0 & 7.0 \\
0 & -2.0 & -3.9 & 0.0 & 1402.0 & -3.9 \\
0 & 3.9 & 5.3 & 7.0 & -3.9 & 24.5
\end{bmatrix} \begin{Bmatrix}
U_{2x} \\ U_{2y} \\ \phi_2 \\ U_{3x} \\ U_{3y} \\ \phi_3
\end{Bmatrix} = \begin{Bmatrix}
-20 \\ 0 \\ 0 \\ 0 \\ 0 \\ 12
\end{Bmatrix} \tag{8.10}$$

The solution of the above system is obtained using MATLAB as follows. Note that the backslash operator "\" is used for Gaussian elimination.

```
» k=K(4:9,4:9)

k =

 1.0e+006 *

   1.0547    0.0000    0.0070   -1.0500         0         0
   0.0000    1.4020    0.0039         0   -0.0020    0.0039
   0.0070    0.0039    0.0245         0   -0.0039    0.0053
  -1.0500         0         0    1.0547    0.0000    0.0070
        0   -0.0020   -0.0039    0.0000    1.4020   -0.0039
        0    0.0039    0.0053    0.0070   -0.0039    0.0245


» f=[-20 ; 0 ; 0 ; 0 ; 0 ; 12]

f =

   -20
     0
     0
     0
     0
    12


» u=k\f

u =

   -0.0038
   -0.0000
    0.0008
   -0.0038
    0.0000
    0.0014
```

It is now clear that the horizontal and vertical displacements at node 2 are −0.0038 m and −0.0 m, respectively, the horizontal and vertical displacements at node 3 are −0.0038 m and 0.0 m, respectively, and the rotations at nodes 2 and 3 are 0.0008 rad (counterclockwise) and 0.0014 rad (counterclockwise), respectively.

Step 6 – Post-processing:

In this step, we obtain the reactions at nodes 1 and 4, and the forces (axial forces, shears and moments) in each plane frame element using MATLAB as follows. First we set up the global nodal displacement vector U, then we calculate the global nodal force vector F.

```
» U=[0 ; 0 ; 0 ; u ; 0 ; 0 ; 0]

U =

         0
         0
         0
   -0.0038
   -0.0000
    0.0008
   -0.0038
    0.0000
    0.0014
         0
         0
         0
```

```
» F=K*U

F =

   12.1897
    8.5865
  -21.0253
  -20.0000
   -0.0000
         0
   -0.0000
         0
   12.0000
    7.8103
   -8.5865
  -16.6286
```

Thus the horizontal reactions at nodes 1 and 4 are forces of 12.1897 kN (to the right) and 7.8103 kN (to the right), respectively, the vertical reactions at nodes 1 and 4 are forces of 8.5865 kN (upward) and 8.5865 kN (downward), respectively, and the moments at nodes 1 and 4 are moments of 21.0253 kN.m (clockwise) and

16.6286 kN.m (clockwise), respectively. It is clear that force equilibrium is satisfied. Next we set up the element nodal displacement vectors u_1, u_2, and u_3, then we calculate the element force vectors f_1, f_2, and f_3 by making calls to the MATLAB function *PlaneFrameElementForces*.

```
» u1=[U(1) ; U(2) ; U(3) ; U(4) ; U(5) ; U(6)]

u1 =

         0
         0
         0
   -0.0038
   -0.0000
    0.0008

» u2=[U(4) ; U(5) ; U(6) ; U(7) ; U(8) ; U(9)]

u2 =

   -0.0038
   -0.0000
    0.0008
   -0.0038
    0.0000
    0.0014

» u3=[U(7) ; U(8) ; U(9) ; U(10) ; U(11) ; U(12)]

u3 =

   -0.0038
    0.0000
    0.0014
         0
         0
         0

» f1=PlaneFrameElementForces(E,A,I,L1,90,u1)

f1 =

    8.5865
  -12.1897
```

```
     -21.0253
      -8.5865
      12.1897
     -15.5438
```

» f2=PlaneFrameElementForces(E,A,I,L2,0,u2)

f2 =

```
      -7.8103
       8.5865
      15.5438
       7.8103
      -8.5865
      18.8023
```

» f3=PlaneFrameElementForces(E,A,I,L3,270,u3)

f3 =

```
      -8.5865
      -7.8103
      -6.8023
       8.5865
       7.8103
     -16.6286
```

Therefore the forces for each element are given above. Element 1 has an axial force of 8.5865 kN, a shear force of – 12.1897 kN and a bending moment of –21.0253 kN.m at its left end while it has an axial force of –8.5865 kN, a shear force of 12.1897 kN and a bending moment of –15.5438 kN.m at its right end. Element 2 has an axial force of –7.8103 kN, a shear force of 8.5865 kN and a bending moment of 15.5438 kN.m at its left end while it has an axial force of 7.8103 kN, a shear force of –8.5865 kN and a bending moment of 18.8023 kN.m at its right end. Element 3 has an axial force of –8.5865 kN, a shear force of –7.8103 kN and a bending moment of –6.8023 kN.m at its left end while it has an axial force of 8.5865 kN, a shear force of 7.8103 kN and a bending moment of –16.6286 kN.m at its right end.

Finally we call the MATLAB functions *PlaneFrameElementAxialDiagram*, *PlaneFrameElementShearDiagram* and *PlaneFrameElementMomentDiagram* to draw the axial force diagram, shear force diagram and bending moment diagram, respectively, for each element. This process is illustrated below.

» `PlaneFrameElementAxialDiagram(f1,L1)`

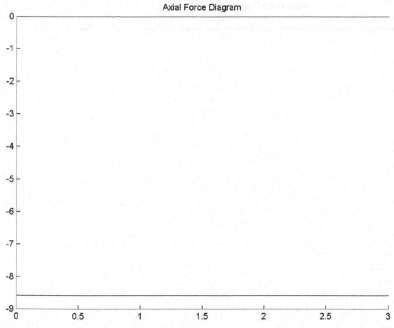

Fig. 8.4. Axial Force Diagram for Element 1

» `PlaneFrameElementAxialDiagram(f2,L2)`

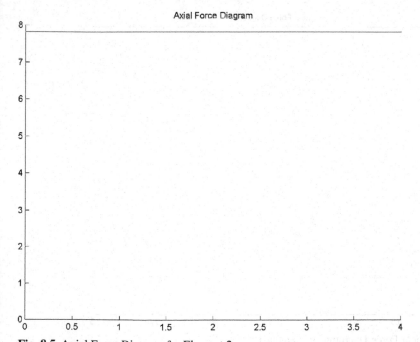

Fig. 8.5. Axial Force Diagram for Element 2

» `PlaneFrameElementAxialDiagram(f3,L3)`

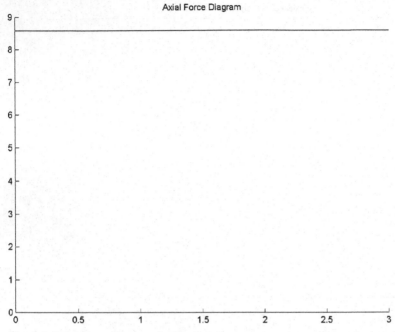

Fig. 8.6. Axial Force Diagram for Element 3

» `PlaneFrameElementShearDiagram(f1,L1)`

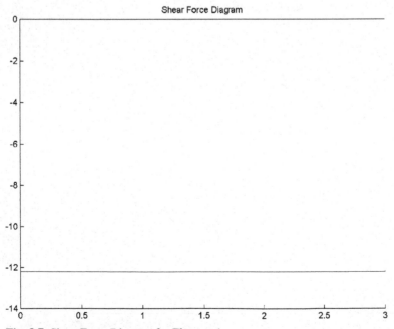

Fig. 8.7. Shear Force Diagram for Element 1

» PlaneFrameElementShearDiagram(f2,L2)

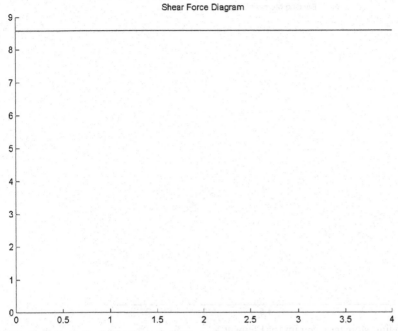

Fig. 8.8. Shear Force Diagram for Element 2

» PlaneFrameElementShearDiagram(f3,L3)

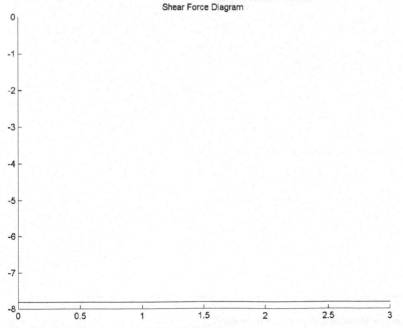

Fig. 8.9. Shear Force Diagram for Element 3

» PlaneFrameElementMomentDiagram(f1,L1)

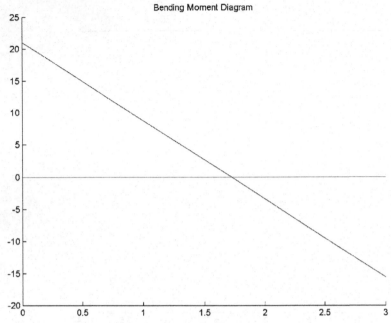

Fig. 8.10. Bending Moment Diagram for Element 1

» PlaneFrameElementMomentDiagram(f2,L2)

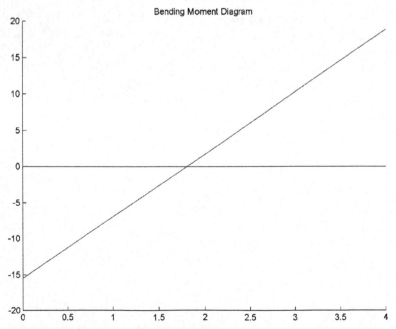

Fig. 8.11. Bending Moment Diagram for Element 2

» `PlaneFrameElementMomentDiagram(f3,L3)`

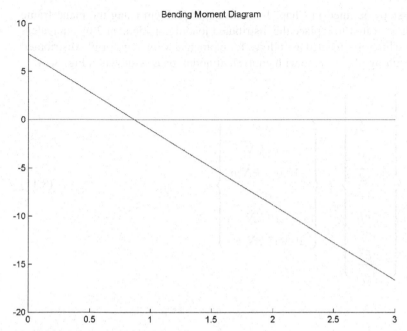

Fig. 8.12. Bending Moment Diagram for Element 3

Example 8.2:

Consider the plane frame shown in Fig. 8.13. Given $E = 200\,\text{GPa}$, $A = 4 \times 10^{-2}\,\text{m}^2$, and $I = 1 \times 10^{-6}\,\text{m}^4$, determine:

1. the global stiffness matrix for the structure.
2. the displacements and rotation at node 2.
3. the reactions at nodes 1 and 3.
4. the axial force, shear force, and bending moment in each element.
5. the axial force diagram for each element.
6. the shear force diagram for each element.
7. the bending moment diagram for each element.

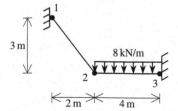

Fig. 8.13. Plane Frame with Distributed Load for Example 8.2

Solution:

Use the six steps outlined in Chap. 1 to solve this problem using the plane frame element. We need first to replace the distributed loading on element 2 by equivalent nodal loads. This is performed as follows for element 2 with a uniformly distributed load. The resulting plane frame with equivalent nodal loads is shown in Fig. 8.14.

$$\{p_2\} = \left\{ \begin{array}{c} 0 \\ -\dfrac{wL}{2} \\ -\dfrac{wL^2}{12} \\ 0 \\ -\dfrac{wL}{2} \\ \dfrac{wL^2}{12} \end{array} \right\} = \left\{ \begin{array}{c} 0 \\ -16\,kN \\ -10.667\,kN.m \\ 0 \\ -16\,kN \\ 10.667\,kN.m \end{array} \right\} \tag{8.11}$$

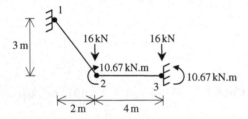

Fig. 8.14. Equivalent Nodal Loads for Example 8.2

Step 1 – Discretizing the Domain:

The domain is already subdivided into two elements and three nodes. The units used in the MATLAB calculations are kN and meter. Table 8.2 shows the element connectivity for this example.

Table 8.2. Element Connectivity for Example 8.2

Element Number	Node i	Node j
1	1	2
2	2	3

Step 2 – Writing the Element Stiffness Matrices:

The two element stiffness matrices k_1 and k_2 are obtained by making calls to the MATLAB function *PlaneFrameElementStiffness*. Each matrix has size 6×6.

```
» E=200e6

E =

   200000000

» A=4e-2

A =

   0.0400

» I=1e-6

I =

   1.0000e-006

» L1=PlaneFrameElementLength(0,3,2,0)

L1 =

   3.6056

» L2=4

L2 =

   4

» theta1=360-atan(3/2)*180/pi

theta1 =

   303.6901

» theta2=0

theta2 =

   0
```

```
» k1=PlaneFrameElementStiffness(E,A,I,L1,theta1)

k1 =

  1.0e+006 *

      0.6827   -1.0240    0.0001   -0.6827    1.0240    0.0001
     -1.0240    1.5361    0.0001    1.0240   -1.5361    0.0001
      0.0001    0.0001    0.0002   -0.0001   -0.0001    0.0001
     -0.6827    1.0240   -0.0001    0.6827   -1.0240   -0.0001
      1.0240   -1.5361   -0.0001   -1.0240    1.5361   -0.0001
      0.0001    0.0001    0.0001   -0.0001   -0.0001    0.0002


» k2=PlaneFrameElementStiffness(E,A,I,L2,theta2)

k2 =

  1.0e+006 *

      2.0000         0         0   -2.0000         0         0
           0    0.0000    0.0001         0   -0.0000    0.0001
           0    0.0001    0.0002         0   -0.0001    0.0001
     -2.0000         0         0    2.0000         0         0
           0   -0.0000   -0.0001         0    0.0000   -0.0001
           0    0.0001    0.0001         0   -0.0001    0.0002
```

Step 3 – Assembling the Global Stiffness Matrix:

Since the structure has three nodes, the size of the global stiffness matrix is 9×9. Therefore to obtain K we first set up a zero matrix of size 9×9 then make two calls to the MATLAB function *PlaneFrameAssemble* since we have two plane frame elements in the structure. Each call to the function will assemble one element. The following are the MATLAB commands:

```
» K=zeros(9,9)

K =

     0     0     0     0     0     0     0     0     0
     0     0     0     0     0     0     0     0     0
     0     0     0     0     0     0     0     0     0
     0     0     0     0     0     0     0     0     0
```

```
        0      0      0      0      0      0      0      0      0
        0      0      0      0      0      0      0      0      0
        0      0      0      0      0      0      0      0      0
        0      0      0      0      0      0      0      0      0
        0      0      0      0      0      0      0      0      0
```

» K=PlaneFrameAssemble(K,k1,1,2)

K =

  1.0e+006 *

  Columns 1 through 7

```
    0.6827  -1.0240   0.0001  -0.6827   1.0240   0.0001        0
   -1.0240   1.5361   0.0001   1.0240  -1.5361   0.0001        0
    0.0001   0.0001   0.0002  -0.0001  -0.0001   0.0001        0
   -0.6827   1.0240  -0.0001   0.6827  -1.0240  -0.0001        0
    1.0240  -1.5361  -0.0001  -1.0240   1.5361  -0.0001        0
    0.0001   0.0001   0.0001  -0.0001  -0.0001   0.0002        0
         0        0        0        0        0        0        0
         0        0        0        0        0        0        0
         0        0        0        0        0        0        0
```

  Columns 8 through 9

```
        0      0
        0      0
        0      0
        0      0
        0      0
        0      0
        0      0
        0      0
        0      0
```

» K=PlaneFrameAssemble(K,k2,2,3)

K =

  1.0e+006 *

```
Columns 1 through 7


    0.6827   -1.0240    0.0001   -0.6827    1.0240    0.0001         0
   -1.0240    1.5361    0.0001    1.0240   -1.5361    0.0001         0
    0.0001    0.0001    0.0002   -0.0001   -0.0001    0.0001         0
   -0.6827    1.0240   -0.0001    2.6827   -1.0240   -0.0001   -2.0000
    1.0240   -1.5361   -0.0001   -1.0240    1.5361    0.0000         0
    0.0001    0.0001    0.0001   -0.0001    0.0000    0.0004         0
         0         0         0   -2.0000         0         0    2.0000
         0         0         0         0   -0.0000   -0.0001         0
         0         0         0         0    0.0001    0.0001         0


Columns 8 through 9


         0                   0
         0                   0
         0                   0
         0                   0
   -0.0000              0.0001
   -0.0001              0.0001
         0                   0
    0.0000             -0.0001
   -0.0001              0.0002
```

Step 4 – Applying the Boundary Conditions:

The matrix (8.2) for this structure is obtained as follows using the global stiffness
matrix obtained in the previous step. Note that we only show the numbers to two
decimal places although the MATLAB calculations are performed using at least four
decimal places.

$$
10^3
\begin{bmatrix}
682.7 & -1024.0 & 0.1 & -682.7 & 1024.0 & 0.1 & 0 & 0 & 0 \\
-1024.0 & 1536.1 & 0.1 & 1024.0 & -1536.1 & 0.1 & 0 & 0 & 0 \\
0.1 & 0.1 & 0.2 & -0.1 & -0.1 & 0.1 & 0 & 0 & 0 \\
-682.7 & 1024.0 & -0.1 & 2682.7 & -1024.0 & -0.1 & -2000.0 & 0 & 0 \\
1024.0 & -1536.1 & -0.1 & -1024.0 & 1536.1 & 0.0 & 0 & -0.0 & 0.1 \\
0.1 & 0.1 & 0.1 & -0.1 & 0.0 & 0.4 & 0 & -0.1 & 0.1 \\
0 & 0 & 0 & -2000.0 & 0 & 0 & 2000.0 & 0 & 0 \\
0 & 0 & 0 & 0 & -0.0 & -0.1 & 0 & 0.0 & -0.1 \\
0 & 0 & 0 & 0 & 0.1 & 0.1 & 0 & -0.1 & 0.2
\end{bmatrix}
$$

$$
\begin{Bmatrix} U_{1x} \\ U_{1y} \\ \phi_1 \\ U_{2x} \\ U_{2y} \\ \phi_2 \\ U_{3x} \\ U_{3y} \\ \phi_3 \end{Bmatrix} = \begin{Bmatrix} F_{1x} \\ F_{1y} \\ M_1 \\ F_{2x} \\ F_{2y} \\ M_2 \\ F_{3x} \\ F_{3y} \\ M_3 \end{Bmatrix}
\tag{8.12}
$$

The boundary conditions for this problem are given as:

$$
U_{1x} = U_{1y} = \phi_1 = U_{3x} = U_{3y} = \phi_3 = 0
$$
$$
F_{2x} = 0, \ F_{2y} = -16, \ M_2 = -10.667
\tag{8.13}
$$

Inserting the above conditions into (8.12) we obtain:

$$
10^3 \begin{bmatrix}
682.7 & -1024.0 & 0.1 & -682.7 & 1024.0 & 0.1 & 0 & 0 & 0 \\
-1024.0 & 1536.1 & 0.1 & 1024.0 & -1536.1 & 0.1 & 0 & 0 & 0 \\
0.1 & 0.1 & 0.2 & -0.1 & -0.1 & 0.1 & 0 & 0 & 0 \\
-682.7 & 1024.0 & -0.1 & 2682.7 & -1024.0 & -0.1 & -2000.0 & 0 & 0 \\
1024.0 & -1536.1 & -0.1 & -1024.0 & 1536.1 & 0.0 & 0 & -0.0 & 0.1 \\
0.1 & 0.1 & 0.1 & -0.1 & 0.0 & 0.4 & 0 & -0.1 & 0.1 \\
0 & 0 & 0 & -2000.0 & 0 & 0 & 2000.0 & 0 & 0 \\
0 & 0 & 0 & 0 & -0.0 & -0.1 & 0 & 0.0 & -0.1 \\
0 & 0 & 0 & 0 & 0.1 & 0.1 & 0 & -0.1 & 0.2
\end{bmatrix}
$$

$$
\begin{Bmatrix} 0 \\ 0 \\ 0 \\ U_{2x} \\ U_{2y} \\ \phi_2 \\ 0 \\ 0 \\ 0 \end{Bmatrix} = \begin{Bmatrix} F_{1x} \\ F_{1y} \\ M_1 \\ 0 \\ -16 \\ -10.667 \\ F_{3x} \\ F_{3y} \\ M_3 \end{Bmatrix}
\tag{8.14}
$$

Step 5 – Solving the Equations:

Solving the system of equations in (8.14) will be performed by partitioning (manually) and Gaussian elimination (with MATLAB). First we partition (8.14) by extracting the submatrix in rows 4 to 6 and columns 4 to 6. Therefore we obtain:

$$10^3 \begin{bmatrix} 2682.7 & -1024.0 & -0.1 \\ -1024.0 & 1536.1 & 0.0 \\ -0.1 & 0.0 & 0.4 \end{bmatrix} \begin{Bmatrix} U_{2x} \\ U_{2y} \\ \phi_2 \end{Bmatrix} = \begin{Bmatrix} 0 \\ -16 \\ -10.667 \end{Bmatrix} \qquad (8.15)$$

The solution of the above system is obtained using MATLAB as follows. Note that the backslash operator "\" is used for Gaussian elimination.

```
» k=K(4:6,4:6)

k =

   1.0e+006 *

              2.6827    -1.0240    -0.0001
             -1.0240     1.5361     0.0000
             -0.0001     0.0000     0.0004


» f=[0 ; -16 ; -10.667]

f =

          0
    -16.0000
    -10.6670

» u=k\f

u =

     -0.0000
     -0.0000
     -0.0253
```

It is now clear that the horizontal and vertical displacements at node 2 are -0.0000 m and -0.0000 m, respectively (both almost zero), while the rotation at node 2 is 0.0253 rad (clockwise).

Step 6 – Post-processing:

In this step, we obtain the reactions at nodes 1 and 3, and the forces (axial forces, shears and moments) in each plane frame element using MATLAB as follows. First we set up the global nodal displacement vector U, then we calculate the global nodal force vector F.

```
» U=[0 ; 0 ; 0 ; u ; 0 ; 0 ; 0]

U =

          0
          0
          0
    -0.0000
    -0.0000
    -0.0253
          0
          0
          0

» F=K*U

F =

    -12.2064
     14.1031
     -2.8039
      0.0000
    -16.0000
    -10.6670
     12.2064
      1.8969
     -2.5295
```

Thus the horizontal and vertical reactions at node 1 are forces of 12.2064 kN (to the left) and 14.1031 kN (upward), respectively, while the horizontal and vertical reactions at node 3 are forces of 12.2064 kN (to the right) and 1.8969 kN (upward), respectively. The moments at nodes 1 and 3 are 2.8039 kN.m (clockwise) and 2.5295 kN.m (clockwise), respectively. It is clear that force equilibrium is satisfied. Next we set up the element nodal displacement vectors u_1 and u_2, then we calculate the element force vectors f_1 and f_2 by making calls to the MATLAB function *PlaneFrameElementForces*.

```
» u1=[U(1) ; U(2) ; U(3) ; U(4) ; U(5) ; U(6)]

u1 =

          0
          0
          0
    -0.0000
    -0.0000
    -0.0253
```

```
» u2=[U(4) ; U(5) ; U(6) ; U(7) ; U(8) ; U(9)]

u2 =

    -0.0000
    -0.0000
    -0.0253
         0
         0
         0


» f1=PlaneFrameElementForces(E,A,I,L1,theta1,u1)

f1 =

    -18.5054
     -2.3333
     -2.8039
     18.5054
      2.3333
     -5.6090


» f2=PlaneFrameElementForces(E,A,I,L2,theta2,u2)

f2 =

    -12.2064
     -1.8969
     -5.0580
     12.2064
      1.8969
     -2.5295
```

Note that the forces for element 2 need to be modified because of the distributed load. In order to obtain the correct forces for element 2 we need to subtract from f2 the vector of equivalent nodal loads given in (8.11). This is performed using MATLAB as follows:

```
» f2=f2-[0 ; -16 ; -10.667 ; 0 ; -16 ; 10.667]

f2 =

    -12.2064
     14.1031
```

```
    5.6090
   12.2064
   17.8969
  -13.1965
```

Therefore the forces for each element are given above. Element 1 has an axial force of -18.5054 kN, a shear force of -2.3333 kN and a bending moment of -2.8039 kN.m at its left end while it has an axial force of 18.5054 kN, a shear force of 2.3333 kN and a bending moment of -5.6090 kN.m at its right end. Element 2 has an axial force of -12.2064 kN, a shear force of 14.1031 kN and a bending moment of 5.6090 kN.m at its left end while it has an axial force of 12.2064 kN, a shear force of 17.8969 kN and a bending moment of -13.1965 kN.m at its right end.

Finally we call the MATLAB functions *PlaneFrameElementAxialDiagram*, *Plane-FrameElementShearDiagram* and *PlaneFrameElementMomentDiagram* to draw the axial force diagram, shear force diagram and bending moment diagram, respectively, for each element. This process is illustrated below.

» `PlaneFrameElementAxialDiagram(f1,L1)`

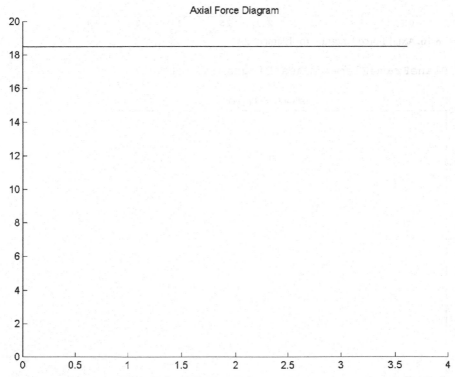

Fig. 8.15. Axial Force Diagram for Element 1

» `PlaneFrameElementAxialDiagram(f2,L2)`

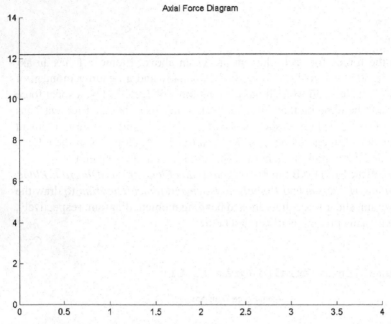

Fig. 8.16. Axial Force Diagram for Element 2

» `PlaneFrameElementShearDiagram(f1,L1)`

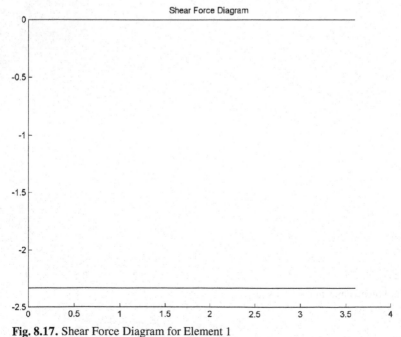

Fig. 8.17. Shear Force Diagram for Element 1

» PlaneFrameElementShearDiagram(f2,L2)

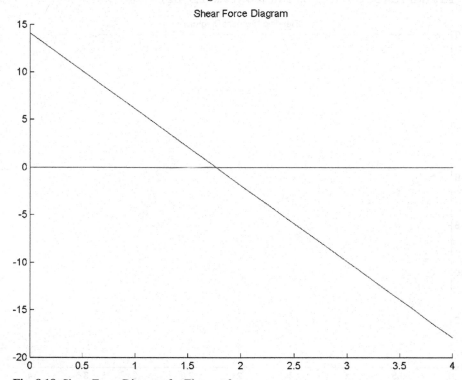

Fig. 8.18. Shear Force Diagram for Element 2

» PlaneFrameElementMomentDiagram(f1,L1)

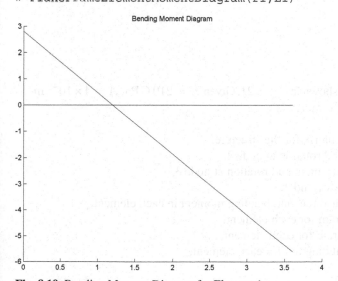

Fig. 8.19. Bending Moment Diagram for Element 1

» `PlaneFrameElementMomentDiagram(f2,L2)`

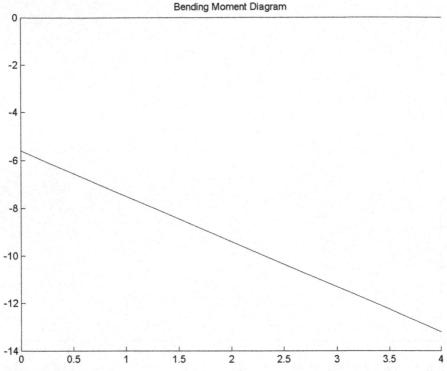

Fig. 8.20. Bending Moment Diagram for Element 2

Problems:

Problem 8.1:

Consider the plane frame shown in Fig. 8.21. Given $E = 210\,\text{GPa}$, $A = 4 \times 10^{-2}\,\text{m}^2$, and $I = 4 \times 10^{-6}\,\text{m}^4$, determine:

1. the global stiffness matrix for the structure.
2. the displacements and rotation at node 2.
3. the horizontal displacement and rotation at node 3.
4. the reactions at nodes 1 and 3.
5. the axial force, shear force, and bending moment in each element.
6. the axial force diagram for each element.
7. the shear force diagram for each element.
8. the bending moment diagram for each element.

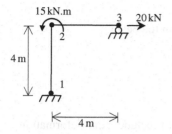

Fig. 8.21. Plane Frame with Two Elements for Problem 8.1

Problem 8.2:

Consider the plane frame shown in Fig. 8.22. Given $E = 210\,\text{GPa}$, $A = 1 \times 10^{-2}\,\text{m}^2$, and $I = 9 \times 10^{-5}\,\text{m}^4$, determine:

1. the global stiffness matrix for the structure.
2. the displacements and rotations at nodes 2 and 3.
3. the reactions at nodes 1 and 4.
4. the axial force, shear force, and bending moment in each element.
5. the axial force diagram for each element.
6. the shear force diagram for each element.
7. the bending moment diagram for each element.

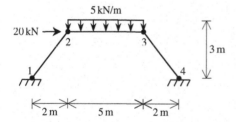

Fig. 8.22. Plane Frame with Distributed Load for Problem 8.2

Problem 8.3:

Consider the structure composed of a beam and a spring as shown in Fig 8.23. Given $E = 70\,\text{GPa}$, $A = 1 \times 10^{-2}\,\text{m}^2$, $I = 1 \times 10^{-5}\,\text{m}^4$, and $k = 5000\,\text{kN/m}$, determine:

1. the global stiffness matrix for the structure.
2. the displacements and rotation at node 1.

3. the reactions at nodes 2 and 3.
4. the axial force, shear force, and bending moment in the beam.
5. the force in the spring.
6. the axial force diagram for the beam.
7. the shear force diagram for the beam.
8. the bending moment diagram for the beam.

(Hint: Use a plane frame element for the beam so as to include axial deformation effects. Also use a plane truss element for the spring so as to include the angle of inclination – in this case determine values for E and A for the spring using the value of k given and the length of the spring).

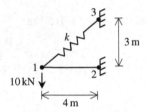

Fig. 8.23. Plane Frame with a Spring for Problem 8.3

9 The Grid Element

9.1
Basic Equations

The grid element is a two-dimensional finite element with both local and global coordinates. The grid element has modulus of elasticity E, shear modulus of elasticity G, moment of inertia I, torsional constant J, and length L. Each grid element has two nodes and is inclined with an angle θ measured counterclockwise from the positive global X axis as shown in Fig. 9.1. Let $C = cos\theta$ and $S = sin\theta$. In this case the element stiffness matrix is given by the following matrix (see [1] and [18]).

$$[k] = [R]^T[k'][R] \tag{9.1}$$

where the rotation matrix $[R]$ and the stiffness matrix $[k']$ are given by:

$$[k'] = \begin{bmatrix} \dfrac{12EI}{L^3} & 0 & \dfrac{6EI}{L^2} & -\dfrac{12EI}{L^3} & 0 & \dfrac{6EI}{L^2} \\ 0 & \dfrac{GJ}{L} & 0 & 0 & -\dfrac{GJ}{L} & 0 \\ \dfrac{6EI}{L^2} & 0 & \dfrac{4EI}{L} & -\dfrac{6EI}{L^2} & 0 & \dfrac{2EI}{L} \\ -\dfrac{12EI}{L^3} & 0 & -\dfrac{6EI}{L^2} & \dfrac{12EI}{L^3} & 0 & -\dfrac{6EI}{L^2} \\ 0 & -\dfrac{GJ}{L} & 0 & 0 & \dfrac{GJ}{L} & 0 \\ \dfrac{6EI}{L^2} & 0 & \dfrac{2EI}{L} & -\dfrac{6EI}{L^2} & 0 & \dfrac{4EI}{L} \end{bmatrix} \tag{9.2}$$

$$[R] = \begin{bmatrix} 1 & 0 & 0 & 0 & 0 & 0 \\ 0 & C & S & 0 & 0 & 0 \\ 0 & -S & C & 0 & 0 & 0 \\ 0 & 0 & 0 & 1 & 0 & 0 \\ 0 & 0 & 0 & 0 & C & S \\ 0 & 0 & 0 & 0 & -S & C \end{bmatrix} \tag{9.3}$$

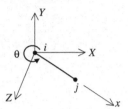

Fig. 9.1. The Grid Element

It is clear that the grid element has six degrees of freedom – three at each node (one displacement and two rotations). Consequently for a structure with n nodes, the global stiffness matrix K will be of size $3n \times 3n$ (since we have three degrees of freedom at each node). The global stiffness matrix K is assembled by making calls to the MATLAB function *GridAssemble* which is written specifically for this purpose. This process will be illustrated in detail in the examples.

Once the global stiffness matrix K is obtained we have the following structure equation:

$$[K]\{U\} = \{F\} \tag{9.4}$$

where U is the global nodal displacement vector and F is the global nodal force vector. At this step the boundary conditions are applied manually to the vectors U and F. Then the matrix (9.4) is solved by partitioning and Gaussian elimination. Finally once the unknown displacements and reactions are found, the nodal force vector is obtained for each element as follows:

$$\{f\} = [k']\,[R]\,\{u\} \tag{9.5}$$

where $\{f\}$ is the 6×1 nodal force vector in the element and u is the 6×1 element displacement vector.

The first element in each vector $\{u\}$ is the transverse displacement while the second and third elements are the rotations about the X and Z axes, respectively, at the first node, while the fourth element in each vector $\{u\}$ is the transverse displacement and the fifth and sixth elements are the rotations about the X and Z axes, respectively, at the second node.

9.2
MATLAB Functions Used

The four MATLAB functions used for the grid element are:

GridElementLength(x_1, y_1, x_2, y_2) – This function returns the element length given the coordinates of the first node (x_1, y_1) and the coordinates of the second node (x_2, y_2).

GridElementStiffness(E, G, I, J, L, *theta*) – This function calculates the element stiffness matrix for each grid element with modulus of elasticity E, shear modulus of elasticity G, moment of inertia I, torsional constant J, and length L. It returns the 6×6 element stiffness matrix k.

GridAssemble(K, k, i, j) – This function assembles the element stiffness matrix k of the grid element joining nodes i and j into the global stiffness matrix K. It returns the $3n \times 3n$ global stiffness matrix K every time an element is assembled.

GridElementForces(E, G, I, J, L, *theta*, u) – This function calculates the element force vector using the modulus of elasticity E, shear modulus of elasticity G, moment of inertia I, torsional constant J, length L, and the element displacement vector u. It returns the 6×1 element force vector f.

The following is a listing of the MATLAB source code for each function:

```
function y = GridElementLength(x1,y1,x2,y2)
%GridElementLength          This function returns the length of the
%                           grid element whose first node has
%                           coordinates (x1,y1) and second node has
%                           coordinates (x2,y2).
y = sqrt((x2-x1) * (x2-x1) + (y2-y1) * (y2-y1));
```

```
function y = GridElementStiffness(E,G,I,J,L,theta)
%GridElementStiffness       This function returns the element
%                           stiffness matrix for a grid
%                           element with modulus of elasticity E,
%                           shear modulus of elasticity G, moment of
%                           inertia I, torsional constant J, length L,
%                           and angle theta (in degrees).
%                           The size of the element stiffness
%                           matrix is 6 x 6.
x = theta*pi/180;
C = cos(x);
S = sin(x);
w1 = 12*E*I/(L*L*L);
w2 = 6*E*I/(L*L);
w3 = G*J/L;
w4 = 4*E*I/L;
w5 = 2*E*I/L;
kprime = [w1 0 w2 -w1 0 w2 ; 0 w3 0 0 -w3 0 ;
   w2 0 w4 -w2 0 w5 ; -w1 0 -w2 w1 0 -w2 ;
   0 -w3 0 0 w3 0 ; w2 0 w5 -w2 0 w4];
R = [1 0 0 0 0 0 ; 0 C S 0 0 0 ; 0 -S C 0 0 0 ;
   0 0 0 1 0 0 ; 0 0 0 0 C S ; 0 0 0 0 -S C];
y = R'*kprime*R;
```

```
function y = GridAssemble(K,k,i,j)
%GridAssemble            This function assembles the element stiffness
%                        matrix k of the grid element with nodes
%                        i and j into the global stiffness matrix K.
%                        This function returns the global stiffness
%                        matrix K after the element stiffness matrix
%                        k is assembled.
K(3*i-2,3*i-2) = K(3*i-2,3*i-2) + k(1,1);
K(3*i-2,3*i-1) = K(3*i-2,3*i-1) + k(1,2);
K(3*i-2,3*i) = K(3*i-2,3*i) + k(1,3);
K(3*i-2,3*j-2) = K(3*i-2,3*j-2) + k(1,4);
K(3*i-2,3*j-1) = K(3*i-2,3*j-1) + k(1,5);
K(3*i-2,3*j) = K(3*i-2,3*j) + k(1,6);
K(3*i-1,3*i-2) = K(3*i-1,3*i-2) + k(2,1);
K(3*i-1,3*i-1) = K(3*i-1,3*i-1) + k(2,2);
K(3*i-1,3*i) = K(3*i-1,3*i) + k(2,3);
K(3*i-1,3*j-2) = K(3*i-1,3*j-2) + k(2,4);
K(3*i-1,3*j-1) = K(3*i-1,3*j-1) + k(2,5);
K(3*i-1,3*j) = K(3*i-1,3*j) + k(2,6);
K(3*i,3*i-2) = K(3*i,3*i-2) + k(3,1);
K(3*i,3*i-1) = K(3*i,3*i-1) + k(3,2);
K(3*i,3*i) = K(3*i,3*i) + k(3,3);
K(3*i,3*j-2) = K(3*i,3*j-2) + k(3,4);
K(3*i,3*j-1) = K(3*i,3*j-1) + k(3,5);
K(3*i,3*j) = K(3*i,3*j) + k(3,6);
K(3*j-2,3*i-2) = K(3*j-2,3*i-2) + k(4,1);
K(3*j-2,3*i-1) = K(3*j-2,3*i-1) + k(4,2);
K(3*j-2,3*i) = K(3*j-2,3*i) + k(4,3);
K(3*j-2,3*j-2) = K(3*j-2,3*j-2) + k(4,4);
K(3*j-2,3*j-1) = K(3*j-2,3*j-1) + k(4,5);
K(3*j-2,3*j) = K(3*j-2,3*j) + k(4,6);
K(3*j-1,3*i-2) = K(3*j-1,3*i-2) + k(5,1);
K(3*j-1,3*i-1) = K(3*j-1,3*i-1) + k(5,2);
K(3*j-1,3*i) = K(3*j-1,3*i) + k(5,3);
K(3*j-1,3*j-2) = K(3*j-1,3*j-2) + k(5,4);
K(3*j-1,3*j-1) = K(3*j-1,3*j-1) + k(5,5);
K(3*j-1,3*j) = K(3*j-1,3*j) + k(5,6);
K(3*j,3*i-2) = K(3*j,3*i-2) + k(6,1);
K(3*j,3*i-1) = K(3*j,3*i-1) + k(6,2);
K(3*j,3*i) = K(3*j,3*i) + k(6,3);
K(3*j,3*j-2) = K(3*j,3*j-2) + k(6,4);
K(3*j,3*j-1) = K(3*j,3*j-1) + k(6,5);
K(3*j,3*j) = K(3*j,3*j) + k(6,6);
y = K;
```

```
function y = GridElementForces(E,G,I,J,L,theta,u)
%GridElementForces   This function returns the element force
%                    vector given the modulus of elasticity E,
%                    the shear modulus of elasticity G, the
%                    moment of inertia I, the torsional constant J,
%                    the length L, the angle theta (in degrees),
%                    and the element nodal displacement vector u.
```

```
x = theta*pi/180;
C = cos(x);
S = sin(x);
w1 = 12*E*I/(L*L*L);
w2 = 6*E*I/(L*L);
w3 = G*J/L;
w4 = 4*E*I/L;
w5 = 2*E*I/L;
kprime = [w1 0 w2 -w1 0 w2 ; 0 w3 0 0 -w3 0 ;
  w2 0 w4 -w2 0 w5 ; -w1 0 -w2 w1 0 -w2 ;
  0 -w3 0 0 w3 0 ; w2 0 w5 -w2 0 w4];
R = [1 0 0 0 0 0 ; 0 C S 0 0 0 ; 0 -S C 0 0 0 ;
  0 0 0 1 0 0 ; 0 0 0 0 C S ; 0 0 0 0 -S C];
y = kprime*R* u;
```

Example 9.1:

Consider the grid shown in Fig. 9.2. Given $E = 210\,\text{GPa}, G = 84\,\text{GPa}$, $I = 20 \times 10^{-5}\,\text{m}^4, J = 5 \times 10^{-5}\,\text{m}^4$, and $P = 20\,\text{kN}$, determine:

1. the global stiffness matrix for the structure.
2. the displacement and rotations at node 1.
3. the reactions at nodes 2, 3, and 4.
4. the forces and moments in each element.

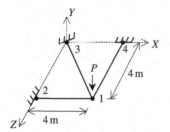

Fig. 9.2. Grid with Three Elements for Example 9.1

Solution:

Use the six steps outlined in Chap. 1 to solve this problem using the grid element.

Step 1 – Discretizing the Domain:

This problem is already discretized. The domain is subdivided into three elements and four nodes. The units used in the MATLAB calculations are kN and meter. Table 9.1 shows the element connectivity for this example.

Table 9.1. Element Connectivity for Example 9.1

Element Number	Node i	Node j
1	1	2
2	1	3
3	1	4

Step 2 – Writing the Element Stiffness Matrices:

The three element stiffness matrices k_1, k_2, and k_3 are obtained by making calls to the MATLAB function *GridElementStiffness*. Each matrix has size 6×6.

```
» E=210e6

E =

   210000000

» G=84e6

G =

   84000000

» I=20e-5

I =

   2.0000e-004

» J=5e-5

J =

   5.0000e-005
```

```
» L1=4

L1 =

    4

» L2=GridElementLength(0,0,4,4)

L2 =

    5.6569

» L3=4

L3 =

    4

» theta1=180

theta1 =

    180

» theta2=180-45

theta2 =

    135

» theta3=90

theta3 =

    90

» k1=GridElementStiffness(E,G,I,J,L1,theta1)

k1 =

    1.0e+004 *
```

```
    0.7875     0.0000    -1.5750    -0.7875     0.0000    -1.5750
    0.0000     0.1050     0.0000     0.0000    -0.1050     0.0000
   -1.5750     0.0000     4.2000     1.5750     0.0000     2.1000
   -0.7875     0.0000     1.5750     0.7875     0.0000     1.5750
    0.0000    -0.1050     0.0000     0.0000     0.1050     0.0000
   -1.5750     0.0000     2.1000     1.5750     0.0000     4.2000
```

» k2=GridElementStiffness(E,G,I,J,L2,theta2)

k2 =

   1.0e+004 *

```
    0.2784    -0.5568    -0.5568    -0.2784    -0.5568    -0.5568
   -0.5568     1.5220     1.4478     0.5568     0.7053     0.7796
   -0.5568     1.4478     1.5220     0.5568     0.7796     0.7053
   -0.2784     0.5568     0.5568     0.2784     0.5568     0.5568
   -0.5568     0.7053     0.7796     0.5568     1.5220     1.4478
   -0.5568     0.7796     0.7053     0.5568     1.4478     1.5220
```

» k3=GridElementStiffness(E,G,I,J,L3,theta3)

k3 =

   1.0e+004 *

```
    0.7875    -1.5750     0.0000    -0.7875    -1.5750     0.0000
   -1.5750     4.2000     0.0000     1.5750     2.1000     0.0000
    0.0000     0.0000     0.1050     0.0000     0.0000    -0.1050
   -0.7875     1.5750     0.0000     0.7875     1.5750     0.0000
   -1.5750     2.1000     0.0000     1.5750     4.2000     0.0000
    0.0000     0.0000    -0.1050     0.0000     0.0000     0.1050
```

Step 3 – Assembling the Global Stiffness Matrix:

Since the structure has four nodes, the size of the global stiffness matrix is 12×12. Therefore to obtain K we first set up a zero matrix of size 12×12 then make three calls to the MATLAB function *GridAssemble* since we have three grid elements in the structure. Each call to the function will assemble one element. The following are the MATLAB commands:

```
» K=zeros(12,12)

K =

     0     0     0     0     0     0     0     0     0     0     0     0
     0     0     0     0     0     0     0     0     0     0     0     0
     0     0     0     0     0     0     0     0     0     0     0     0
     0     0     0     0     0     0     0     0     0     0     0     0
     0     0     0     0     0     0     0     0     0     0     0     0
     0     0     0     0     0     0     0     0     0     0     0     0
     0     0     0     0     0     0     0     0     0     0     0     0
     0     0     0     0     0     0     0     0     0     0     0     0
     0     0     0     0     0     0     0     0     0     0     0     0
     0     0     0     0     0     0     0     0     0     0     0     0
     0     0     0     0     0     0     0     0     0     0     0     0
     0     0     0     0     0     0     0     0     0     0     0     0

» K=GridAssemble(K,k1,1,2)

K =

   1.0e+004 *

   Columns 1 through 7

    0.7875    0.0000   -1.5750   -0.7875    0.0000   -1.5750   0
    0.0000    0.1050    0.0000    0.0000   -0.1050    0.0000   0
   -1.5750    0.0000    4.2000    1.5750    0.0000    2.1000   0
   -0.7875    0.0000    1.5750    0.7875    0.0000    1.5750   0
    0.0000   -0.1050    0.0000    0.0000    0.1050    0.0000   0
   -1.5750    0.0000    2.1000    1.5750    0.0000    4.2000   0
         0         0         0         0         0         0   0
         0         0         0         0         0         0   0
         0         0         0         0         0         0   0
         0         0         0         0         0         0   0
         0         0         0         0         0         0   0
         0         0         0         0         0         0   0

   Columns 8 through 12

         0         0         0         0         0
         0         0         0         0         0
         0         0         0         0         0
         0         0         0         0         0
```

```
    0          0          0          0          0
    0          0          0          0          0
    0          0          0          0          0
    0          0          0          0          0
    0          0          0          0          0
    0          0          0          0          0
    0          0          0          0          0
    0          0          0          0          0
```

» K=GridAssemble(K,k2,1,3)

K =

   1.0e+004 *

   Columns 1 through 7

```
    1.0659  -0.5568  -2.1318  -0.7875   0.0000  -1.5750  -0.2784
   -0.5568   1.6270   1.4478   0.0000  -0.1050   0.0000   0.5568
   -2.1318   1.4478   5.7220   1.5750   0.0000   2.1000   0.5568
   -0.7875   0.0000   1.5750   0.7875   0.0000   1.5750        0
    0.0000  -0.1050   0.0000   0.0000   0.1050   0.0000        0
   -1.5750   0.0000   2.1000   1.5750   0.0000   4.2000        0
   -0.2784   0.5568   0.5568        0        0        0   0.2784
   -0.5568   0.7053   0.7796        0        0        0   0.5568
   -0.5568   0.7796   0.7053        0        0        0   0.5568
         0        0        0        0        0        0        0
         0        0        0        0        0        0        0
         0        0        0        0        0        0        0
```

   Columns 8 through 12

```
   -0.5568     -0.5568      0      0      0
    0.7053      0.7796      0      0      0
    0.7796      0.7053      0      0      0
         0           0      0      0      0
         0           0      0      0      0
         0           0      0      0      0
    0.5568      0.5568      0      0      0
    1.5220      1.4478      0      0      0
    1.4478      1.5220      0      0      0
         0           0      0      0      0
         0           0      0      0      0
         0           0      0      0      0
```

```
» K=GridAssemble(K,k3,1,4)

K =

   1.0e+004 *

   Columns 1 through 7


    1.8534   -2.1318   -2.1318   -0.7875    0.0000   -1.5750   -0.2784
   -2.1318    5.8270    1.4478    0.0000   -0.1050    0.0000    0.5568
   -2.1318    1.4478    5.8270    1.5750    0.0000    2.1000    0.5568
   -0.7875    0.0000    1.5750    0.7875    0.0000    1.5750         0
    0.0000   -0.1050    0.0000    0.0000    0.1050    0.0000         0
   -1.5750    0.0000    2.1000    1.5750    0.0000    4.2000         0
   -0.2784    0.5568    0.5568         0         0         0    0.2784
   -0.5568    0.7053    0.7796         0         0         0    0.5568
   -0.5568    0.7796    0.7053         0         0         0    0.5568
   -0.7875    1.5750    0.0000         0         0         0         0
   -1.5750    2.1000    0.0000         0         0         0         0
    0.0000    0.0000   -0.1050         0         0         0         0


   Columns 8 through 12


   -0.5568   -0.5568   -0.7875   -1.5750    0.0000
    0.7053    0.7796    1.5750    2.1000    0.0000
    0.7796    0.7053    0.0000    0.0000   -0.1050
         0         0         0         0         0
         0         0         0         0         0
         0         0         0         0         0
    0.5568    0.5568         0         0         0
    1.5220    1.4478         0         0         0
    1.4478    1.5220         0         0         0
         0         0    0.7875    1.5750    0.0000
         0         0    1.5750    4.2000    0.0000
         0         0    0.0000    0.0000    0.1050
```

Step 4 – Applying the Boundary Conditions:

The matrix (9.4) for this structure is obtained as follows using the global stiffness matrix obtained in the previous step:

$$
10^3
\begin{bmatrix}
18.53 & -21.32 & -21.32 & -7.88 & 0.0 & -15.75 & -2.78 & -5.57 & -5.57 & -7.88 & -15.75 & 0.0 \\
-21.32 & 58.27 & 14.48 & 0.0 & -1.05 & 0.0 & 5.57 & 7.05 & 7.80 & 15.75 & 21.00 & 0.0 \\
-21.32 & 14.48 & 58.27 & 15.75 & 0.0 & 21.00 & 5.57 & 7.80 & 7.05 & 0.0 & 0.0 & -1.05 \\
-7.88 & 0.0 & 15.75 & 7.88 & 0.0 & 15.75 & 0 & 0 & 0 & 0 & 0 & 0 \\
0.0 & -1.05 & 0.0 & 0.0 & 1.05 & 0.0 & 0 & 0 & 0 & 0 & 0 & 0 \\
-15.75 & 0.0 & 21.00 & 15.75 & 0.0 & 42.00 & 0 & 0 & 0 & 0 & 0 & 0 \\
-2.78 & 5.57 & 5.57 & 0 & 0 & 0 & 2.78 & 5.57 & 5.57 & 0 & 0 & 0 \\
-5.57 & 7.05 & 7.80 & 0 & 0 & 0 & 5.57 & 15.22 & 14.48 & 0 & 0 & 0 \\
-5.57 & 7.80 & 7.05 & 0 & 0 & 0 & 5.57 & 14.48 & 15.22 & 0 & 0 & 0 \\
-7.88 & 15.75 & 0.0 & 0 & 0 & 0 & 0 & 0 & 0 & 7.88 & 15.75 & 0.0 \\
-15.75 & 21.00 & 0.0 & 0 & 0 & 0 & 0 & 0 & 0 & 15.75 & 42.00 & 0.0 \\
0.0 & 0.0 & -1.05 & 0 & 0 & 0 & 0 & 0 & 0 & 0.0 & 0.0 & 1.05
\end{bmatrix}
$$

$$
\begin{Bmatrix}
U_{1y} \\
\phi_{1x} \\
\phi_{1z} \\
U_{2y} \\
\phi_{2x} \\
\phi_{2z} \\
U_{3y} \\
\phi_{3x} \\
\phi_{3z} \\
U_{4y} \\
\phi_{4x} \\
\phi_{4z}
\end{Bmatrix}
=
\begin{Bmatrix}
F_{1y} \\
M_{1x} \\
M_{1z} \\
F_{2y} \\
M_{2x} \\
M_{2z} \\
F_{3y} \\
M_{3x} \\
M_{3z} \\
F_{4y} \\
M_{4x} \\
M_{4z}
\end{Bmatrix}
\qquad (9.6)
$$

The boundary conditions for this problem are given as:

$$
F_{1y} = -20, \quad M_{1x} = M_{1z} = 0
$$
$$
U_{2y} = \phi_{2x} = \phi_{2z} = U_{3y} = \phi_{3x} = \phi_{3z} = U_{4y} = \phi_{4x} = \phi_{4z} = 0 \qquad (9.7)
$$

Inserting the above conditions into (9.6) we obtain:

$$
10^3
\begin{bmatrix}
18.53 & -21.32 & -21.32 & -7.88 & 0.0 & -15.75 & -2.78 & -5.57 & -5.57 & -7.88 & -15.75 & 0.0 \\
-21.32 & 58.27 & 14.48 & 0.0 & -1.05 & 0.0 & 5.57 & 7.05 & 7.80 & 15.75 & 21.00 & 0.0 \\
-21.32 & 14.48 & 58.27 & 15.75 & 0.0 & 21.00 & 5.57 & 7.80 & 7.05 & 0.0 & 0.0 & -1.05 \\
-7.88 & 0.0 & 15.75 & 7.88 & 0.0 & 15.75 & 0 & 0 & 0 & 0 & 0 & 0 \\
0.0 & -1.05 & 0.0 & 0.0 & 1.05 & 0.0 & 0 & 0 & 0 & 0 & 0 & 0 \\
-15.75 & 0.0 & 21.00 & 15.75 & 0.0 & 42.00 & 0 & 0 & 0 & 0 & 0 & 0 \\
-2.78 & 5.57 & 5.57 & 0 & 0 & 0 & 2.78 & 5.57 & 5.57 & 0 & 0 & 0 \\
-5.57 & 7.05 & 7.80 & 0 & 0 & 0 & 5.57 & 15.22 & 14.48 & 0 & 0 & 0 \\
-5.57 & 7.80 & 7.05 & 0 & 0 & 0 & 5.57 & 14.48 & 15.22 & 0\,0 & 0 & \\
-7.88 & 15.75 & 0.0 & 0 & 0 & 0 & 0 & 0 & 0 & 7.88 & 15.75 & 0.0 \\
-15.75 & 21.00 & 0.0 & 0 & 0 & 0 & 0 & 0 & 0 & 15.75 & 42.00 & 0.0 \\
0.0 & 0.0 & -1.05 & 0 & 0 & 0 & 0 & 0 & 0 & 0.0 & 0.0 & 1.05
\end{bmatrix}
$$

$$
\begin{Bmatrix} U_{1y} \\ \phi_{1x} \\ \phi_{1z} \\ 0 \\ 0 \\ 0 \\ 0 \\ 0 \\ 0 \\ 0 \\ 0 \\ 0 \end{Bmatrix} = \begin{Bmatrix} -20 \\ 0 \\ 0 \\ F_{2y} \\ M_{2x} \\ M_{2z} \\ F_{3y} \\ M_{3x} \\ M_{3z} \\ F_{4y} \\ M_{4x} \\ M_{4z} \end{Bmatrix}
\tag{9.8}
$$

Step 5 – Solving the Equations:

Solving the system of equations in (9.8) will be performed by partitioning (manually) and Gaussian elimination (with MATLAB). First we partition equation (9.8) by extracting the submatrix in rows 1 to 3 and columns 1 to 3. Therefore we obtain:

$$
10^3 \begin{bmatrix} 18.53 & -21.32 & -21.32 \\ -21.32 & 58.27 & 14.48 \\ -21.32 & 14.48 & 58.27 \end{bmatrix} \begin{Bmatrix} U_{1y} \\ \phi_{1x} \\ \phi_{1z} \end{Bmatrix} = \begin{Bmatrix} -20 \\ 0 \\ 0 \end{Bmatrix}
\tag{9.9}
$$

The solution of the above system is obtained using MATLAB as follows. Note that the backslash operator "\" is used for Gaussian elimination.

```
» k=K(1:3,1:3)

k =

   1.0e+004 *

     1.8534     -2.1318     -2.1318
    -2.1318      5.8270      1.4478
    -2.1318      1.4478      5.8270

» f=[-20 ; 0 ; 0]

f =

    -20
      0
      0

» u=k\f
```

u =

   -0.0033
   -0.0010
   -0.0010

It is now clear that the vertical displacement at node 1 is –0.0033 m and the rotations at node 1 along the X and Z axes are –0.0010 rad and –0.0010 rad, respectively. The symmetry in the results in this problem is clear.

Step 6 – Post-processing:

In this step, we obtain the reactions at nodes 2, 3, and 4, and the forces and moments in each grid element using MATLAB as follows. First we set up the global nodal displacement vector U, then we calculate the global nodal force vector F.

» U=[u ; 0 ; 0 ; 0 ; 0 ; 0 ; 0 ; 0 ; 0 ; 0]

U =

   -0.0033
   -0.0010
   -0.0010
      0
      0
      0
      0
      0
      0
      0
      0
      0

» F=K*U

F =

  -20.0000
    0.0000
    0.0000
   10.7937
    1.0189
   31.7764
   -1.5874
    4.0299

```
   4.0299
  10.7937
  31.7764
   1.0189
```

Thus the vertical reaction and moments at node 2 are 10.7937 kN, 1.0189 kN.m, and 31.7764 kN.m. The vertical reaction and moments at node 3 are -1.5874 kN, 4.0299 kN.m, and 4.0299 kN.m. The vertical reaction and moments at node 4 are 10.7937 kN, 31.7764 kN.m, and 1.0189 kN.m. It is clear that force equilibrium is satisfied. Next we set up the element nodal displacement vectors u_1, u_2, and u_3, then we calculate the element force vectors f_1, f_2, and f_3 by making calls to the MATLAB function *GridElementForces*.

```
» u1=[U(1)  ;  U(2)  ;  U(3)  ;  U(4)  ;  U(5)  ;  U(6)]

u1 =

   -0.0033
   -0.0010
   -0.0010
         0
         0
         0

» u2=[U(1)  ;  U(2)  ;  U(3)  ;  U(7)  ;  U(8)  ;  U(9)]

u2 =

   -0.0033
   -0.0010
   -0.0010
         0
         0
         0

» u3=[U(1)  ;  U(2)  ;  U(3)  ;  U(10)  ;  U(11)  ;  U(12)]

u3 =

   -0.0033
   -0.0010
   -0.0010
         0
         0
         0
```

```
» f1=GridElementForces(E,G,I,J,L1,theta1,u1)

f1 =

    -10.7937
      1.0189
    -11.3984
     10.7937
     -1.0189
    -31.7764

» f2=GridElementForces(E,G,I,J,L2,theta2,u2)


f2 =

      1.5874
           0
     14.6788
     -1.5874
           0
     -5.6992

» f3=GridElementForces(E,G,I,J,L3,theta3,u3)

f3 =

    -10.7937
     -1.0189
    -11.3984
     10.7937
      1.0189
    -31.7764
```

Therefore the forces for each element are given above. Element 1 has a force of −10.7937 kN and moments of 1.0189 kN.m and −11.3984 kN.m at its left end while it has a force of 10.7937 kN and moments of −1.0189 kN.m and −31.7764 kN.m at its right end. Element 2 has a force of 1.5874 kN and moments of 0 kN.m and 14.6788 kN.m at its left end while it has a force of −1.5874 kN and moments of 0 kN.m and −5.6992 kN.m at its right end. Element 3 has a force of −10.7937 kN and moments of −1.0189 kN.m and −11.3984 kN.m at its left end while it has a force of 10.7937 kN and moments of 1.0189 kN.m and −31.7764 kN.m at its right end.

Problems:

Problem 9.1:

Consider the grid shown in Fig. 9.3. Given $E = 210\,\text{GPa}$, $G = 84\,\text{GPa}$, $I = 20 \times 10^{-5}\,\text{m}^4$, and $J = 5 \times 10^{-5}\,\text{m}^4$, determine:

1. the global stiffness matrix for the structure.
2. the displacement and rotations at node 1.
3. the reactions at nodes 2, and 3.
4. the forces and moments in each element.

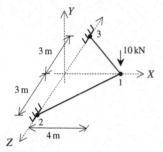

Fig. 9.3. Gird with Two Elements for Problem 9.1

10 The Space Frame Element

10.1
Basic Equations

The space frame element is a three-dimensional finite element with both local and global coordinates. The space frame element has modulus of elasticity E, shear modulus of elasticity G, cross-sectional area A, moments of inertia I_x and I_y, polar moment of inertia J, and length L. Each space frame element has two nodes and is inclined with angles $\theta_{Xx}\theta_{Yx}$ and θ_{Zx} measured from the global X, Y, and Z axes, respectively, to the local x axis as shown in Fig. 10.1. In this case the element stiffness matrix is given by the following matrix (see [1] and [18]).

$$[k] = [R]^T[k'][R] \tag{10.1}$$

where the rotation matrix $[R]$ and the stiffness matrix $[k']$ are given by:

$$[k'] =
\begin{bmatrix}
\frac{EA}{L} & 0 & 0 & 0 & 0 & 0 & -\frac{EA}{L} & 0 & 0 & 0 & 0 & 0 \\
0 & \frac{12EI_z}{L^3} & 0 & 0 & 0 & \frac{6EI_z}{L^2} & 0 & -\frac{12EI_z}{L^3} & 0 & 0 & 0 & \frac{6EI_z}{L^2} \\
0 & 0 & \frac{12EI_y}{L^3} & 0 & -\frac{6EI_y}{L^2} & 0 & 0 & 0 & -\frac{12EI_y}{L^3} & 0 & -\frac{6EI_y}{L^2} & 0 \\
0 & 0 & 0 & \frac{GJ}{L} & 0 & 0 & 0 & 0 & 0 & -\frac{GJ}{L} & 0 & 0 \\
0 & 0 & -\frac{6EI_y}{L^2} & 0 & \frac{4EI_y}{L} & 0 & 0 & 0 & \frac{6EI_y}{L^2} & 0 & \frac{2EI_y}{L} & 0 \\
0 & \frac{6EI_z}{L^2} & 0 & 0 & 0 & \frac{4EI_z}{L} & 0 & -\frac{6EI_z}{L^2} & 0 & 0 & 0 & \frac{2EI_z}{L} \\
-\frac{EA}{L} & 0 & 0 & 0 & 0 & 0 & \frac{EA}{L} & 0 & 0 & 0 & 0 & 0 \\
0 & -\frac{12EI_z}{L^3} & 0 & 0 & 0 & -\frac{6EI_z}{L^2} & 0 & \frac{12EI_z}{L^3} & 0 & 0 & 0 & -\frac{6EI_z}{L^2} \\
0 & 0 & -\frac{12EI_y}{L^3} & 0 & \frac{6EI_y}{L^2} & 0 & 0 & 0 & \frac{12EI_y}{L^3} & 0 & \frac{6EI_y}{L^2} & 0 \\
0 & 0 & 0 & -\frac{GJ}{L} & 0 & 0 & 0 & 0 & 0 & \frac{GJ}{L} & 0 & 0 \\
0 & 0 & -\frac{6EI_y}{L^2} & 0 & \frac{2EI_y}{L} & 0 & 0 & 0 & \frac{6EI_y}{L^2} & 0 & \frac{4EI_y}{L} & 0 \\
0 & \frac{6EI_z}{L^2} & 0 & 0 & 0 & \frac{2EI_z}{L} & 0 & -\frac{6EI_z}{L^2} & 0 & 0 & 0 & \frac{4EI_z}{L}
\end{bmatrix}$$

$$\tag{10.2}$$

$$[R] = \begin{bmatrix} [r] & 0 & 0 & 0 \\ 0 & [r] & 0 & 0 \\ 0 & 0 & [r] & 0 \\ 0 & 0 & 0 & [r] \end{bmatrix} \tag{10.3}$$

where r is the 3×3 matrix of direction cosines given as follows:

$$[r] = \begin{bmatrix} C_{Xx} & C_{Yx} & C_{Zx} \\ C_{Xy} & C_{Yy} & C_{Zy} \\ C_{Xz} & C_{Yz} & C_{Zz} \end{bmatrix} \tag{10.4}$$

where $C_{Xx} = \cos\theta_{Xx}$, $C_{Yx} = \cos\theta_{Yx}$, ..., etc.

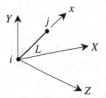

Fig. 10.1. The Space Frame Element

It is clear that the space frame element has twelve degrees of freedom – six at each node (three displacements and three rotations). Consequently for a structure with n nodes, the global stiffness matrix K will be of size $6n \times 6n$ (since we have six degrees of freedom at each node). The global stiffness matrix K is assembled by making calls to the MATLAB function *SpaceFrameAssemble* which is written specifically for this purpose. This process will be illustrated in detail in the examples.

Once the global stiffness matrix K is obtained we have the following structure equation:

$$[K]\{U\} = \{F\} \tag{10.5}$$

where U is the global nodal displacement vector and F is the global nodal force vector. At this step the boundary conditions are applied manually to the vectors U and F. Then the matrix (10.5) is solved by partitioning and Gaussian elimination. Finally once the unknown displacements and reactions are found, the nodal force vector is obtained for each element as follows:

$$\{f\} = [k'][R]\{u\} \tag{10.6}$$

where $\{f\}$ is the 12×1 nodal force vector in the element and u is the 12×1 element displacement vector.

The first, second, and third elements in each vector $\{u\}$ are the three displacements while the fourth, fifth, and sixth elements are the three rotations, respectively, at the first node, while the seventh, eighth, and ninth elements in each vector $\{u\}$ are the three displacements while the tenth, eleventh, and twelvth elements are the three rotations, respectively, at the second node.

10.2
MATLAB Functions Used

The ten MATLAB functions used for the space frame element are:

SpaceFrameElementLength$(x_1, y_1, z_1, x_2, y_2, z_2)$ – This function returns the element length given the coordinates of the first node (x_1, y_1, z_1) and the coordinates of the second node (x_2, y_2, z_2).

SpaceFrameElementStiffness$(E, G, A, I_y, I_z, J, x_1, y_1, z_1, x_2, y_2, z_2)$ – This function calculates the element stiffness matrix for each space frame element with modulus of elasticity E, shear modulus of elasticity G, cross-sectional area A, moments of inertia I_x and I_y, polar moment of inertia J, coordinates (x_1, y_1, z_1) for the first node, and coordinates (x_2, y_2, z_2) for the second node. It returns the 12×12 element stiffness matrix k.

SpaceFrameAssemble(K, k, i, j) – This function assembles the element stiffness matrix k of the space frame element joining nodes i and j into the global stiffness matrix K. It returns the $6n \times 6n$ global stiffness matrix K every time an element is assembled.

SpaceFrameElementForces$(E, G, A, I_y, I_z, J, x_1, y_1, z_1, x_2, y_2, z_2, u)$ – This function calculates the element force vector using the modulus of elasticity E, shear modulus of elasticity G, cross-sectional area A, moments of inertia I_x and I_y, polar moment of inertia J, coordinates (x_1, y_1, z_1) for the first node, coordinates (x_2, y_2, z_2) for the second node, and the element displacement vector u. It returns the 12×1 element force vector f.

SpaceFrameElementAxialDiagram(f, L) – This function plots the axial force diagram for the element with nodal force vector f and length L.

SpaceFrameElementShearZDiagram(f, L) – This function plots the shear force (along the Z axis) diagram for the element with nodal force vector f and length L.

SpaceFrameElementShearYDiagram(f, L) – This function plots the shear force (along the Y axis) diagram for the element with nodal force vector f and length L.

SpaceFrameElementTorsionDiagram(f, L) – This function plots the torsional moment diagram for the element with nodal force vector f and length L.

SpaceFrameElementMomentZDiagram(f, L) – This function plots the bending moment (along the Z axis) diagram for the element with nodal force vector *f* and length *L*.

SpaceFrameElementMomentYDiagram(f, L) – This function plots the bending moment (along the Y axis) diagram for the element with nodal force vector *f* and length *L*.

The following is a listing of the MATLAB source code for each function:

```
function y = SpaceFrameElementLength(x1,y1,z1,x2,y2,z2)
%SpaceFrameElementLength    This function returns the length of the
%                           space frame element whose first node has
%                           coordinates (x1,y1,z1) and second node has
%                           coordinates (x2,y2,z2).
y = sqrt((x2-x1)*(x2-x1) + (y2-y1)*(y2-y1) + (z2-z1)*(z2-z1));
```

```
function y =
         SpaceFrameElementStiffness(E,G,A,Iy,Iz,J,x1,y1,z1,x2,y2,z2)
%SpaceFrameElementStiffness    This function returns the element
%                              stiffness matrix for a space frame
%                              element with modulus of elasticity E,
%                              shear modulus of elasticity G, cross-
%                              sectional area A, moments of inertia
%                              Iy and Iz, torsional constant J,
%                              coordinates (x1,y1,z1) for the first
%                              node and coordinates (x2,y2,z2) for the
%                              second node.
%                              The size of the element stiffness
%                              matrix is 12 x 12.
L = sqrt((x2-x1)*(x2-x1) + (y2-y1)*(y2-y1) + (z2-z1)*(z2-z1));
w1 = E*A/L;
w2 = 12*E*Iz/(L*L*L);
w3 = 6*E*Iz/(L*L);
w4 = 4*E*Iz/L;
w5 = 2*E*Iz/L;
w6 = 12*E*Iy/(L*L*L);
w7 = 6*E*Iy/(L*L);
w8 = 4*E*Iy/L;
w9 = 2*E*Iy/L;
w10 = G*J/L;
kprime = [w1 0 0 0 0 0 -w1 0 0 0 0 0 ;
   0 w2 0 0 0 w3 0 -w2 0 0 0 w3 ;
   0 0 w6 0 -w7 0 0 0 -w6 0 -w7 0 ;
   0 0 0 w10 0 0 0 0 0 -w10 0 0 ;
   0 0 -w7 0 w8 0 0 0 w7 0 w9 0 ;
   0 w3 0 0 0 w4 0 -w3 0 0 0 w5 ;
   -w1 0 0 0 0 0 w1 0 0 0 0 0 ;
   0 -w2 0 0 0 -w3 0 w2 0 0 0 -w3 ;
   0 0 -w6 0 w7 0 0 0 w6 0 w7 0 ;
```

```
     0 0 0 -w10 0 0 0 0 0 w10 0 0 ;
     0 0 -w7 0 w9 0 0 0 w7 0 w8 0 ;
     0 w3 0 0 0 w5 0 -w3 0 0 0 w4];
if x1 == x2 & y1 == y2
   if z2 > z1
     Lambda = [0 0 1 ; 0 1 0 ; -1 0 0];
   else
     Lambda = [0 0 -1 ; 0 1 0 ; 1 0 0];
   end
else
   CXx = (x2-x1)/L;
   CYx = (y2-y1)/L;
   CZx = (z2-z1)/L;
   D = sqrt(CXx*CXx + CYx*CYx);
   CXy = -CYx/D;
   CYy = CXx/D;
   CZy = 0;
   CXz = -CXx*CZx/D;
   CYz = -CYx*CZx/D;
   CZz = D;
   Lambda = [CXx CYx CZx ; CXy CYy CZy ; CXz CYz CZz];
end
R = [Lambda zeros(3) zeros(3) zeros(3) ;
     zeros(3) Lambda zeros(3) zeros(3) ;
     zeros(3) zeros(3) Lambda zeros(3) ;
     zeros(3) zeros(3) zeros(3) Lambda];
y = R'*kprime*R;
```

```
function y = SpaceFrameAssemble(K,k,i,j)
%SpaceFrameAssemble      This function assembles the element stiffness
%                        matrix k of the space frame element with nodes
%                        i and j into the global stiffness matrix K.
%                        This function returns the global stiffness
%                        matrix K after the element stiffness matrix
%                        k is assembled.
K(6*i-5,6*i-5) = K(6*i-5,6*i-5) + k(1,1);
K(6*i-5,6*i-4) = K(6*i-5,6*i-4) + k(1,2);
K(6*i-5,6*i-3) = K(6*i-5,6*i-3) + k(1,3);
K(6*i-5,6*i-2) = K(6*i-5,6*i-2) + k(1,4);
K(6*i-5,6*i-1) = K(6*i-5,6*i-1) + k(1,5);
K(6*i-5,6*i) = K(6*i-5,6*i) + k(1,6);
K(6*i-5,6*j-5) = K(6*i-5,6*j-5) + k(1,7);
K(6*i-5,6*j-4) = K(6*i-5,6*j-4) + k(1,8);
K(6*i-5,6*j-3) = K(6*i-5,6*j-3) + k(1,9);
K(6*i-5,6*j-2) = K(6*i-5,6*j-2) + k(1,10);
K(6*i-5,6*j-1) = K(6*i-5,6*j-1) + k(1,11);
K(6*i-5,6*j) = K(6*i-5,6*j) + k(1,12);
K(6*i-4,6*i-5) = K(6*i-4,6*i-5) + k(2,1);
K(6*i-4,6*i-4) = K(6*i-4,6*i-4) + k(2,2);
K(6*i-4,6*i-3) = K(6*i-4,6*i-3) + k(2,3);
K(6*i-4,6*i-2) = K(6*i-4,6*i-2) + k(2,4);
```

```
K(6*i-4,6*i-1) = K(6*i-4,6*i-1) + k(2,5);
K(6*i-4,6*i) = K(6*i-4,6*i) + k(2,6);
K(6*i-4,6*j-5) = K(6*i-4,6*j-5) + k(2,7);
K(6*i-4,6*j-4) = K(6*i-4,6*j-4) + k(2,8);
K(6*i-4,6*j-3) = K(6*i-4,6*j-3) + k(2,9);
K(6*i-4,6*j-2) = K(6*i-4,6*j-2) + k(2,10);
K(6*i-4,6*j-1) = K(6*i-4,6*j-1) + k(2,11);
K(6*i-4,6*j) = K(6*i-4,6*j) + k(2,12);
K(6*i-3,6*i-5) = K(6*i-3,6*i-5) + k(3,1);
K(6*i-3,6*i-4) = K(6*i-3,6*i-4) + k(3,2);
K(6*i-3,6*i-3) = K(6*i-3,6*i-3) + k(3,3);
K(6*i-3,6*i-2) = K(6*i-3,6*i-2) + k(3,4);
K(6*i-3,6*i-1) = K(6*i-3,6*i-1) + k(3,5);
K(6*i-3,6*i) = K(6*i-3,6*i) + k(3,6);
K(6*i-3,6*j-5) = K(6*i-3,6*j-5) + k(3,7);
K(6*i-3,6*j-4) = K(6*i-3,6*j-4) + k(3,8);
K(6*i-3,6*j-3) = K(6*i-3,6*j-3) + k(3,9);
K(6*i-3,6*j-2) = K(6*i-3,6*j-2) + k(3,10);
K(6*i-3,6*j-1) = K(6*i-3,6*j-1) + k(3,11);
K(6*i-3,6*j) = K(6*i-3,6*j) + k(3,12);
K(6*i-2,6*i-5) = K(6*i-2,6*i-5) + k(4,1);
K(6*i-2,6*i-4) = K(6*i-2,6*i-4) + k(4,2);
K(6*i-2,6*i-3) = K(6*i-2,6*i-3) + k(4,3);
K(6*i-2,6*i-2) = K(6*i-2,6*i-2) + k(4,4);
K(6*i-2,6*i-1) = K(6*i-2,6*i-1) + k(4,5);
K(6*i-2,6*i) = K(6*i-2,6*i) + k(4,6);
K(6*i-2,6*j-5) = K(6*i-2,6*j-5) + k(4,7);
K(6*i-2,6*j-4) = K(6*i-2,6*j-4) + k(4,8);
K(6*i-2,6*j-3) = K(6*i-2,6*j-3) + k(4,9);
K(6*i-2,6*j-2) = K(6*i-2,6*j-2) + k(4,10);
K(6*i-2,6*j-1) = K(6*i-2,6*j-1) + k(4,11);
K(6*i-2,6*j) = K(6*i-2,6*j) + k(4,12);
K(6*i-1,6*i-5) = K(6*i-1,6*i-5) + k(5,1);
K(6*i-1,6*i-4) = K(6*i-1,6*i-4) + k(5,2);
K(6*i-1,6*i-3) = K(6*i-1,6*i-3) + k(5,3);
K(6*i-1,6*i-2) = K(6*i-1,6*i-2) + k(5,4);
K(6*i-1,6*i-1) = K(6*i-1,6*i-1) + k(5,5);
K(6*i-1,6*i) = K(6*i-1,6*i) + k(5,6);
K(6*i-1,6*j-5) = K(6*i-1,6*j-5) + k(5,7);
K(6*i-1,6*j-4) = K(6*i-1,6*j-4) + k(5,8);
K(6*i-1,6*j-3) = K(6*i-1,6*j-3) + k(5,9);
K(6*i-1,6*j-2) = K(6*i-1,6*j-2) + k(5,10);
K(6*i-1,6*j-1) = K(6*i-1,6*j-1) + k(5,11);
K(6*i-1,6*j) = K(6*i-1,6*j) + k(5,12);
K(6*i,6*i-5) = K(6*i,6*i-5) + k(6,1);
K(6*i,6*i-4) = K(6*i,6*i-4) + k(6,2);
K(6*i,6*i-3) = K(6*i,6*i-3) + k(6,3);
K(6*i,6*i-2) = K(6*i,6*i-2) + k(6,4);
K(6*i,6*i-1) = K(6*i,6*i-1) + k(6,5);
K(6*i,6*i) = K(6*i,6*i) + k(6,6);
K(6*i,6*j-5) = K(6*i,6*j-5) + k(6,7);
K(6*i,6*j-4) = K(6*i,6*j-4) + k(6,8);
K(6*i,6*j-3) = K(6*i,6*j-3) + k(6,9);
```

```
K(6*i,6*j-2) = K(6*i,6*j-2) + k(6,10);
K(6*i,6*j-1) = K(6*i,6*j-1) + k(6,11);
K(6*i,6*j) = K(6*i,6*j) + k(6,12);
K(6*j-5,6*i-5) = K(6*j-5,6*i-5) + k(7,1);
K(6*j-5,6*i-4) = K(6*j-5,6*i-4) + k(7,2);
K(6*j-5,6*i-3) = K(6*j-5,6*i-3) + k(7,3);
K(6*j-5,6*i-2) = K(6*j-5,6*i-2) + k(7,4);
K(6*j-5,6*i-1) = K(6*j-5,6*i-1) + k(7,5);
K(6*j-5,6*i) = K(6*j-5,6*i) + k(7,6);
K(6*j-5,6*j-5) = K(6*j-5,6*j-5) + k(7,7);
K(6*j-5,6*j-4) = K(6*j-5,6*j-4) + k(7,8);
K(6*j-5,6*j-3) = K(6*j-5,6*j-3) + k(7,9);
K(6*j-5,6*j-2) = K(6*j-5,6*j-2) + k(7,10);
K(6*j-5,6*j-1) = K(6*j-5,6*j-1) + k(7,11);
K(6*j-5,6*j) = K(6*j-5,6*j) + k(7,12);
K(6*j-4,6*i-5) = K(6*j-4,6*i-5) + k(8,1);
K(6*j-4,6*i-4) = K(6*j-4,6*i-4) + k(8,2);
K(6*j-4,6*i-3) = K(6*j-4,6*i-3) + k(8,3);
K(6*j-4,6*i-2) = K(6*j-4,6*i-2) + k(8,4);
K(6*j-4,6*i-1) = K(6*j-4,6*i-1) + k(8,5);
K(6*j-4,6*i) = K(6*j-4,6*i) + k(8,6);
K(6*j-4,6*j-5) = K(6*j-4,6*j-5) + k(8,7);
K(6*j-4,6*j-4) = K(6*j-4,6*j-4) + k(8,8);
K(6*j-4,6*j-3) = K(6*j-4,6*j-3) + k(8,9);
K(6*j-4,6*j-2) = K(6*j-4,6*j-2) + k(8,10);
K(6*j-4,6*j-1) = K(6*j-4,6*j-1) + k(8,11);
K(6*j-4,6*j) = K(6*j-4,6*j) + k(8,12);
K(6*j-3,6*i-5) = K(6*j-3,6*i-5) + k(9,1);
K(6*j-3,6*i-4) = K(6*j-3,6*i-4) + k(9,2);
K(6*j-3,6*i-3) = K(6*j-3,6*i-3) + k(9,3);
K(6*j-3,6*i-2) = K(6*j-3,6*i-2) + k(9,4);
K(6*j-3,6*i-1) = K(6*j-3,6*i-1) + k(9,5);
K(6*j-3,6*i) = K(6*j-3,6*i) + k(9,6);
K(6*j-3,6*j-5) = K(6*j-3,6*j-5) + k(9,7);
K(6*j-3,6*j-4) = K(6*j-3,6*j-4) + k(9,8);
K(6*j-3,6*j-3) = K(6*j-3,6*j-3) + k(9,9);
K(6*j-3,6*j-2) = K(6*j-3,6*j-2) + k(9,10);
K(6*j-3,6*j-1) = K(6*j-3,6*j-1) + k(9,11);
K(6*j-3,6*j) = K(6*j-3,6*j) + k(9,12);
K(6*j-2,6*i-5) = K(6*j-2,6*i-5) + k(10,1);
K(6*j-2,6*i-4) = K(6*j-2,6*i-4) + k(10,2);
K(6*j-2,6*i-3) = K(6*j-2,6*i-3) + k(10,3);
K(6*j-2,6*i-2) = K(6*j-2,6*i-2) + k(10,4);
K(6*j-2,6*i-1) = K(6*j-2,6*i-1) + k(10,5);
K(6*j-2,6*i) = K(6*j-2,6*i) + k(10,6);
K(6*j-2,6*j-5) = K(6*j-2,6*j-5) + k(10,7);
K(6*j-2,6*j-4) = K(6*j-2,6*j-4) + k(10,8);
K(6*j-2,6*j-3) = K(6*j-2,6*j-3) + k(10,9);
K(6*j-2,6*j-2) = K(6*j-2,6*j-2) + k(10,10);
K(6*j-2,6*j-1) = K(6*j-2,6*j-1) + k(10,11);
K(6*j-2,6*j) = K(6*j-2,6*j) + k(10,12);
K(6*j-1,6*i-5) = K(6*j-1,6*i-5) + k(11,1);
K(6*j-1,6*i-4) = K(6*j-1,6*i-4) + k(11,2);
K(6*j-1,6*i-3) = K(6*j-1,6*i-3) + k(11,3);
K(6*j-1,6*i-2) = K(6*j-1,6*i-2) + k(11,4);
```

```
K(6*j-1,6*i-1) = K(6*j-1,6*i-1) + k(11,5);
K(6*j-1,6*i) = K(6*j-1,6*i) + k(11,6);
K(6*j-1,6*j-5) = K(6*j-1,6*j-5) + k(11,7);
K(6*j-1,6*j-4) = K(6*j-1,6*j-4) + k(11,8);
K(6*j-1,6*j-3) = K(6*j-1,6*j-3) + k(11,9);
K(6*j-1,6*j-2) = K(6*j-1,6*j-2) + k(11,10);
K(6*j-1,6*j-1) = K(6*j-1,6*j-1) + k(11,11);
K(6*j-1,6*j) = K(6*j-1,6*j) + k(11,12);
K(6*j,6*i-5) = K(6*j,6*i-5) + k(12,1);
K(6*j,6*i-4) = K(6*j,6*i-4) + k(12,2);
K(6*j,6*i-3) = K(6*j,6*i-3) + k(12,3);
K(6*j,6*i-2) = K(6*j,6*i-2) + k(12,4);
K(6*j,6*i-1) = K(6*j,6*i-1) + k(12,5);
K(6*j,6*i) = K(6*j,6*i) + k(12,6);
K(6*j,6*j-5) = K(6*j,6*j-5) + k(12,7);
K(6*j,6*j-4) = K(6*j,6*j-4) + k(12,8);
K(6*j,6*j-3) = K(6*j,6*j-3) + k(12,9);
K(6*j,6*j-2) = K(6*j,6*j-2) + k(12,10);
K(6*j,6*j-1) = K(6*j,6*j-1) + k(12,11);
K(6*j,6*j) = K(6*j,6*j) + k(12,12);
y = K;
```

```
function y =
            SpaceFrameElementForces(E,G,A,Iy,Iz,J,x1,y1,z1,x2,y2,z2,u)
%SpaceFrameElementForces    This function returns the element force
%                           vector given the modulus of elasticity E,
%                           the shear modulus of elasticity G, the
%                           cross-sectional area A, moments of inertia
%                           Iy and Iz, the torsional constant J,
%                           the coordinates (x1,y1,z1) of the first
%                           node, the coordinates (x2,y2,z2) of the
%                           second node, and the element nodal
%                           displacement vector u.
L = sqrt((x2-x1)*(x2-x1) + (y2-y1)*(y2-y1) + (z2-z1)*(z2-z1));
w1 = E*A/L;
w2 = 12*E*Iz/(L*L*L);
w3 = 6*E*Iz/(L*L);
w4 = 4*E*Iz/L;
w5 = 2*E*Iz/L;
w6 = 12*E*Iy/(L*L*L);
w7 = 6*E*Iy/(L*L);
w8 = 4*E*Iy/L;
w9 = 2*E*Iy/L;
w10 = G*J/L;
kprime = [w1 0 0 0 0 0 -w1 0 0 0 0 0 ;
   0 w2 0 0 0 w3 0 -w2 0 0 0 w3 ;
   0 0 w6 0 -w7 0 0 0 -w6 0 -w7 0 ;
   0 0 0 w10 0 0 0 0 0 -w10 0 0 ;
   0 0 -w7 0 w8 0 0 0 w7 0 w9 0 ;
   0 w3 0 0 0 w4 0 -w3 0 0 0 w5 ;
   -w1 0 0 0 0 0 w1 0 0 0 0 0 ;
   0 -w2 0 0 0 -w3 0 w2 0 0 0 -w3 ;
   0 0 -w6 0 w7 0 0 0 w6 0 w7 0 ;
   0 0 0 -w10 0 0 0 0 0 w10 0 0 ;
   0 0 -w7 0 w9 0 0 0 w7 0 w8 0 ;
   0 w3 0 0 0 w5 0 -w3 0 0 0 w4];
```

```
if x1 == x2 & y1 == y2
   if z2 > z1
      Lambda = [0 0 1 ; 0 1 0 ; -1 0 0];
   else
      Lambda = [0 0 -1 ; 0 1 0 ; 1 0 0];
   end
else
  CXx = (x2-x1)/L;
  CYx = (y2-y1)/L;
  CZx = (z2-z1)/L;
  D = sqrt(CXx*CXx + CYx*CYx);
  CXy = -CYx/D;
  CYy = CXx/D;
  CZy = 0;
  CXz = -CXx*CZx/D;
  CYz = -CYx*CZx/D;
  CZz = D;
  Lambda = [CXx CYx CZx ; CXy CYy CZy ; CXz CYz CZz];
end
R = [Lambda zeros(3) zeros(3) zeros(3) ;
  zeros(3) Lambda zeros(3) zeros(3) ;
  zeros(3) zeros(3) Lambda zeros(3) ;
  zeros(3) zeros(3) zeros(3) Lambda];
y = kprime*R* u;
```

```
function y = SpaceFrameElementAxialDiagram(f, L)
%SpaceFrameElementAxialDiagram   This function plots the axial force
%                                diagram for the space frame element
%                                with nodal force vector f and length
%                                L.
x = [0 ; L];
z = [-f(1) ; f(7)];
hold on;
title('Axial Force Diagram');
plot(x,z);
y1 = [0 ; 0];
plot(x,y1,'k')
```

```
function y = SpaceFrameElementShearYDiagram(f, L)
%SpaceFrameElementShearYDiagram   This function plots the shear force
%                                 diagram for the space frame element
%                                 with nodal force vector f and
%                                 length L.
x = [0 ; L];
z = [f(2) ; -f(8)];
hold on;
title('Shear Force Diagram in Y Direction');
plot(x,z);
y1 = [0 ; 0];
plot(x,y1,'k')
```

```
function y = SpaceFrameElementShearZDiagram(f, L)
%SpaceFrameElementShearZDiagram    This function plots the shear force
%                                  diagram for the space frame element
%                                  with nodal force vector f and
%                                  length L.
x = [0 ; L];
z = [f(3) ; -f(9)];
hold on;
title('Shear Force Diagram in Z Direction');
plot(x,z);
y1 = [0 ; 0];
plot(x,y1,'k')
```

```
function y = SpaceFrameElementTorsionDiagram(f, L)
%SpaceFrameElementTorsionDiagram    This function plots the torsion
%                                   diagram for the space frame
%                                   element with nodal force vector f
%                                   and length L.
x = [0 ; L];
z = [f(4) ; -f(10)];
hold on;
title('Torsion Diagram');
plot(x,z);
y1 = [0 ; 0];
plot(x,y1,'k')
```

```
function y = SpaceFrameElementMomentYDiagram(f, L)
%SpaceFrameElementMomentYDiagram    This function plots the bending
%                                   moment diagram for the space frame
%                                   element with nodal force vector f
%                                   and length L.
x = [0 ; L];
z = [f(5) ; -f(11)];
hold on;
title('Bending Moment Diagram along Y Axis');
plot(x,z);
y1 = [0 ; 0];
plot(x,y1,'k')
```

```
function y = SpaceFrameElementMomentZDiagram(f, L)
%SpaceFrameElementMomentZDiagram    This function plots the bending
%                                   moment diagram for the space frame
%                                   element with nodal force vector f
%                                   and length L.
x = [0 ; L];
z = [f(6) ; -f(12)];
hold on;
title('Bending Moment Diagram along Z Axis');
plot(x,z);
y1 = [0 ; 0];
plot(x,y1,'k')
```

Example 10.1:

Consider the space frame shown in Fig. 10.2. Given $E = 210\,\text{GPa}$, $G = 84\,\text{GPa}$, $A = 2 \times 10^{-2}\,\text{m}^2$, $I_y = 10 \times 10^{-5}\,\text{m}^4$, $I_z = 20 \times 10^{-5}\,\text{m}^4$, and $J = 5 \times 10^{-5}\,\text{m}^4$, determine:

1. the global stiffness matrix for the structure.
2. the displacements and rotations at node 1.
3. the reactions at nodes 2, 3, and 4.
4. the forces (axial, shears, torsion, bending moments) in each element.

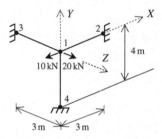

Fig. 10.2. Space Frame with Three Elements for Example 10.1

Solution:

Use the six steps outlined in Chap. 1 to solve this problem using the space frame element.

Step 1 – Discretizing the Domain:

This problem is already discretized. The domain is subdivided into three elements and four nodes. The units used in the MATLAB calculations are kN and meter. Table 10.1 shows the element connectivity for this example.

Table 10.1. Element Connectivity for Example 10.1

Element Number	Node i	Node j
1	1	2
2	1	3
3	1	4

Step 2 – Writing the Element Stiffness Matrices:

The three element stiffness matrices k_1, k_2, and k_3 are obtained by making calls to the MATLAB function *SpaceFrameElementStiffness*. Each matrix has size 12×12.

```
» E=210e6

E =

   210000000

» G=84e6

G =

   84000000

» A=2e-2

A =

   0.0200

» Iy=10e-5

Iy =

  1.0000e-004

» Iz=20e-5

Iz =

  2.0000e-004

» J=5e-5

J =

  5.0000e-005

» k1=SpaceFrameElementStiffness(E,G,A,Iy,Iz,J,0,0,0,3,0,0)

k1 =

  1.0e+006  *
```

Columns 1 through 7

```
 1.4000        0        0        0        0        0 -1.4000
      0   0.0187        0        0        0   0.0280        0
      0        0   0.0093        0  -0.0140        0        0
      0        0        0   0.0014        0        0        0
      0        0  -0.0140        0   0.0280        0        0
      0   0.0280        0        0        0   0.0560        0
-1.4000        0        0        0        0        0  1.4000
      0  -0.0187        0        0        0  -0.0280        0
      0        0  -0.0093        0   0.0140        0        0
      0        0        0  -0.0014        0        0        0
      0        0  -0.0140        0   0.0140        0        0
      0   0.0280        0        0        0   0.0280        0
```

Columns 8 through 12

```
      0        0        0        0        0
-0.0187        0        0        0   0.0280
      0  -0.0093        0  -0.0140        0
      0        0  -0.0014        0        0
      0   0.0140        0   0.0140        0
-0.0280        0        0        0   0.0280
      0        0        0        0        0
 0.0187        0        0        0  -0.0280
      0   0.0093        0   0.0140        0
      0        0   0.0014        0        0
      0   0.0140        0   0.0280        0
-0.0280        0        0        0   0.0560
```

» k2=SpaceFrameElementStiffness(E,G,A,Iy,Iz,J,0,0,0,0,0,-3)

k2 =

  1.0e+006 *

Columns 1 through 7

```
0.0093        0        0        0  -0.0140        0  -0.0093
     0   0.0187        0   0.0280        0        0        0
     0        0   1.4000        0        0        0        0
     0   0.0280        0   0.0560        0        0        0
```

```
-0.0140          0          0          0  0.0280          0  0.0140
      0          0          0          0          0   0.0014        0
-0.0093          0          0          0  0.0140          0  0.0093
      0   -0.0187          0   -0.0280          0          0        0
      0          0   -1.4000          0          0          0        0
      0    0.0280          0    0.0280          0          0        0
-0.0140          0          0          0   0.0140          0  0.0140
      0          0          0          0          0  -0.0014        0
```

Columns 8 through 12

```
      0          0          0   -0.0140          0
-0.0187          0    0.0280          0          0
      0   -1.4000          0          0          0
-0.0280          0    0.0280          0          0
      0          0          0    0.0140          0
      0          0          0          0   -0.0014
      0          0          0    0.0140          0
 0.0187          0   -0.0280          0          0
      0    1.4000          0          0          0
-0.0280          0    0.0560          0          0
      0          0          0    0.0280          0
      0          0          0          0    0.0014
```

» k3=SpaceFrameElementStiffness(E,G,A,Iy,Iz,J,0,0,0,0,-4,0)

k3 =

   1.0e+006 *

Columns 1 through 7

```
 0.0079          0          0          0          0   0.0158  -0.0079
      0    1.0500          0          0          0          0        0
      0          0    0.0039  -0.0079          0          0        0
      0          0   -0.0079   0.0210          0          0        0
      0          0          0          0   0.0010          0        0
 0.0158          0          0          0          0   0.0420  -0.0158
-0.0079          0          0          0          0  -0.0158   0.0079
      0   -1.0500          0          0          0          0        0
```

```
        0       0   -0.0039    0.0079         0            0            0
        0       0   -0.0079    0.0105         0            0            0
        0       0        0        0      -0.0010          0            0
   0.0158       0        0        0           0       0.0210     -0.0158
```

Columns 8 through 12

```
        0              0            0            0       0.0158
  -1.0500              0            0            0            0
        0        -0.0039      -0.0079          0            0
        0         0.0079       0.0105          0            0
        0              0            0      -0.0010          0
        0              0            0            0       0.0210
        0              0            0            0      -0.0158
   1.0500              0            0            0            0
        0         0.0039       0.0079          0            0
        0         0.0079       0.0210          0            0
        0              0            0       0.0010          0
        0              0            0            0       0.0420
```

Step 3 – Assembling the Global Stiffness Matrix:

Since the structure has four nodes, the size of the global stiffness matrix is 24×24. Therefore to obtain K we first set up a zero matrix of size 24×24 then make three calls to the MATLAB function *SpaceFrameAssemble* since we have three space frame elements in the structure. Each call to the function will assemble one element. The following are the MATLAB commands where it is noted that the only the final result is shown.

» K=zeros(24,24);

» K=SpaceFrameAssemble(K,k1,1,2);

» K=SpaceFrameAssemble(K,k2,1,3);

» K=SpaceFrameAssemble(K,k3,1,4)

K =

  1.0e+006 *

Columns 1 through 7

```
 1.4172        0         0         0  -0.0140    0.0158  -1.4000
      0   1.0873         0    0.0280         0    0.0280        0
      0        0    1.4133   -0.0079  -0.0140         0        0
      0   0.0280   -0.0079    0.0784         0         0        0
-0.0140        0   -0.0140         0    0.0570         0        0
 0.0158   0.0280         0         0         0    0.0994        0
-1.4000        0         0         0         0         0   1.4000
      0  -0.0187         0         0         0   -0.0280        0
      0        0   -0.0093         0    0.0140         0        0
      0        0         0   -0.0014         0         0        0
      0        0   -0.0140         0    0.0140         0        0
      0   0.0280         0         0         0    0.0280        0
-0.0093        0         0         0    0.0140         0        0
      0  -0.0187         0   -0.0280         0         0        0
      0        0   -1.4000         0         0         0        0
      0   0.0280         0    0.0280         0         0        0
-0.0140        0         0         0    0.0140         0        0
      0        0         0         0         0   -0.0014        0
-0.0079        0         0         0         0   -0.0158        0
      0  -1.0500         0         0         0         0        0
      0        0   -0.0039    0.0079         0         0        0
      0        0   -0.0079    0.0105         0         0        0
      0        0         0         0   -0.0010         0        0
 0.0158        0         0         0         0    0.0210        0
```

Columns 8 through 14

```
      0        0         0         0         0   -0.0093        0
-0.0187        0         0         0    0.0280         0  -0.0187
      0  -0.0093         0   -0.0140         0         0        0
      0        0   -0.0014         0         0         0  -0.0280
      0   0.0140         0    0.0140         0    0.0140        0
-0.0280        0         0         0    0.0280         0        0
      0        0         0         0         0         0        0
 0.0187        0         0         0   -0.0280         0        0
      0   0.0093         0    0.0140         0         0        0
      0        0    0.0014         0         0         0        0
      0   0.0140         0    0.0280         0         0        0
```

```
 -0.0280        0        0        0   0.0560           0           0
       0        0        0        0        0      0.0093           0
       0        0        0        0        0           0      0.0187
       0        0        0        0        0           0           0
       0        0        0        0        0           0     -0.0280
       0        0        0        0        0      0.0140           0
       0        0        0        0        0           0           0
       0        0        0        0        0           0           0
       0        0        0        0        0           0           0
       0        0        0        0        0           0           0
       0        0        0        0        0           0           0
       0        0        0        0        0           0           0
       0        0        0        0        0           0           0
       0        0        0        0        0           0           0
```

Columns 15 through 21

```
       0        0  -0.0140        0  -0.0079           0           0
       0   0.0280        0        0        0     -1.0500           0
 -1.4000        0        0        0        0           0     -0.0039
       0   0.0280        0        0        0           0      0.0079
       0        0   0.0140        0        0           0           0
       0        0        0  -0.0014  -0.0158           0           0
       0        0        0        0        0           0           0
       0        0        0        0        0           0           0
       0        0        0        0        0           0           0
       0        0        0        0        0           0           0
       0        0        0        0        0           0           0
       0        0   0.0140        0        0           0           0
       0  -0.0280        0        0        0           0           0
  1.4000        0        0        0        0           0           0
       0   0.0560        0        0        0           0           0
       0        0   0.0280        0        0           0           0
       0        0        0   0.0014        0           0           0
       0        0        0        0   0.0079           0           0
       0        0        0        0        0      1.0500           0
       0        0        0        0        0           0      0.0039
       0        0        0        0        0           0      0.0079
       0        0        0        0        0           0           0
       0        0        0        0  -0.0158           0           0
```

Columns 22 through 24

0	0	0.0158
0	0	0
-0.0079	0	0
0.0105	0	0
0	-0.0010	0
0	0	0.0210
0	0	0
0	0	0
0	0	0
0	0	0
0	0	0
0	0	0
0	0	0
0	0	0
0	0	0
0	0	0
0	0	0
0	0	-0.0158
0	0	0
0.0079	0	0
0.0210	0	0
0	0.0010	0
0	0	0.0420

Step 4 – Applying the Boundary Conditions:

The matrix (10.5) for this structure is obtained using the global stiffness matrix shown above (last martrix shown above). This matrix equation will not be shown explicitly below because $[K]$ is very large (of size 24×24). We will proceed directly to the boundary conditions.

The boundary conditions for this problem are given as:

$$F_{1x} = -10, \ F_{1y} = 0, F_{1z} = 20, \ M_{1x} = M_{1y} = M_{1z} = 0$$
$$U_{2x} = U_{2y} = U_{2z} = \phi_{2x} = \phi_{2y} = \phi_{2z} = 0$$
$$U_{3x} = U_{3y} = U_{3z} = \phi_{3x} = \phi_{3y} = \phi_{3z} = 0$$
$$U_{4x} = U_{4y} = U_{4z} = \phi_{4x} = \phi_{4y} = \phi_{4z} = 0 \tag{10.7}$$

We next insert the above conditions into the global stiffness equation (not shown) and proceed to the solution step below.

Step 5 – Solving the Equations:

Solving the resulting system of equations will be performed by partitioning (manually) and Gaussian elimination (with MATLAB). First we partition the equation by extracting the submatrix in rows 1 to 6 and columns 1 to 6. Therefore we obtain:

$$10^6 \begin{bmatrix} 1.4172 & 0 & 0 & 0 & -0.0140 & 0.0158 \\ 0 & 1.0873 & 0 & 0.0280 & 0 & 0.0280 \\ 0 & 0 & 1.4133 & -0.0079 & -0.0140 & 0 \\ 0 & 0.0280 & -0.0079 & 0.0784 & 0 & 0 \\ -0.0140 & 0 & -0.0140 & 0 & 0.0570 & 0 \\ 0.0158 & 0.0280 & 0 & 0 & 0 & 0.0994 \end{bmatrix} \begin{Bmatrix} U_{1x} \\ U_{1y} \\ U_{1z} \\ \phi_{1x} \\ \phi_{1y} \\ \phi_{1z} \end{Bmatrix} = \begin{Bmatrix} -10 \\ 0 \\ 20 \\ 0 \\ 0 \\ 0 \end{Bmatrix} \qquad (10.8)$$

The solution of the above system is obtained using MATLAB as follows. Note that the backslash operator "\" is used for Gaussian elimination.

```
» k=K(1:6,1:6)

k =

   1.0e+006 *

      1.4172        0        0        0  -0.0140   0.0158
           0   1.0873        0   0.0280        0   0.0280
           0        0   1.4133  -0.0079  -0.0140        0
           0   0.0280  -0.0079   0.0784        0        0
     -0.0140        0  -0.0140        0   0.0570        0
      0.0158   0.0280        0        0        0   0.0994

» f=[-10 ; 0 ; 20 ; 0 ; 0 ; 0]

f =

    -10
      0
     20
      0
      0
      0
```

```
» u=k\f

u =

   1.0e-004 *

   -0.0705
   -0.0007
    0.1418
    0.0145
    0.0175
    0.0114
```

It is now clear that the three displacements at node 1 are -0.0705×10^{-4} m, -0.0007×10^{-4} m, and 0.1418×10^{-4} m along the X, Y, and Z axes, respectively. Also the three rotations at node 1 are 0.0145×10^{-4} rad, 0.0175×10^{-4} rad, and 0.0114×10^{-4} rad along the X, Y, and Z axes, respectively.

Step 6 – Post-processing:

In this step, we obtain the reactions at nodes 2, 3, and 4, and the forces and moments in each space frame element using MATLAB as follows. First we set up the global nodal displacement vector U, then we calculate the global nodal force vector F.

```
» U=[u ; 0;0;0;0;0;0;0;0;0;0;0;0;0;0;0;0;0;0]

U =

   1.0e-004 *

    -0.0705
    -0.0007
     0.1418
     0.0145
     0.0175
     0.0114
          0
          0
          0
          0
          0
          0
          0
          0
          0
```

```
                    0
                    0
                    0
                    0
                    0
                    0
                    0
                    0
                    0
```

» F=K*U

F =

```
  -10.0000
         0
   20.0000
    0.0000
    0.0000
    0.0000
    9.8721
   -0.0306
   -0.1078
   -0.0020
   -0.1740
    0.0299
    0.0903
   -0.0393
  -19.8477
    0.0387
    0.1232
   -0.0016
    0.0376
    0.0699
   -0.0444
   -0.0964
   -0.0018
   -0.0872
```

Next we set up the element nodal displacement vectors u_1, u_2, and u_3, then we calculate the element force vectors f_1, f_2, and f_3 by making calls to the MATLAB function *SpaceFrameElementForces*.

» u1=[U(1);U(2);U(3);U(4);U(5);U(6);
       U(7);U(8);U(9);U(10);U(11);U(12)]

```
u1 =
   1.0e-004 *

   -0.0705
   -0.0007
    0.1418
    0.0145
    0.0175
    0.0114
         0
         0
         0
         0
         0
         0
```

» u2=[U(1);U(2);U(3);U(4);U(5);U(6);U(13);
      U(14);U(15);U(16);U(17);U(18)]

```
u2 =

   1.0e-004 *

   -0.0705
   -0.0007
    0.1418
    0.0145
    0.0175
    0.0114
         0
         0
         0
         0
         0
         0
```

» u3=[U(1);U(2);U(3);U(4);U(5);U(6);U(19);
      U(20);U(21);U(22);U(23);U(24)]

```
u3 =

   1.0e-004 *

   -0.0705
   -0.0007
```

```
        0.1418
        0.0145
        0.0175
        0.0114
             0
             0
             0
             0
             0
             0
```

»f1=SpaceFrameElementForces(E,G,A,Iy,Iz,J,0,0,0,3,0,0,u1)

f1 =

```
   -9.8721
    0.0306
    0.1078
    0.0020
   -0.1495
    0.0618
    9.8721
   -0.0306
   -0.1078
   -0.0020
   -0.1740
    0.0299
```

»f2=SpaceFrameElementForces(E,G,A,Iy,Iz,J,0,0,0,0,0,-3,u2)

f2 =

```
  -19.8477
    0.0393
   -0.0903
   -0.0016
    0.1477
    0.0792
   19.8477
   -0.0393
    0.0903
    0.0016
    0.1232
    0.0387
```

```
»f3=SpaceFrameElementForces(E,G,A,Iy,Iz,J,0,0,0,0,-4,0,u3)
```

f3 =

```
     0.0699
    -0.0376
     0.0444
    -0.0018
    -0.0812
    -0.0633
    -0.0699
     0.0376
    -0.0444
     0.0018
    -0.0964
    -0.0872
```

Problems:

Problem 10.1:

Consider the space frame shown in Fig. 10.3. Given $E = 210\,\text{GPa}$, $G = 84\,\text{GPa}$, $A = 2 \times 10^{-2}\,\text{m}^2$, $I_y = 10 \times 10^{-5}\,\text{m}^4$, $I_z = 20 \times 10^{-5}\,\text{m}^4$, and $J = 5 \times 10^{-5}\,\text{m}^4$, determine:

1. the global stiffness matrix for the structure.
2. the displacements and rotations at nodes 5, 6, 7, and 8.
3. the reactions at nodes 1, 2, 3, and 4.
4. the forces (axial, shears, torsion, bending moments) in each element.

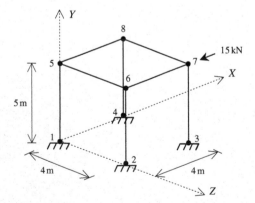

Fig. 10.3. Space Frame with Eight Elements for Problem 10.1

11 The Linear Triangular Element

11.1
Basic Equations

The linear triangular element is a two-dimensional finite element with both local and global coordinates. It is characterized by linear shape functions. This element can be used for plane stress or plane strain problems in elasticity. It is also called the constant strain triangle. The linear triangular element has modulus of elasticity E, Poisson's ratio ν, and thickness t. Each linear triangle has three nodes with two in-plane degrees of freedom at each node as shown in Fig. 11.1. The global coordinates of the three nodes are denoted by (x_i, y_i), (x_j, y_j), and (x_m, y_m). The order of the nodes for each element is important – they should be listed in a counterclockwise direction starting from any node. The area of each triangle should be positive – you can actually check this by using the MATLAB function *LinearTriangleElementArea* which is written specifically for this purpose. In this case the element stiffness matrix is given by (see [1]).

$$[k] = tA[B]^T[D][B] \tag{11.1}$$

where A is the area of the element given by

$$2A = x_i(y_j - y_m) + x_j(y_m - y_i) + x_m(y_i - y_j) \tag{11.2}$$

and the matrix $[B]$ is given by

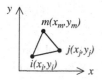

Fig. 11.1. The Linear Triangular Element

$$[B] = \frac{1}{2A} \begin{bmatrix} \beta_i & 0 & \beta_j & 0 & \beta_m & 0 \\ 0 & \gamma_i & 0 & \gamma_j & 0 & \gamma_m \\ \gamma_i & \beta_i & \gamma_j & \beta_j & \gamma_m & \beta_m \end{bmatrix} \tag{11.3}$$

where $\beta_i, \beta_j, \beta_m, \gamma_i, \gamma_j$, and γ_m are given by

$$\begin{aligned} \beta_i &= y_j - y_m \\ \beta_j &= y_m - y_i \\ \beta_m &= y_i - y_j \\ \gamma_i &= x_m - x_j \\ \gamma_j &= x_i - x_m \\ \gamma_m &= x_j - x_i \end{aligned} \tag{11.4}$$

For cases of plane stress the matrix $[D]$ is given by

$$[D] = \frac{E}{1 - \nu^2} \begin{bmatrix} 1 & \nu & 0 \\ \nu & 1 & 0 \\ 0 & 0 & \frac{1-\nu}{2} \end{bmatrix} \tag{11.5}$$

For cases of plane strain the matrix $[D]$ is given by

$$[D] = \frac{E}{(1+\nu)(1-2\nu)} \begin{bmatrix} 1-\nu & \nu & 0 \\ \nu & 1-\nu & 0 \\ 0 & 0 & \frac{1-2\nu}{2} \end{bmatrix} \tag{11.6}$$

It is clear that the linear triangular element has six degrees of freedom – two at each node. Consequently for a structure with n nodes, the global stiffness matrix K will be of size $2n \times 2n$ (since we have two degrees of freedom at each node). The global stiffness matrix K is assembled by making calls to the MATLAB function *LinearTriangleAssemble* which is written specifically for this purpose. This process will be illustrated in detail in the examples.

Once the global stiffness matrix K is obtained we have the following structure equation:

$$[K]\{U\} = \{F\} \tag{11.7}$$

where U is the global nodal displacement vector and F is the global nodal force vector. At this step the boundary conditions are applied manually to the vectors U and

F. Then the matrix (11.7) is solved by partitioning and Gaussian elimination. Finally once the unknown displacements and reactions are found, the stress vector is obtained for each element as follows:

$$\{\sigma\} = [D][B]\{u\} \tag{11.8}$$

where σ is the stress vector in the element (of size 3×1) and u is the 6×1 element displacement vector. The vector σ is written for each element as $\{\sigma\} = [\sigma_x \sigma_y \tau_{xy}]^{\mathrm{T}}$.

11.2
MATLAB Functions Used

The five MATLAB functions used for the linear triangular element are:

LinearTriangleElementArea$(x_i, y_i, x_j, y_j, x_m, y_m)$ – This function returns the element area given the coordinates of the first node (x_i, y_i), the coordinates of the second node (x_j, y_j), and the coordinates of the third node (x_m, y_m).

LinearTriangleElementStiffness$(E, NU, t, x_i, y_i, x_j, y_j, x_m, y_m, p)$ – This function calculates the element stiffness matrix for each linear triangle with modulus of elasticity E, Poisson's ratio NU, thickness t, and coordinates (x_i, y_i) for the first node, (x_j, y_j) for the second node, and (x_m, y_m) for the third node. Use $p = 1$ for cases of plane stress and $p = 2$ for cases of plane strain. It returns the 6×6 element stiffness matrix k.

LinearTriangleAssemble(K, k, i, j, m) – This function assembles the element stiffness matrix k of the linear triangle joining nodes i, j, and m into the global stiffness matrix K. It returns the $2n \times 2n$ global stiffness matrix K every time an element is assembled.

LinearTriangleElementStresses$(E, NU, x_i, y_i, x_j, y_j, x_m, y_m, p, u)$ – This function calculates the element stresses using the modulus of elasticity E, Poisson's ratio NU, the coordinates (x_i, y_i) for the first node, (x_j, y_j) for the second node, and (x_m, y_m) for the third node, and the element displacement vector u. Use $p = 1$ for cases of plane stress and $p = 2$ for cases of plane strain. It returns the stress vector for the element.

LinearTriangleElementPStresses$(sigma)$ – This function calculates the element principal stresses using the element stress vector *sigma*. It returns a 3×1 vector in the form $[sigma1\ sigma2\ theta]^T$ where *sigma1* and *sigma2* are the principal stresses for the element and *theta* is the principal angle.

The following is a listing of the MATLAB source code for each function:

```
function y = LinearTriangleElementArea(xi,yi,xj,yj,xm,ym)
%LinearTriangleElementArea         This function returns the area of the
%                                  linear triangular element whose first
%                                  node has coordinates (xi,yi), second
%                                  node has coordinates (xj,yj), and
%                                  third node has coordinates (xm,ym).
y = (xi*(yj-ym) + xj*(ym-yi) + xm*(yi-yj))/2;
```

```
function y = LinearTriangleElementStiffness(E,NU,t,xi,yi,xj,yj,xm,ym,p)
%LinearTriangleElementStiffness  This function returns the element
%                                  stiffness matrix for a linear
%                                  triangular element with modulus of
%                                  elasticity E, Poisson's ratio NU,
%                                  thickness t, coordinates of the
%                                  first node (xi,yi), coordinates of
%                                  the second node (xj,yj), and
%                                  coordinates of the third node
%                                  (xm,ym). Use p = 1 for cases of
%                                  plane stress, and p = 2 for cases
%                                  of plane strain.
%                                  The size of the element stiffness
%                                  matrix is 6 x 6.
A = (xi*(yj-ym) + xj*(ym-yi) + xm*(yi-yj))/2;
betai = yj-ym;
betaj = ym-yi;
betam = yi-yj;
gammai = xm-xj;
gammaj = xi-xm;
gammam = xj-xi;
B = [betai 0 betaj 0 betam 0 ;
  0 gammai 0 gammaj 0 gammam ;
  gammai betai gammaj betaj gammam betam]/(2*A);
if p == 1
   D = (E/(1-NU*NU))*[1 NU 0 ; NU 1 0 ; 0 0 (1-NU)/2];
elseif p == 2
   D = (E/(1+NU)/(1-2*NU))*[1-NU NU 0 ; NU 1-NU 0 ; 0 0 (1-
2*NU)/2];
end
y = t*A*B'*D*B;
```

```
function y = LinearTriangleAssemble(K,k,i,j,m)
%LinearTriangleAssemble    This function assembles the element
%                          stiffness matrix k of the linear
%                          triangular element with nodes i, j,
%                          and m into the global stiffness matrix K.
%                          This function returns the global stiffness
%                          matrix K after the element stiffness matrix
%                          k is assembled.
K(2*i-1,2*i-1) = K(2*i-1,2*i-1) + k(1,1);
K(2*i-1,2*i) = K(2*i-1,2*i) + k(1, 2);
K(2*i-1,2*j-1) = K(2*i-1,2*j-1) + k(1,3);
```

```
K(2*i-1,2*j) = K(2*i-1,2*j) + k(1,4);
K(2*i-1,2*m-1) = K(2*i-1,2*m-1) + k(1,5);
K(2*i-1,2*m) = K(2*i-1,2*m) + k(1,6);
K(2*i,2*i-1) = K(2*i,2*i-1) + k(2,1);
K(2*i,2*i) = K(2*i,2*i) + k(2,2);
K(2*i,2*j-1) = K(2*i,2*j-1) + k(2,3);
K(2*i,2*j) = K(2*i,2*j) + k(2,4);
K(2*i,2*m-1) = K(2*i,2*m-1) + k(2,5);
K(2*i,2*m) = K(2*i,2*m) + k(2,6);
K(2*j-1,2*i-1) = K(2*j-1,2*i-1) + k(3,1);
K(2*j-1,2*i) = K(2*j-1,2*i) + k(3,2);
K(2*j-1,2*j-1) = K(2*j-1,2*j-1) + k(3,3);
K(2*j-1,2*j) = K(2*j-1,2*j) + k(3,4);
K(2*j-1,2*m-1) = K(2*j-1,2*m-1) + k(3,5);
K(2*j-1,2*m) = K(2*j-1,2*m) + k(3,6);
K(2*j,2*i-1) = K(2*j,2*i-1) + k(4,1);
K(2*j,2*i) = K(2*j,2*i) + k(4,2);
K(2*j,2*j-1) = K(2*j,2*j-1) + k(4,3);
K(2*j,2*j) = K(2*j,2*j) + k(4,4);
K(2*j,2*m-1) = K(2*j,2*m-1) + k(4,5);
K(2*j,2*m) = K(2*j,2*m) + k(4,6);
K(2*m-1,2*i-1) = K(2*m-1,2*i-1) + k(5,1);
K(2*m-1,2*i) = K(2*m-1,2*i) + k(5,2);
K(2*m-1,2*j-1) = K(2*m-1,2*j-1) + k(5,3);
K(2*m-1,2*j) = K(2*m-1,2*j) + k(5,4);
K(2*m-1,2*m-1) = K(2*m-1,2*m-1) + k(5,5);
K(2*m-1,2*m) = K(2*m-1,2*m) + k(5,6);
K(2*m,2*i-1) = K(2*m,2*i-1) + k(6,1);
K(2*m,2*i) = K(2*m,2*i) + k(6,2);
K(2*m,2*j-1) = K(2*m,2*j-1) + k(6,3);
K(2*m,2*j) = K(2*m,2*j) + k(6,4);
K(2*m,2*m-1) = K(2*m,2*m-1) + k(6,5);
K(2*m,2*m) = K(2*m,2*m) + k(6,6);
y = K;
```

```
function y = LinearTriangleElementStresses(E,NU,xi,yi,xj,yj,xm,ym,p,u)
%LinearTriangleElementStressesThis function returns the element
%                              stress vector for a linear
%                              triangular element with modulus of
%                              elasticity E, Poisson's ratio NU,
%                              coordinates of the
%                              first node (xi,yi), coordinates of
%                              the second node (xj,yj),
%                              coordinates of the third node
%                              (xm,ym), and element displacement
%                              vector u. Use p = 1 for cases of
%                              plane stress, and p = 2 for cases
%                              of plane strain.
%                              The size of the element stress
%                              vector is 3 x 1.
A = (xi*(yj-ym) + xj*(ym-yi) + xm*(yi-yj))/2;
betai = yj-ym;
betaj = ym-yi;
betam = yi-yj;
gammai = xm-xj;
```

```
gammaj = xi-xm;
gammam = xj-xi;
B = [betai 0 betaj 0 betam 0 ;
   0 gammai 0 gammaj 0 gammam ;
   gammai betai gammaj betaj gammam betam]/(2*A);
if p == 1
   D = (E/(1-NU*NU))*[1 NU 0 ; NU 1 0 ; 0 0 (1-NU)/2];
elseif p == 2
   D = (E/(1+NU)/(1-2*NU))*[1-NU NU 0 ; NU 1-NU 0 ; 0 0 (1-
2*NU)/2];
end
y = D*B*u;
```

```
function y = LinearTriangleElementPStresses(sigma)
%LinearTriangleElementPStresses      This function returns the element
%                                    principal stresses and their
%                                    angle given the element
%                                    stress vector.
R = (sigma(1) + sigma(2))/2;
Q = ((sigma(1) - sigma(2))/2)^2 + sigma(3)*sigma(3);
M = 2*sigma(3)/(sigma(1) - sigma(2));
s1 = R + sqrt(Q);
s2 = R - sqrt(Q);
theta = (atan(M)/2)*180/pi;
y = [s1 ; s2 ; theta];
```

Example 11.1:

Consider the thin plate subjected to a uniformly distributed load as shown in Fig. 11.2. The plate is discretized using two linear triangular elements as shown in Fig. 11.3. Given $E = 210\,\text{GPa}$, $\nu = 0.3$, $t = 0.025\,\text{m}$, and $w = 3000\,\text{kN/m}^2$, determine:

1. the global stiffness matrix for the structure.
2. the horizontal and vertical displacements at nodes 2 and 3.
3. the reactions at nodes 1 and 4.
4. the stresses in each element.
5. the principal stresses and principal angle for each element.

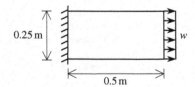

0.25 m

w

0.5 m

Fig. 11.2. Thin Plate for Example 11.1

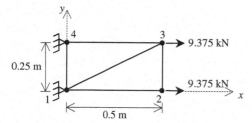

Fig. 11.3. Discretization of Thin Plate Using Two Linear Triangles

Solution

Use the six steps outlined in Chap. 1 to solve this problem using the linear triangular element.

Step 1 – Discretizing the Domain:

We subdivide the plate into two elements only for illustration purposes. More elements must be used in order to obtain reliable results. Thus the domain is subdivided into two elements and four nodes as shown in Fig. 11.3. The total force due to the distributed load is divided equally between nodes 2 and 3. Since the plate is thin, a case of plane stress is assumed. The units used in the MATLAB calculations are kN and meter. Table 11.1 shows the element connectivity for this example.

Table 11.1. Element Connectivity for Example 11.1

Element Number	Node i	Node j	Node m
1	1	3	4
2	1	2	3

Step 2 – Writing the Element Stiffness Matrices:

The two element stiffness matrices k_1 and k_2 are obtained by making calls to the MATLAB function *LinearTriangleElementStiffness*. Each matrix has size 6×6.

```
» E=210e6

E =

   210000000

» NU=0.3
```

NU =

   0.3000

» t=0.025

t =

   0.0250

» k1=LinearTriangleElementStiffness(E,NU,t,0,0,0.5,
       0.25,0,0.25,1)

k1 =

   1.0e+006 *

```
    2.0192         0         0   -1.0096   -2.0192    1.0096
         0    5.7692   -0.8654         0    0.8654   -5.7692
         0   -0.8654    1.4423         0   -1.4423    0.8654
   -1.0096         0         0    0.5048    1.0096   -0.5048
   -2.0192    0.8654   -1.4423    1.0096    3.4615   -1.8750
    1.0096   -5.7692    0.8654   -0.5048   -1.8750    6.2740
```

» k2=LinearTriangleElementStiffness(E,NU,t,0,0,0.5,
       0,0.5,0.25,1)

k2 =

   1.0e+006 *

```
    1.4423         0   -1.4423    0.8654         0   -0.8654
         0    0.5048    1.0096   -0.5048   -1.0096         0
   -1.4423    1.0096    3.4615   -1.8750   -2.0192    0.8654
    0.8654   -0.5048   -1.8750    6.2740    1.0096   -5.7692
         0   -1.0096   -2.0192    1.0096    2.0192         0
   -0.8654         0    0.8654   -5.7692         0    5.7692
```

Step 3 – Assembling the Global Stiffness Matrix:

Since the structure has four nodes, the size of the global stiffness matrix is 8×8. Therefore to obtain K we first set up a zero matrix of size 8×8 then make two calls to the MATLAB function *LinearTriangleAssemble* since we have two elements in the structure. Each call to the function will assemble one element. The following are the MATLAB commands:

```
» K=zeros(8,8)

K =

     0    0    0    0    0    0    0    0
     0    0    0    0    0    0    0    0
     0    0    0    0    0    0    0    0
     0    0    0    0    0    0    0    0
     0    0    0    0    0    0    0    0
     0    0    0    0    0    0    0    0
     0    0    0    0    0    0    0    0
     0    0    0    0    0    0    0    0

» K=LinearTriangleAssemble(K,k1,1,3,4)

K =

   1.0e+006 *

  Columns 1 through 7

    2.0192         0    0    0         0   -1.0096   -2.0192
         0    5.7692    0    0   -0.8654         0    0.8654
         0         0    0    0         0         0         0
         0         0    0    0         0         0         0
         0   -0.8654    0    0    1.4423         0   -1.4423
   -1.0096         0    0    0         0    0.5048    1.0096
   -2.0192    0.8654    0    0   -1.4423    1.0096    3.4615
    1.0096   -5.7692    0    0    0.8654   -0.5048   -1.8750

  Column 8

    1.0096
   -5.7692
         0
         0
    0.8654
   -0.5048
   -1.8750
    6.2740
```

```
» K=LinearTriangleAssemble(K,k2,1,2,3)

K =

   1.0e+006 *

   Columns 1 through 7

     3.4615          0 -1.4423   0.8654          0 -1.8750 -2.0192
          0    6.2740   1.0096  -0.5048  -1.8750          0   0.8654
    -1.4423    1.0096   3.4615  -1.8750  -2.0192   0.8654         0
     0.8654   -0.5048  -1.8750   6.2740   1.0096  -5.7692         0
          0   -1.8750  -2.0192   1.0096   3.4615          0 -1.4423
    -1.8750          0   0.8654  -5.7692          0   6.2740   1.0096
    -2.0192    0.8654        0         0  -1.4423   1.0096   3.4615
     1.0096   -5.7692        0         0   0.8654  -0.5048  -1.8750

   Column 8

     1.0096
    -5.7692
          0
          0
     0.8654
    -0.5048
    -1.8750
     6.2740
```

Step 4 – Applying the Boundary Conditions:

The matrix (11.7) for this structure is obtained as follows using the global stiffness matrix obtained in the previous step:

$$
10^6 \begin{bmatrix}
 3.46 & 0   & -1.44 & 0.87 & 0   & -1.88 & -2.02 & 1.01 \\
 0   & 6.27 & 1.01 & -0.50 & -1.88 & 0   & 0.87 & -5.77 \\
-1.44 & 1.01 & 3.46 & -1.88 & -2.02 & 0.87 & 0   & 0   \\
 0.87 & -0.50 & -1.88 & 6.27 & 1.01 & -5.77 & 0   & 0   \\
 0   & -1.88 & -2.02 & 1.01 & 3.46 & 0   & -1.44 & 0.87 \\
-1.88 & 0   & 0.87 & -5.77 & 0   & 6.27 & 1.01 & -0.50 \\
-2.02 & 0.87 & 0   & 0   & -1.44 & 1.01 & 3.46 & -1.88 \\
 1.01 & -5.77 & 0   & 0   & 0.87 & -0.50 & -1.88 & 6.27
\end{bmatrix}
$$

$$
\begin{Bmatrix}
U_{1x} \\
U_{1y} \\
U_{2x} \\
U_{2y} \\
U_{3x} \\
U_{3y} \\
U_{4x} \\
U_{4y}
\end{Bmatrix}
=
\begin{Bmatrix}
F_{1x} \\
F_{1y} \\
F_{2x} \\
F_{2y} \\
F_{3x} \\
F_{3y} \\
F_{4x} \\
F_{4y}
\end{Bmatrix}
\tag{11.9}
$$

The boundary conditions for this problem are given as:

$$
U_{1x} = U_{1y} = U_{4x} = U_{4y} = 0
$$
$$
F_{2x} = 9.375, \ F_{2y} = 0, \ F_{3x} = 9.375, \ F_{3y} = 0 \tag{11.10}
$$

Inserting the above conditions into (11.9) we obtain:

$$
10^6
\begin{bmatrix}
3.46 & 0 & -1.44 & 0.87 & 0 & -1.88 & -2.02 & 1.01 \\
0 & 6.27 & 1.01 & -0.50 & -1.88 & 0 & 0.87 & -5.77 \\
-1.44 & 1.01 & 3.46 & -1.88 & -2.02 & 0.87 & 0 & 0 \\
0.87 & -0.50 & -1.88 & 6.27 & 1.01 & -5.77 & 0 & 0 \\
0 & -1.88 & -2.02 & 1.01 & 3.46 & 0 & -1.44 & 0.87 \\
-1.88 & 0 & 0.87 & -5.77 & 0 & 6.27 & 1.01 & -0.50 \\
-2.02 & 0.87 & 0 & 0 & -1.44 & 1.01 & 3.46 & -1.88 \\
1.01 & -5.77 & 0 & 0 & 0.87 & -0.50 & -1.88 & 6.27
\end{bmatrix}
$$
$$
\begin{Bmatrix}
0 \\
0 \\
U_{2x} \\
U_{2y} \\
U_{3x} \\
U_{3y} \\
0 \\
0
\end{Bmatrix}
=
\begin{Bmatrix}
F_{1x} \\
F_{1y} \\
9.375 \\
0 \\
9.375 \\
0 \\
F_{4x} \\
F_{4y}
\end{Bmatrix}
\tag{11.11}
$$

Step 5 – Solving the Equations:

Solving the system of equations in (11.11) will be performed by partitioning (manually) and Gaussian elimination (with MATLAB). First we partition (11.11) by extracting the submatrix in rows 3 to 6 and columns 3 to 6. Therefore we obtain:

$$
10^6
\begin{bmatrix}
3.46 & -1.88 & -2.02 & 0.87 \\
-1.88 & 6.27 & 1.01 & -5.77 \\
-2.02 & 1.01 & 3.46 & 0 \\
0.87 & -5.77 & 0 & 6.27
\end{bmatrix}
\begin{Bmatrix}
U_{2x} \\
U_{2y} \\
U_{3x} \\
U_{3y}
\end{Bmatrix}
=
\begin{Bmatrix}
9.375 \\
0 \\
9.375 \\
0
\end{Bmatrix}
\tag{11.12}
$$

The solution of the above system is obtained using MATLAB as follows. Note that the backslash operator "\" is used for Gaussian elimination.

```
» k=K(3:6,3:6)

k =

   1.0e+006 *


      3.4615     -1.8750     -2.0192      0.8654
     -1.8750      6.2740      1.0096     -5.7692
     -2.0192      1.0096      3.4615           0
      0.8654     -5.7692           0      6.2740


» f=[9.375 ; 0 ; 9.375 ; 0]

f =

      9.3750
           0
      9.3750
           0

» u=k\f

u =

   1.0e-005 *

      0.7111
      0.1115
      0.6531
      0.0045
```

It is now clear that the horizontal and vertical displacements at node 2 are 0.7111 m and 0.1115 m, respectively, and the horizontal and vertical displacements at node 3 are 0.6531 m and 0.0045 m, respectively. When a larger number of elements is used we expect to get the same result for the horizontal displacements at nodes 2 and 3.

Step 6 – Post-processing:

In this step, we obtain the reactions at nodes 1 and 4, and the stresses in each element using MATLAB as follows. First we set up the global nodal displacement vector U, then we calculate the global nodal force vector F.

```
» U=[0 ; 0 ; u ; 0 ; 0]

U =

   1.0e-005 *

            0
            0
       0.7111
       0.1115
       0.6531
       0.0045
            0
            0

» F=K*U

F =

      -9.3750
      -5.6295
       9.3750
            0
       9.3750
       0.0000
      -9.3750
       5.6295
```

Thus the horizontal and vertical reactions at node 1 are forces of 9.375 kN (directed to the left) and 5.6295 kN (directed downwards). The horizontal and vertical reactions at node 4 are forces of 9.375 kN (directed to the left) and 5.6295 kN (directed upwards). Obviously force equilibrium is satisfied for this problem. Next we set up the element nodal displacement vectors u_1 and u_2 then we calculate the element stresses *sigma1* and *sigma2* by making calls to the MATLAB function *LinearTriangleElementStresses*.

```
» u1=[U(1) ; U(2) ; U(5) ; U(6) ; U(7) ; U(8)]

u1 =

   1.0e-005 *

            0
            0
       0.6531
```

```
        0.0045
            0
            0
```

» u2=[U(1) ; U(2) ; U(3) ; U(4) ; U(5) ; U(6)]

u2 =

   1.0e-005 *

            0
            0
       0.7111
       0.1115
       0.6531
       0.0045

» sigma1=LinearTriangleElementStresses(E,NU,0,0,0.5,0.25,
          0,0.25,1,u1)

sigma1 =

   1.0e+003 *

       3.0144
       0.9043
       0.0072

» sigma2=LinearTriangleElementStresses(E,NU,0,0,0.5,0,
          0.5,0.25,1,u2)

sigma2 =

   1.0e+003 *

       2.9856
      -0.0036
      -0.0072

Thus it is clear that the stresses in element 1 are $\sigma_x = 3.0144$ MPa (tensile), $\sigma_y = 0.9043$ MPa (tensile), and $\tau_{xy} = 0.0072$ MPa (positive). The stresses in element 2 are $\sigma_x = 2.9856$ MPa (tensile), $\sigma_y = 0.0036$ MPa (compressive), and $\tau_{xy} = 0.0072$ MPa (negative). It is clear that the stresses in the x-direction approach closely the correct value of 3 MPa (tensile). Next we calculate the principal stresses and principal angle for each element by making calls to the MATLAB function *LinearTriangleElementPStresses*.

```
» s1=LinearTriangleElementPStresses(sigma1)

s1 =

   1.0e+003 *

      3.0144
      0.9043
      0.0002
» s2=LinearTriangleElementPStresses(sigma2)

s2 =

   1.0e+003 *

      2.9856
     -0.0036
     -0.0001
```

Thus it is clear that the principal stresses in element 1 are $\sigma_1 = 3.0144\,\text{MPa}$ (tensile), $\sigma_2 = 0.9043\,\text{MPa}$ (tensile), while the principal angle $\theta_p = 0.2°$. The principal stresses in element 2 are $\sigma_1 = 2.9856\,\text{MPa}$ (tensile), $\sigma_2 = 0.0036\,\text{MPa}$ (compressive), while the principal angle $\theta_p = -0.1°$.

Example 11.2:

Consider the thin plate subjected to both a uniformly distributed load and a concentrated load as shown in Fig. 11.4. The plate is discretized using twelve linear triangles as shown in Fig. 11.5. Given $E = 210\,\text{GPa}$, $\nu = 0.3$, $t = 0.025\,\text{m}$, $w = 100\,\text{kN/m}$, and $P = 12.5\,\text{kN}$, determine:

1. the global stiffness matrix for the structure.
2. the horizontal and vertical displacements at each node.
3. the reactions at nodes 1, 4, and 7.

Solution:

Use the six steps outlined in Chap. 1 to solve this problem using the linear triangular element.

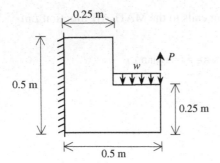

Fig. 11.4. Thin Plate with a Distributed Load and a Concentrated Load for Example 11.2

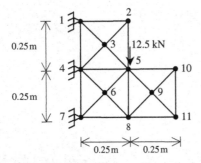

Fig. 11.5. Discretization of Thin Plate Using Twelve Linear Triangles

Step 1 – Discretizing the Domain:

We subdivide the plate into twelve elements and eleven nodes as shown in Fig. 11.5. The total force due to the distributed load is divided equally between nodes 5 and 10. However, the resultant applied force at node 10 cancels out and we are left with a concentrated force of 12.5 kN applied vertically downwards at node 5. Since the plate is thin, a case of plane stress is assumed. The units used in the MATLAB calculations are kN and meter. Table 11.2 shows the element connectivity for this example.

Table 11.2. Element Connectivity for Example 11.2

Element Number	Node i	Node j	Node m
1	1	3	2
2	1	4	3
3	3	5	2
4	3	4	5
5	4	6	5
6	4	7	6
7	5	6	8
8	6	7	8
9	5	8	9
10	5	9	10
11	8	11	9
12	9	11	10

Step 2 – Writing the Element Stiffness Matrices:

The twelve element stiffness matrices are obtained by making calls to the MATLAB function *LinearTriangleElementStiffness*. Each matrix has size 6×6.

```
» E=210e6

E =

   210000000

» NU=0.3

NU =

    0.3000

» t=0.025

t =

    0.0250

» k1=LinearTriangleElementStiffness(E,NU,t,0,0.5,0.125,
         0.375,0.25,0.5,1)

k1 =

   1.0e+006 *

    1.9471   -0.9375   -1.0096    0.8654   -0.9375    0.0721
   -0.9375    1.9471    1.0096   -2.8846   -0.0721    0.9375
   -1.0096    1.0096    2.0192         0   -1.0096   -1.0096
    0.8654   -2.8846         0    5.7692   -0.8654   -2.8846
   -0.9375   -0.0721   -1.0096   -0.8654    1.9471    0.9375
    0.0721    0.9375   -1.0096   -2.8846    0.9375    1.9471

» k2=LinearTriangleElementStiffness(E,NU,t,0,0.5,0,0.25,
         0.125,0.375,1)

k2 =

   1.0e+006 *
```

```
     1.9471    -0.9375     0.9375    -0.0721    -2.8846     1.0096
    -0.9375     1.9471     0.0721    -0.9375     0.8654    -1.0096
     0.9375     0.0721     1.9471     0.9375    -2.8846    -1.0096
    -0.0721    -0.9375     0.9375     1.9471    -0.8654    -1.0096
    -2.8846     0.8654    -2.8846    -0.8654     5.7692          0
     1.0096    -1.0096    -1.0096    -1.0096          0     2.0192
```

» k3=LinearTriangleElementStiffness(E,NU,t,0.125,0.375,
        0.25,0.25,0.25,0.5,1)

k3 =

   1.0e+006 *

```
     5.7692          0    -2.8846     0.8654    -2.8846    -0.8654
          0     2.0192     1.0096    -1.0096    -1.0096    -1.0096
    -2.8846     1.0096     1.9471    -0.9375     0.9375    -0.0721
     0.8654    -1.0096    -0.9375     1.9471     0.0721    -0.9375
    -2.8846    -1.0096     0.9375     0.0721     1.9471     0.9375
    -0.8654    -1.0096    -0.0721    -0.9375     0.9375     1.9471
```

» k4=LinearTriangleElementStiffness(E,NU,t,0.125,0.375,
        0,0.25,0.25,0.25,1)

k4 =

   1.0e+006 *

```
     2.0192          0    -1.0096    -1.0096    -1.0096     1.0096
          0     5.7692    -0.8654    -2.8846     0.8654    -2.8846
    -1.0096    -0.8654     1.9471     0.9375    -0.9375    -0.0721
    -1.0096    -2.8846     0.9375     1.9471     0.0721     0.9375
    -1.0096     0.8654    -0.9375     0.0721     1.9471    -0.9375
     1.0096    -2.8846    -0.0721     0.9375    -0.9375     1.9471
```

» k5=LinearTriangleElementStiffness(E,NU,t,0,0.25,0.125,
        0.125,0.25,0.25,1)
k5 =

   1.0e+006 *

```
      1.9471   -0.9375   -1.0096    0.8654   -0.9375    0.0721
     -0.9375    1.9471    1.0096   -2.8846   -0.0721    0.9375
     -1.0096    1.0096    2.0192         0   -1.0096   -1.0096
      0.8654   -2.8846         0    5.7692   -0.8654   -2.8846
     -0.9375   -0.0721   -1.0096   -0.8654    1.9471    0.9375
      0.0721    0.9375   -1.0096   -2.8846    0.9375    1.9471
```

» k6=LinearTriangleElementStiffness(E,NU,t,0,0.25,0,0,
        0.125,0.125,1)

k6 =

    1.0e+006 *

```
      1.9471   -0.9375    0.9375   -0.0721   -2.8846    1.0096
     -0.9375    1.9471    0.0721   -0.9375    0.8654   -1.0096
      0.9375    0.0721    1.9471    0.9375   -2.8846   -1.0096
     -0.0721   -0.9375    0.9375    1.9471   -0.8654   -1.0096
     -2.8846    0.8654   -2.8846   -0.8654    5.7692         0
      1.0096   -1.0096   -1.0096   -1.0096         0    2.0192
```

» k7=LinearTriangleElementStiffness(E,NU,t,0.25,0.25,
        0.125,0.125,0.25,0,1)

k7 =

    1.0e+006 *

```
      1.9471    0.9375   -2.8846   -1.0096    0.9375    0.0721
      0.9375    1.9471   -0.8654   -1.0096   -0.0721   -0.9375
     -2.8846   -0.8654    5.7692         0   -2.8846    0.8654
     -1.0096   -1.0096         0    2.0192    1.0096   -1.0096
      0.9375   -0.0721   -2.8846    1.0096    1.9471   -0.9375
      0.0721   -0.9375    0.8654   -1.0096   -0.9375    1.9471
```

» k8=LinearTriangleElementStiffness(E,NU,t,0.125,0.125,
        0,0,0.25,0,1)

k8 =

    1.0e+006 *

```
  2.0192         0   -1.0096   -1.0096   -1.0096    1.0096
       0    5.7692   -0.8654   -2.8846    0.8654   -2.8846
 -1.0096   -0.8654    1.9471    0.9375   -0.9375   -0.0721
 -1.0096   -2.8846    0.9375    1.9471    0.0721    0.9375
 -1.0096    0.8654   -0.9375    0.0721    1.9471   -0.9375
  1.0096   -2.8846   -0.0721    0.9375   -0.9375    1.9471
```

» k9=LinearTriangleElementStiffness(E,NU,t,0.25,0.25,0.25,
        0,0.375,0.125,1)

k9 =

   1.0e+006 *

```
  1.9471   -0.9375    0.9375   -0.0721   -2.8846    1.0096
 -0.9375    1.9471    0.0721   -0.9375    0.8654   -1.0096
  0.9375    0.0721    1.9471    0.9375   -2.8846   -1.0096
 -0.0721   -0.9375    0.9375    1.9471   -0.8654   -1.0096
 -2.8846    0.8654   -2.8846   -0.8654    5.7692         0
  1.0096   -1.0096   -1.0096   -1.0096         0    2.0192
```

» k10=LinearTriangleElementStiffness(E,NU,t,0.25,0.25,
        0.375,0.125,0.5,0.25,1)

k10 =

   1.0e+006 *

```
  1.9471   -0.9375   -1.0096    0.8654   -0.9375    0.0721
 -0.9375    1.9471    1.0096   -2.8846   -0.0721    0.9375
 -1.0096    1.0096    2.0192         0   -1.0096   -1.0096
  0.8654   -2.8846         0    5.7692   -0.8654   -2.8846
 -0.9375   -0.0721   -1.0096   -0.8654    1.9471    0.9375
  0.0721    0.9375   -1.0096   -2.8846    0.9375    1.9471
```

» k11=LinearTriangleElementStiffness(E,NU,t,0.25,0,0.5,
        0,0.375,0.125,1)

k11 =

   1.0e+006 *

```
   1.9471      0.9375     -0.9375     -0.0721     -1.0096     -0.8654
   0.9375      1.9471      0.0721      0.9375     -1.0096     -2.8846
  -0.9375      0.0721      1.9471     -0.9375     -1.0096      0.8654
  -0.0721      0.9375     -0.9375      1.9471      1.0096     -2.8846
  -1.0096     -1.0096     -1.0096      1.0096      2.0192           0
  -0.8654     -2.8846      0.8654     -2.8846           0      5.7692
```

» k12=LinearTriangleElementStiffness(E,NU,t,0.375,0.125,
       0.5,0,0.5,0.25,1)

k12 =

  1.0e+006 *

```
   5.7692           0     -2.8846      0.8654     -2.8846     -0.8654
        0      2.0192      1.0096     -1.0096     -1.0096     -1.0096
  -2.8846      1.0096      1.9471     -0.9375      0.9375     -0.0721
   0.8654     -1.0096     -0.9375      1.9471      0.0721     -0.9375
  -2.8846     -1.0096      0.9375      0.0721      1.9471      0.9375
  -0.8654     -1.0096     -0.0721     -0.9375      0.9375      1.9471
```

Step 3 – Assembling the Global Stiffness Matrix:

Since the structure has eleven nodes, the size of the global stiffness matrix is 22×22.
Therefore to obtain K we first set up a zero matrix of size 22×22 then make twelve calls
to the MATLAB function *LinearTriangleAssemble* since we have twelve elements in
the structure. Each call to the function will assemble one element. The following are
the MATLAB commands – note that the result for the global stiffness matrix is not
shown at each step except at the last step.

» K=zeros(22,22);

» K=LinearTriangleAssemble(K,k1,1,3,2);

» K=LinearTriangleAssemble(K,k2,1,4,3);

» K=LinearTriangleAssemble(K,k3,3,5,2);

» K=LinearTriangleAssemble(K,k4,3,4,5);

» K=LinearTriangleAssemble(K,k5,4,6,5);

```
» K=LinearTriangleAssemble(K,k6,4,7,6);

» K=LinearTriangleAssemble(K,k7,5,6,8);

» K=LinearTriangleAssemble(K,k8,6,7,8);

» K=LinearTriangleAssemble(K,k9,5,8,9);

» K=LinearTriangleAssemble(K,k10,5,9,10);

» K=LinearTriangleAssemble(K,k11,8,11,9);

» K=LinearTriangleAssemble(K,k12,9,11,10)
```

K =

   1.0e+007 *

Columns 1 through 7

```
  0.3894  -0.1875  -0.0938   0.0072  -0.3894   0.1875   0.0938
 -0.1875   0.3894  -0.0072   0.0938   0.1875  -0.3894   0.0072
 -0.0938  -0.0072   0.3894   0.1875  -0.3894  -0.1875        0
  0.0072   0.0938   0.1875   0.3894  -0.1875  -0.3894        0
 -0.3894   0.1875  -0.3894  -0.1875   1.5577        0  -0.3894
  0.1875  -0.3894  -0.1875  -0.3894        0   1.5577  -0.1875
  0.0938   0.0072        0        0  -0.3894  -0.1875   0.7788
 -0.0072  -0.0938        0        0  -0.1875  -0.3894        0
       0        0   0.0938  -0.0072  -0.3894   0.1875  -0.1875
       0        0   0.0072  -0.0938   0.1875  -0.3894        0
       0        0        0        0        0        0  -0.3894
       0        0        0        0        0        0   0.1875
       0        0        0        0        0        0   0.0938
       0        0        0        0        0        0  -0.0072
       0        0        0        0        0        0        0
       0        0        0        0        0        0        0
       0        0        0        0        0        0        0
       0        0        0        0        0        0        0
       0        0        0        0        0        0        0
       0        0        0        0        0        0        0
       0        0        0        0        0        0        0
```

Columns 8 through 14

-0.0072	0	0	0	0	0	0
-0.0938	0	0	0	0	0	0
0	0.0938	0.0072	0	0	0	0
0	-0.0072	-0.0938	0	0	0	0
-0.1875	-0.3894	0.1875	0	0	0	0
-0.3894	0.1875	-0.3894	0	0	0	0
0	-0.1875	0	-0.3894	0.1875	0.0938	-0.0072
0.7788	0	0.1875	0.1875	-0.3894	0.0072	-0.0938
0	1.1683	-0.1875	-0.3894	-0.1875	0	0
0.1875	-0.1875	1.1683	-0.1875	-0.3894	0	0
0.1875	-0.3894	-0.1875	1.5577	0	-0.3894	-0.1875
-0.3894	-0.1875	-0.3894	0	1.5577	-0.1875	-0.3894
0.0072	0	0	-0.3894	-0.1875	0.3894	0.1875
-0.0938	0	0	-0.1875	-0.3894	0.1875	0.3894
0	0.1875	0	-0.3894	0.1875	-0.0938	0.0072
0	0	-0.1875	0.1875	-0.3894	-0.0072	0.0938
0	-0.3894	0.1875	0	0	0	0
0	0.1875	-0.3894	0	0	0	0
0	-0.0938	-0.0072	0	0	0	0
0	0.0072	0.0938	0	0	0	0
0	0	0	0	0	0	0
0	0	0	0	0	0	0

Columns 15 through 21

0	0	0	0	0	0	0
0	0	0	0	0	0	0
0	0	0	0	0	0	0
0	0	0	0	0	0	0
0	0	0	0	0	0	0
0	0	0	0	0	0	0
0	0	0	0	0	0	0
0	0	0	0	0	0	0
0.1875	0	-0.3894	0.1875	-0.0938	0.0072	0
0	-0.1875	0.1875	-0.3894	-0.0072	0.0938	0
-0.3894	0.1875	0	0	0	0	0
0.1875	-0.3894	0	0	0	0	0
-0.0938	-0.0072	0	0	0	0	0
0.0072	0.0938	0	0	0	0	0
0.7788	0	-0.3894	-0.1875	0	0	-0.0938
0	0.7788	-0.1875	-0.3894	0	0	0.0072
-0.3894	-0.1875	1.5577	0	-0.3894	-0.1875	-0.3894

```
    0    0.7788  -0.1875  -0.3894        0           0    0.0072
-0.3894  -0.1875   1.5577         0  -0.3894  -0.1875  -0.3894
-0.1875  -0.3894        0    1.5577  -0.1875  -0.3894   0.1875
    0          0  -0.3894  -0.1875   0.3894   0.1875   0.0938
    0          0  -0.1875  -0.3894   0.1875   0.3894  -0.0072
-0.0938   0.0072  -0.3894   0.1875   0.0938  -0.0072   0.3894
-0.0072   0.0938   0.1875  -0.3894   0.0072  -0.0938  -0.1875
```

```
Column 22

    0
    0
    0
    0
    0
    0
    0
    0
    0
    0
    0
    0
    0
    0
-0.0072
 0.0938
 0.1875
-0.3894
 0.0072
-0.0938
-0.1875
 0.3894
```

Step 4 – Applying the Boundary Conditions:

The matrix (11.7) for this structure can be obtained (but is not written explicitly here for lack of space) using the global stiffness matrix obtained in the previous step. The boundary conditions for this problem are given as:

$$U_{1x} = U_{1y} = U_{4x} = U_{4y} = U_{7x} = U_{7y} = 0$$

$$F_{2x} = F_{2y} = F_{3x} = F_{3y} = F_{6x} = F_{6y} = F_{8x} = F_{8y} = F_{9x} = F_{9y} \quad (11.13)$$

$$= F_{10x} = F_{10y} = F_{11x} = F_{11y} = 0$$

$$F_{5x} = 0, \ F_{5y} = -12.5 \quad (11.14)$$

Inserting the above conditions into the matrix equation for this structure we can proceed to the next step. The final equation to be solved in not shown here explicitly because it is too large (of size 22 × 22) and will not fit here.

Step 5 – Solving the Equations:

Solving the system of equations for this structure will be performed by partitioning (manually) and Gaussian elimination (with MATLAB). First we partition the matrix equation by extracting the submatrices in rows 3 to 6, 9 to 12, 15 to 22, and columns 3 to 6, 9 to 12, 15 to 22. We then obtain a reduced matrix equation of size 16 × 16 which is not shown here explicitly because it is too large. However, it is shown in the MATLAB output below. The solution of the resulting system is obtained using MATLAB as follows. Note that the backslash operator "\" is used for Gaussian elimination.

```
» k=[K(3:6,3:6)  K(3:6,9:12)  K(3:6,15:22)  ;  K(9:12,3:6)
K(9:12,9:12)  K(9:12,15:22)  ;  K(15:22,3:6)  K(15:22,9:12)
K(15:22,15:22)]

k =

   1.0e+007 *

Columns 1 through 7


   0.3894    0.1875   -0.3894   -0.1875    0.0938    0.0072         0
   0.1875    0.3894   -0.1875   -0.3894   -0.0072   -0.0938         0
  -0.3894   -0.1875    1.5577         0   -0.3894    0.1875         0
  -0.1875   -0.3894         0    1.5577    0.1875   -0.3894         0
   0.0938   -0.0072   -0.3894    0.1875    1.1683   -0.1875   -0.3894
   0.0072   -0.0938    0.1875   -0.3894   -0.1875    1.1683   -0.1875
        0         0         0         0   -0.3894   -0.1875    1.5577
        0         0         0         0   -0.1875   -0.3894         0
        0         0         0         0    0.1875         0   -0.3894
        0         0         0         0         0   -0.1875    0.1875
        0         0         0         0   -0.3894    0.1875         0
        0         0         0         0    0.1875   -0.3894         0
        0         0         0         0   -0.0938   -0.0072         0
        0         0         0         0    0.0072    0.0938         0
        0         0         0         0         0         0         0
        0         0         0         0         0         0         0
```

Columns 8 through 14

```
        0         0          0         0          0         0          0
        0         0          0         0          0         0          0
        0         0          0         0          0         0          0
        0         0          0         0          0         0          0
  -0.1875    0.1875          0   -0.3894    0.1875   -0.0938    0.0072
  -0.3894         0    -0.1875    0.1875   -0.3894   -0.0072    0.0938
        0   -0.3894    0.1875          0          0         0          0
   1.5577    0.1875   -0.3894          0          0         0          0
   0.1875    0.7788          0   -0.3894   -0.1875         0          0
  -0.3894         0    0.7788   -0.1875   -0.3894         0          0
        0   -0.3894   -0.1875    1.5577          0   -0.3894   -0.1875
        0   -0.1875   -0.3894          0    1.5577   -0.1875   -0.3894
        0         0          0   -0.3894   -0.1875    0.3894    0.1875
        0         0          0   -0.1875   -0.3894    0.1875    0.3894
        0   -0.0938    0.0072   -0.3894    0.1875    0.0938   -0.0072
        0   -0.0072    0.0938    0.1875   -0.3894    0.0072   -0.0938
```

Columns 15 through 16

```
        0              0
        0              0
        0              0
        0              0
        0              0
        0              0
        0              0
        0              0
  -0.0938        -0.0072
   0.0072         0.0938
  -0.3894         0.1875
   0.1875        -0.3894
   0.0938         0.0072
  -0.0072        -0.0938
   0.3894        -0.1875
  -0.1875         0.3894
```

» f=[0;0;0;0;0;-12.5;0;0;0;0;0;0;0;0;0;0]

f =

```
        0
        0
        0
```

```
                        0
                        0
             -12.5000
                        0
                        0
                        0
                        0
                        0
                        0
                        0
                        0
                        0
                        0

» u=k\f

u =

   1.0e-005 *

      0.1158
     -0.3010
      0.0432
     -0.1657
      0.0013
     -0.4170
     -0.0409
     -0.1762
     -0.1340
     -0.3528
     -0.0530
     -0.4526
      0.0083
     -0.5114
     -0.1270
     -0.5290
```

The horizontal and vertical displacements are shown clearly above for the eight free nodes in the structure.

Step 6 – Post-processing:

In this step, we obtain the reactions at nodes 1, 4, and 7 using MATLAB as follows. We set up the global nodal displacement vector U, then we calculate the global nodal force vector F.

» U=[0;0;u(1:4);0;0;u(5:8);0;0;u(9:16)]

U =

    1.0e-005 *

                0
                0
           0.1158
          -0.3010
           0.0432
          -0.1657
                0
                0
           0.0013
          -0.4170
          -0.0409
          -0.1762
                0
                0
          -0.1340
          -0.3528
          -0.0530
          -0.4526
           0.0083
          -0.5114
          -0.1270
          -0.5290

» F=K*U

F =

          -6.0939
           4.3591
           0.0000
           0.0000
           0.0000
           0.0000
          -0.3122
           3.9181
           0.0000
         -12.5000
           0.0000
           0.0000

```
6.4061
4.2228
      0
 0.0000
 0.0000
 0.0000
      0
 0.0000
 0.0000
 0.0000
```

The reactions are shown clearly above. The stresses and principal stresses in each element are not required in this problem but they can be easily obtained for each one of the twelve elements by making successive calls to the MATLAB functions *LinearTriangleElementStresses* and *LinearTriangleElementPStresses*.

Problems:

Problem 11.1:

Consider the thin plate problem solved in Example 11.1. Solve the problem again using four linear triangular elements instead of two elements as shown in Fig. 11.6. Compare your answers for the displacements at nodes 2 and 3 with the answers obtained in the example. Compare also the stresses obtained for the four elements with those obtained for the two elements in the example.

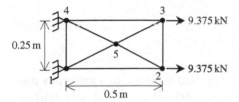

Fig. 11.6. Discretization of Thin Plate Using Four Linear Triangles

Problem 11.2:

Consider the thin plate with a hole in the middle that is subjected to a concentrated load at the corner as shown in Fig. 11.7. Given $E = 70\,\text{GPa}$, $\nu = 0.25$, $t = 0.02\,\text{m}$, and $P = 20\,\text{kN}$, use the finite element mesh (composed of linear triangles) shown in Fig. 11.8 to determine:

1. the global stiffness matrix for the structure.
2. the horizontal and vertical displacements at nodes 4, 6, 7, 10, 11 and 16.
3. the reactions at nodes 1, 5, 9, and 13.

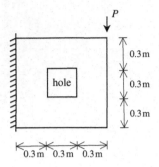

Fig. 11.7. Thin Plate with a Hole

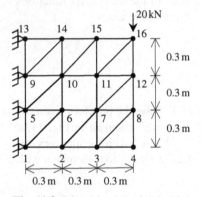

Fig. 11.8. Discretization of Thin Plate with a Hole Using 16 Linear Triangles

Problem 11.3:

Consider the thin plate supported on two springs and subjected to a uniformly distributed load as shown in Fig. 11.9. Given $E = 200\,\text{GPa}$, $\nu = 0.3$, $t = 0.01\,\text{m}$, $k = 4000\,\text{kN/m}$, and $w = 5000\,\text{kN/m}^2$, determine:

1. the global stiffness matrix for the structure.
2. the horizontal and vertical displacements at nodes 1, 2, 3, and 4.
3. the reactions at nodes 5 and 6.
4. the stresses in each element.
5. the principal stresses and principal angle for each element.
6. the force in each spring.

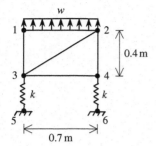

Fig. 11.9. Thin Plate Supported on Two Springs

12 The Quadratic Triangular Element

12.1
Basic Equations

The quadratic triangular element is a two-dimensional finite element with both local and global coordinates. It is characterized by quadratic shape functions. This element can be used for plane stress or plane strain problems in elasticity. It is also called the linear strain triangle. The quadratic triangular element has modulus of elasticity E, Poisson's ratio ν, and thickness t. Each quadratic triangle has six nodes with two in-plane degrees of freedom at each node as shown in Fig. 12.1. The global coordinates of the six nodes are denoted by (x_1, y_1), (x_2, y_2), (x_3, y_3), (x_4, y_4), (x_5, y_5), and (x_6, y_6). The order of the nodes for each element is important – they should be listed in a counterclockwise direction starting from the corner nodes then the midside nodes. The area of each triangle should be positive – you can actually check this by using the MATLAB function *QuadTriangleElementArea* which is written specifically for this purpose. In this case the element stiffness matrix is not written explicitly but calculated through symbolic differentiation and integration with the aid of the MATLAB Symbolic Math Toolbox. The six shape functions for this element are listed explicitly as follows (see [1] and [14]).

$$N_1 = \frac{(x_{23}(y - y_3) - y_{23}(x - x_3))(x_{46}(y - y_6) - y_{46}(x - x_6))}{(x_{23}y_{13} - y_{23}x_{13})(x_{46}y_{16} - y_{46}x_{16})}$$

$$N_2 = \frac{(x_{31}(y - y_1) - y_{31}(x - x_1))(x_{54}(y - y_4) - y_{54}(x - x_4))}{(x_{31}y_{21} - y_{31}x_{21})(x_{54}y_{24} - y_{54}x_{24})}$$

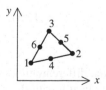

Fig. 12.1. The Quadratic Triangular Element

$$N_3 = \frac{(x_{21}(y - y_1) - y_{21}(x - x_1))(x_{56}(y - y_6) - y_{56}(x - x_6))}{(x_{21}y_{31} - y_{21}x_{31})(x_{56}y_{36} - y_{56}x_{36})}$$

$$N_4 = \frac{(x_{31}(y - y_1) - y_{31}(x - x_1))(x_{23}(y - y_3) - y_{23}(x - x_3))}{(x_{31}y_{41} - y_{31}x_{41})(x_{23}y_{43} - y_{23}x_{43})}$$

$$N_5 = \frac{(x_{31}(y - y_1) - y_{31}(x - x_1))(x_{21}(y - y_1) - y_{21}(x - x_1))}{(x_{31}y_{51} - y_{31}x_{51})(x_{21}y_{51} - y_{21}x_{51})}$$

$$N_6 = \frac{(x_{21}(y - y_1) - y_{21}(x - x_1))(x_{23}(y - y_3) - y_{23}(x - x_3))}{(x_{21}y_{61} - y_{21}x_{61})(x_{23}y_{63} - y_{23}x_{63})} \tag{12.1}$$

where $x_{ij} = x_i - x_j$ and $y_{ij} = y_i - y_j$.

The $[B]$ matrix is given as follows for this element:

$$[B] = \begin{bmatrix} \frac{\partial N_1}{\partial x} & 0 & \frac{\partial N_2}{\partial x} & 0 & \frac{\partial N_3}{\partial x} & 0 & \frac{\partial N_4}{\partial x} & 0 & \frac{\partial N_5}{\partial x} & 0 & \frac{\partial N_6}{\partial x} & 0 \\ 0 & \frac{\partial N_1}{\partial y} & 0 & \frac{\partial N_2}{\partial y} & 0 & \frac{\partial N_3}{\partial y} & 0 & \frac{\partial N_4}{\partial y} & 0 & \frac{\partial N_5}{\partial y} & 0 & \frac{\partial N_6}{\partial y} \\ \frac{\partial N_1}{\partial y} & \frac{\partial N_1}{\partial x} & \frac{\partial N_2}{\partial y} & \frac{\partial N_2}{\partial x} & \frac{\partial N_3}{\partial y} & \frac{\partial N_3}{\partial x} & \frac{\partial N_4}{\partial y} & \frac{\partial N_4}{\partial x} & \frac{\partial N_5}{\partial y} & \frac{\partial N_5}{\partial x} & \frac{\partial N_6}{\partial y} & \frac{\partial N_6}{\partial x} \end{bmatrix} \tag{12.2}$$

For cases of plane stress the matrix $[D]$ is given by

$$[D] = \frac{E}{1 - \nu^2} \begin{bmatrix} 1 & \nu & 0 \\ \nu & 1 & 0 \\ 0 & 0 & \frac{1 - \nu}{2} \end{bmatrix} \tag{12.3}$$

For cases of plane strain the matrix $[D]$ is given by

$$[D] = \frac{E}{(1 + \nu)(1 - 2\nu)} \begin{bmatrix} 1 - \nu & \nu & 0 \\ \nu & 1 - \nu & 0 \\ 0 & 0 & \frac{1 - 2\nu}{2} \end{bmatrix} \tag{12.4}$$

The element stiffness matrix for the quadratic triangular element is written in terms of a double integral as follows:

$$[k] = t \iint_A [B]^T [D][B] \, dx \, dy \tag{12.5}$$

where A is the area and t is the thickness of the element. The partial differentiation of (12.2) and the double integration of (12.5) are carried out symbolically with the aid of the MATLAB Symbolic Math Toolbox. See the details of the MATLAB code for the function *QuadTriangleElementStiffness* which calculates the element stiffness matrix for this element. The reader should note the calculation of this matrix will be somewhat slow due to the symbolic computations involved.

It is clear that the quadratic triangular element has twelve degrees of freedom – two at each node. Consequently for a structure with n nodes, the global stiffness matrix K will be of size $2n \times 2n$ (since we have two degrees of freedom at each node). The global stiffness matrix K is assembled by making calls to the MATLAB function *QuadTriangleAssemble* which is written specifically for this purpose. This process will be illustrated in detail in the examples.

Once the global stiffness matrix K is obtained we have the following structure equation:

$$[K]\{U\} = \{F\} \tag{12.6}$$

where U is the global nodal displacement vector and F is the global nodal force vector. At this step the boundary conditions are applied manually to the vectors U and F. Then the matrix (12.6) is solved by partitioning and Gaussian elimination. Finally once the unknown displacements and reactions are found, the stress vector is obtained for each element as follows:

$$\{\sigma\} = [D][B]\{u\} \tag{12.7}$$

where σ is the stress vector in the element (of size 3×1) and u is the 12×1 element displacement vector. The vector σ is written for each element as $\{\sigma\} = [\sigma_x \sigma_y \tau_{xy}]^T$. It should be noted that in this case this vector is a linear function of x and y. Usually numerical results are obtained at the centroid of the element. The MATLAB function *QuadTriangleElementStresses* gives two results – the general linear stress functions in x and y, and the numerical values of the stresses at the centroid of the element.

12.2
MATLAB Functions Used

The five MATLAB functions used for the quadratic triangular element are:

QuadTriangleElementArea$(x_1, y_1, x_2, y_2, x_3, y_3)$ – This function returns the element area given the coordinates of the first node (x_1, y_1), the coordinates of the second node (x_2, y_2), and the coordinates of the third node (x_3, y_3).

QuadTriangleElementStiffness$(E, NU, t, x_1, y_1, x_2, y_2, x_3, y_3, p)$ – This function calculates the element stiffness matrix for each linear triangle with modulus of elasticity E, Poisson's ratio NU, thickness t, and coordinates (x_1, y_1) for the first node, (x_2, y_2) for the second node, and (x_3, y_3) for the third node. Use $p = 1$ for cases of plane stress and $p = 2$ for cases of plane strain. It returns the 12×12 element stiffness matrix k.

QuadTriangleAssemble(K, k, i, j, m, p, q, r) – This function assembles the element stiffness matrix k of the linear triangle joining nodes i, j, m, p, q, and r into the global stiffness matrix K. It returns the $2n \times 2n$ global stiffness matrix K every time an element is assembled.

QuadTriangleElementStresses(E, NU, x_1, y_1, x_2, y_2, x_3, y_3, p, u) – This function calculates the element stresses using the modulus of elasticity E, Poisson's ratio NU, the coordinates (x_1, y_1) for the first node, (x_2, y_2) for the second node, and (x_3, y_3) for the third node, and the element displacement vector u. Use $p = 1$ for cases of plane stress and $p = 2$ for cases of plane strain. It returns the stress vector for the element.

QuadTriangleElementPStresses($sigma$) – This function calculates the element principal stresses using the element stress vector $sigma$. It returns a 3×1 vector in the form $[sigma1 \ sigma2 \ theta]^{\mathrm{T}}$ where $sigma1$ and $sigma2$ are the principal stresses for the element and $theta$ is the principal angle.

The following is a listing of the MATLAB source code for each function:

```
function y = QuadTriangleElementArea(x1,y1,x2,y2,x3,y3)
%QuadTriangleElementArea      This function returns the area of the
%                             quadratic triangular element whose first
%                             node has coordinates (x1,y1), second
%                             node has coordinates (x2,y2), and
%                             third node has coordinates (x3,y3).
y = (x1*(y2-y3) + x2*(y3-y1) + x3*(y1-y2))/2;
```

```
function w =
QuadTriangleElementStiffness(E,NU,t,x1,y1,x2,y2,x3,y3,p)
%QuadTriangleElementStiffness      This function returns the element
%                                  stiffness matrix for a quadratic
%                                  triangular element with modulus
%                                  of elasticity E, Poisson's ratio
%                                  NU, thickness t, coordinates of
%                                  the node 1 (x1,y1), coordinates
%                                  of node 2 (x2,y2), and
%                                  coordinates of node 3
%                                  (x3,y3). Use p = 1 for cases of
%                                  plane stress, and p = 2 for
%                                  cases of plane strain.
%                                  The size of the element
%                                  stiffness matrix is 12 x 12.
```

```
syms x y;
x4 = (x1 + x2)/2;
y4 = (y1 + y2)/2;
x5 = (x2 + x3)/2;
y5 = (y2 + y3)/2;
x6 = (x1 + x3)/2;
y6 = (y1 + y3)/2;
x21 = x2 - x1;
y21 = y2 - y1;
x23 = x2 - x3;
y23 = y2 - y3;
x46 = x4 - x6;
y46 = y4 - y6;
x13 = x1 - x3;
y13 = y1 - y3;
x16 = x1 - x6;
y16 = y1 - y6;
x31 = x3 - x1;
y31 = y3 - y1;
x54 = x5 - x4;
y54 = y5 - y4;
x24 = x2 - x4;
y24 = y2 - y4;
x56 = x5 - x6;
y56 = y5 - y6;
x36 = x3 - x6;
y36 = y3 - y6;
x41 = x4 - x1;
y41 = y4 - y1;
x43 = x4 - x3;
y43 = y4 - y3;
x51 = x5 - x1;
y51 = y5 - y1;
x61 = x6 - x1;
y61 = y6 - y1;
x63 = x6 - x3;
y63 = y6 - y3;
N1 =(x23*(y-y3)-y23*(x-x3))*(x46*(y-y6)-y46*(x-x6))/
    ((x23*y13-y23*x13)*(x46*y16-y46*x16));
N2 =(x31*(y-y1)-y31*(x-x1))*(x54*(y-y4)-y54*(x-x4))/
    ((x31*y21-y31*x21)*(x54*y24-y54*x24));
N3 =(x21*(y-y1)-y21*(x-x1))*(x56*(y-y6)-y56*(x-x6))/
    ((x21*y31-y21*x31)*(x56*y36-y56*x36));
N4 =(x31*(y-y1)-y31*(x-x1))*(x23*(y-y3)-y23*(x-x3))/
    ((x31*y41-y31*x41)*(x23*y43-y23*x43));
N5 =(x31*(y-y1)-y31*(x-x1))*(x21*(y-y1)-y21*(x-x1))/
    ((x31*y51-y31*x51)*(x21*y51-y21*x51));
N6 =(x21*(y-y1)-y21*(x-x1))*(x23*(y-y3)-y23*(x-x3))/
    (x21*y61-y21*x61)*(x23*y63-y23*x63));
N1x = diff(N1,x);
N1y = diff(N1,y);
N2x = diff(N2,x);
N2y = diff(N2,y);
N3x = diff(N3,x);
```

```
N3y = diff(N3,y);
N4x = diff(N4,x);
N4y = diff(N4,y);
N5x = diff(N5,x);
N5y = diff(N5,y);
N6x = diff(N6,x);
N6y = diff(N6,y);
B = [N1x, 0, N2x, 0, N3x, 0, N4x, 0, N5x, 0, N6x, 0 ;
     0, N1y, 0, N2y, 0, N3y, 0, N4y, 0, N5y, 0, N6y;
     N1y, N1x, N2y, N2x, N3y, N3x, N4y, N4x, N5y, N5x, N6y, N6x];
if p == 1
    D = (E/(1-NU*NU))*[1, NU, 0 ; NU, 1, 0 ; 0, 0, (1-NU)/2];
elseif p == 2
    D = (E/(1+NU)/(1-2*NU))*[1-NU, NU, 0 ; NU, 1-NU, 0 ; 0, 0,
        (1-2*NU)/2];
end
BD = transpose(B)*D*B;
l1 = y1 + (x-x1)*(y2-y1)/(x2-x1);
l2 = y1 + (x-x1)*(y3-y1)/(x3-x1);
l3 = y2 + (x-x2)*(y3-y2)/(x3-x2);
r1 = int(int(BD, y, l1, l2), x, x1, x3);
r2 = int(int(BD, y, l1, l3), x, x3, x2);
z = t*(r1+r2);
w = double(z);
```

```
function y = QuadTriangleAssemble(K,k,i,j,m,p,q,r)
%QuadTriangleAssemble          This function assembles the element
%                              stiffness matrix k of the quadratic
%                              triangular element with nodes i, j,
%                              m, p, q, and r into the global
%                              stiffness matrix K.
%                              This function returns the global
%                              stiffness matrix K after the element
%                              stiffness matrix k is assembled.
K(2*i-1,2*i-1) = K(2*i-1,2*i-1) + k(1,1);
K(2*i-1,2*i) = K(2*i-1,2*i) + k(1,2);
K(2*i-1,2*j-1) = K(2*i-1,2*j-1) + k(1,3);
K(2*i-1,2*j) = K(2*i-1,2*j) + k(1,4);
K(2*i-1,2*m-1) = K(2*i-1,2*m-1) + k(1,5);
K(2*i-1,2*m) = K(2*i-1,2*m) + k(1,6);
K(2*i-1,2*p-1) = K(2*i-1,2*p-1) + k(1,7);
K(2*i-1,2*p) = K(2*i-1,2*p) + k(1,8);
K(2*i-1,2*q-1) = K(2*i-1,2*q-1) + k(1,9);
K(2*i-1,2*q) = K(2*i-1,2*q) + k(1,10);
K(2*i-1,2*r-1) = K(2*i-1,2*r-1) + k(1,11);
K(2*i-1,2*r) = K(2*i-1,2*r) + k(1,12);
K(2*i,2*i-1) = K(2*i,2*i-1) + k(2,1);
K(2*i,2*i) = K(2*i,2*i) + k(2,2);
K(2*i,2*j-1) = K(2*i,2*j-1) + k(2,3);
K(2*i,2*j) = K(2*i,2*j) + k(2,4);
K(2*i,2*m-1) = K(2*i,2*m-1) + k(2,5);
K(2*i,2*m) = K(2*i,2*m) + k(2,6);
K(2*i,2*p-1) = K(2*i,2*p-1) + k(2,7);
K(2*i,2*p) = K(2*i,2*p) + k(2,8);
K(2*i,2*q-1) = K(2*i,2*q-1) + k(2,9);
```

```
K(2*i,2*q) = K(2*i,2*q) + k(2,10);
K(2*i,2*r-1) = K(2*i,2*r-1) + k(2,11);
K(2*i,2*r) = K(2*i,2*r) + k(2,12);
K(2*j-1,2*i-1) = K(2*j-1,2*i-1) + k(3,1);
K(2*j-1,2*i) = K(2*j-1,2*i) + k(3,2);
K(2*j-1,2*j-1) = K(2*j-1,2*j-1) + k(3,3);
K(2*j-1,2*j) = K(2*j-1,2*j) + k(3,4);
K(2*j-1,2*m-1) = K(2*j-1,2*m-1) + k(3,5);
K(2*j-1,2*m) = K(2*j-1,2*m) + k(3,6);
K(2*j-1,2*p-1) = K(2*j-1,2*p-1) + k(3,7);
K(2*j-1,2*p) = K(2*j-1,2*p) + k(3,8);
K(2*j-1,2*q-1) = K(2*j-1,2*q-1) + k(3,9);
K(2*j-1,2*q) = K(2*j-1,2*q) + k(3,10);
K(2*j-1,2*r-1) = K(2*j-1,2*r-1) + k(3,11);
K(2*j-1,2*r) = K(2*j-1,2*r) + k(3,12);
K(2*j,2*i-1) = K(2*j,2*i-1) + k(4,1);
K(2*j,2*i) = K(2*j,2*i) + k(4,2);
K(2*j,2*j-1) = K(2*j,2*j-1) + k(4,3);
K(2*j,2*j) = K(2*j,2*j) + k(4,4);
K(2*j,2*m-1) = K(2*j,2*m-1) + k(4,5);
K(2*j,2*m) = K(2*j,2*m) + k(4,6);
K(2*j,2*p-1) = K(2*j,2*p-1) + k(4,7);
K(2*j,2*p) = K(2*j,2*p) + k(4,8);
K(2*j,2*q-1) = K(2*j,2*q-1) + k(4,9);
K(2*j,2*q) = K(2*j,2*q) + k(4,10);
K(2*j,2*r-1) = K(2*j,2*r-1) + k(4,11);
K(2*j,2*r) = K(2*j,2*r) + k(4,12);
K(2*m-1,2*i-1) = K(2*m-1,2*i-1) + k(5,1);
K(2*m-1,2*i) = K(2*m-1,2*i) + k(5,2);
K(2*m-1,2*j-1) = K(2*m-1,2*j-1) + k(5,3);
K(2*m-1,2*j) = K(2*m-1,2*j) + k(5,4);
K(2*m-1,2*m-1) = K(2*m-1,2*m-1) + k(5,5);
K(2*m-1,2*m) = K(2*m-1,2*m) + k(5,6);
K(2*m-1,2*p-1) = K(2*m-1,2*p-1) + k(5,7);
K(2*m-1,2*p) = K(2*m-1,2*p) + k(5,8);
K(2*m-1,2*q-1) = K(2*m-1,2*q-1) + k(5,9);
K(2*m-1,2*q) = K(2*m-1,2*q) + k(5,10);
K(2*m-1,2*r-1) = K(2*m-1,2*r-1) + k(5,11);
K(2*m-1,2*r) = K(2*m-1,2*r) + k(5,12);
K(2*m,2*i-1) = K(2*m,2*i-1) + k(6,1);
K(2*m,2*i) = K(2*m,2*i) + k(6,2);
K(2*m,2*j-1) = K(2*m,2*j-1) + k(6,3);
K(2*m,2*j) = K(2*m,2*j) + k(6,4);
K(2*m,2*m-1) = K(2*m,2*m-1) + k(6,5);
K(2*m,2*m) = K(2*m,2*m) + k(6,6);
K(2*m,2*p-1) = K(2*m,2*p-1) + k(6,7);
K(2*m,2*p) = K(2*m,2*p) + k(6,8);
K(2*m,2*q-1) = K(2*m,2*q-1) + k(6,9);
K(2*m,2*q) = K(2*m,2*q) + k(6,10);
K(2*m,2*r-1) = K(2*m,2*r-1) + k(6,11);
K(2*m,2*r) = K(2*m,2*r) + k(6,12);
K(2*p-1,2*i-1) = K(2*p-1,2*i-1) + k(7,1);
K(2*p-1,2*i) = K(2*p-1,2*i) + k(7,2);
K(2*p-1,2*j-1) = K(2*p-1,2*j-1) + k(7,3);
```

```
K(2*p-1,2*j) = K(2*p-1,2*j) + k(7,4);
K(2*p-1,2*m-1) = K(2*p-1,2*m-1) + k(7,5);
K(2*p-1,2*m) = K(2*p-1,2*m) + k(7,6);
K(2*p-1,2*p-1) = K(2*p-1,2*p-1) + k(7,7);
K(2*p-1,2*p) = K(2*p-1,2*p) + k(7,8);
K(2*p-1,2*q-1) = K(2*p-1,2*q-1) + k(7,9);
K(2*p-1,2*q) = K(2*p-1,2*q) + k(7,10);
K(2*p-1,2*r-1) = K(2*p-1,2*r-1) + k(7,11);
K(2*p-1,2*r) = K(2*p-1,2*r) + k(7,12);
K(2*p,2*i-1) = K(2*p,2*i-1) + k(8,1);
K(2*p,2*i) = K(2*p,2*i) + k(8,2);
K(2*p,2*j-1) = K(2*p,2*j-1) + k(8,3);
K(2*p,2*j) = K(2*p,2*j) + k(8,4);
K(2*p,2*m-1) = K(2*p,2*m-1) + k(8,5);
K(2*p,2*m) = K(2*p,2*m) + k(8,6);
K(2*p,2*p-1) = K(2*p,2*p-1) + k(8,7);
K(2*p,2*p) = K(2*p,2*p) + k(8,8);
K(2*p,2*q-1) = K(2*p,2*q-1) + k(8,9);
K(2*p,2*q) = K(2*p,2*q) + k(8,10);
K(2*p,2*r-1) = K(2*p,2*r-1) + k(8,11);
K(2*p,2*r) = K(2*p,2*r) + k(8,12);
K(2*q-1,2*i-1) = K(2*q-1,2*i-1) + k(9,1);
K(2*q-1,2*i) = K(2*q-1,2*i) + k(9,2);
K(2*q-1,2*j-1) = K(2*q-1,2*j-1) + k(9,3);
K(2*q-1,2*j) = K(2*q-1,2*j) + k(9,4);
K(2*q-1,2*m-1) = K(2*q-1,2*m-1) + k(9,5);
K(2*q-1,2*m) = K(2*q-1,2*m) + k(9,6);
K(2*q-1,2*p-1) = K(2*q-1,2*p-1) + k(9,7);
K(2*q-1,2*p) = K(2*q-1,2*p) + k(9,8);
K(2*q-1,2*q-1) = K(2*q-1,2*q-1) + k(9,9);
K(2*q-1,2*q) = K(2*q-1,2*q) + k(9,10);
K(2*q-1,2*r-1) = K(2*q-1,2*r-1) + k(9,11);
K(2*q-1,2*r) = K(2*q-1,2*r) + k(9,12);
K(2*q,2*i-1) = K(2*q,2*i-1) + k(10,1);
K(2*q,2*i) = K(2*q,2*i) + k(10,2);
K(2*q,2*j-1) = K(2*q,2*j-1) + k(10,3);
K(2*q,2*j) = K(2*q,2*j) + k(10,4);
K(2*q,2*m-1) = K(2*q,2*m-1) + k(10,5);
K(2*q,2*m) = K(2*q,2*m) + k(10,6);
K(2*q,2*p-1) = K(2*q,2*p-1) + k(10,7);
K(2*q,2*p) = K(2*q,2*p) + k(10,8);
K(2*q,2*q-1) = K(2*q,2*q-1) + k(10,9);
K(2*q,2*q) = K(2*q,2*q) + k(10,10);
K(2*q,2*r-1) = K(2*q,2*r-1) + k(10,11);
K(2*q,2*r) = K(2*q,2*r) + k(10,12);
K(2*r-1,2*i-1) = K(2*r-1,2*i-1) + k(11,1);
K(2*r-1,2*i) = K(2*r-1,2*i) + k(11,2);
K(2*r-1,2*j-1) = K(2*r-1,2*j-1) + k(11,3);
K(2*r-1,2*j) = K(2*r-1,2*j) + k(11,4);
K(2*r-1,2*m-1) = K(2*r-1,2*m-1) + k(11,5);
K(2*r-1,2*m) = K(2*r-1,2*m) + k(11,6);
K(2*r-1,2*p-1) = K(2*r-1,2*p-1) + k(11,7);
K(2*r-1,2*p) = K(2*r-1,2*p) + k(11,8);
K(2*r-1,2*q-1) = K(2*r-1,2*q-1) + k(11,9);
```

```
K(2*r-1,2*q) = K(2*r-1,2*q) + k(11,10);
K(2*r-1,2*r-1) = K(2*r-1,2*r-1) + k(11,11);
K(2*r-1,2*r) = K(2*r-1,2*r) + k(11,12);
K(2*r,2*i-1) = K(2*r,2*i-1) + k(12,1);
K(2*r,2*i) = K(2*r,2*i) + k(12,2);
K(2*r,2*j-1) = K(2*r,2*j-1) + k(12,3);
K(2*r,2*j) = K(2*r,2*j) + k(12,4);
K(2*r,2*m-1) = K(2*r,2*m-1) + k(12,5);
K(2*r,2*m) = K(2*r,2*m) + k(12,6);
K(2*r,2*p-1) = K(2*r,2*p-1) + k(12,7);
K(2*r,2*p) = K(2*r,2*p) + k(12,8);
K(2*r,2*q-1) = K(2*r,2*q-1) + k(12,9);
K(2*r,2*q) = K(2*r,2*q) + k(12,10);
K(2*r,2*r-1) = K(2*r,2*r-1) + k(12,11);
K(2*r,2*r) = K(2*r,2*r) + k(12,12);
y = K;
```

```
function w =
QuadTriangleElementSresses(E,NU,x1,y1,x2,y2,x3,y3,p,u)
%QuadTriangleElementStresses     This function returns the element
%                                stresses for a quadratic
%                                triangular element with modulus of
%                                elasticity E, Poisson's ratio NU,
%                                coordinates of the
%                                node 1 (x1,y1), coordinates of
%                                node 2 (x2,y2),
%                                coordinates of node 3
%                                (x3,y3), and element displacement
%                                vector u. Use p = 1 for cases of
%                                plane stress, and p = 2 for cases
%                                of plane strain.
%                                The size of the element stiffness
%                                matrix is 12 x 12.
syms x y;
x4 = (x1 + x2)/2;
y4 = (y1 + y2)/2;
x5 = (x2 + x3)/2;
y5 = (y2 + y3)/2;
x6 = (x1 + x3)/2;
y6 = (y1 + y3)/2;
x21 = x2 - x1;
y21 = y2 - y1;
x23 = x2 - x3;
y23 = y2 - y3;
x46 = x4 - x6;
y46 = y4 - y6;
x13 = x1 - x3;
y13 = y1 - y3;
x16 = x1 - x6;
y16 = y1 - y6;
x31 = x3 - x1;
y31 = y3 - y1;
x54 = x5 - x4;
y54 = y5 - y4;
x24 = x2 - x4;
```

```
y24 = y2 - y4;
x56 = x5 - x6;
y56 = y5 - y6;
x36 = x3 - x6;
y36 = y3 - y6;
x41 = x4 - x1;
y41 = y4 - y1;
x43 = x4 - x3;
y43 = y4 - y3;
x51 = x5 - x1;
y51 = y5 - y1;
x61 = x6 - x1;
y61 = y6 - y1;
x63 = x6 - x3;
y63 = y6 - y3;
N1 = (x23*(y-y3)-y23*(x-x3))*(x46*(y-y6)-y46*(x-x6))/
     ((x23*y13-y23*x13)*(x46*y16-y46*x16));
N2 = (x31*(y-y1)-y31*(x-x1))*(x54*(y-y4)-y54*(x-x4))/
     ((x31*y21-y31*x21)*(x54*y24-y54*x24));
N3 = (x21*(y-y1)-y21*(x-x1))*(x56*(y-y6)-y56*(x-x6))/
     ((x21*y31-y21*x31)*(x56*y36-y56*x36));
N4 = (x31*(y-y1)-y31*(x-x1))*(x23*(y-y3)-y23*(x-x3))/
     ((x31*y41-y31*x41)*(x23*y43-y23*x43));
N5 = (x31*(y-y1)-y31*(x-x1))*(x21*(y-y1)-y21*(x-x1))/
     ((x31*y51-y31*x51)*(x21*y51-y21*x51));
N6 = (x21*(y-y1)-y21*(x-x1))*(x23*(y-y3)-y23*(x-x3))/
     ((x21*y61-y21*x61)*(x23*y63-y23*x63));
N1x = diff(N1,x);
N1y = diff(N1,y);
N2x = diff(N2,x);
N2y = diff(N2,y);
N3x = diff(N3,x);
N3y = diff(N3,y);
N4x = diff(N4,x);
N4y = diff(N4,y);
N5x = diff(N5,x);
N5y = diff(N5,y);
N6x = diff(N6,x);
N6y = diff(N6,y);
B = [N1x, 0, N2x, 0, N3x, 0, N4x, 0, N5x, 0, N6x, 0 ;
    0, N1y, 0, N2y, 0, N3y, 0, N4y, 0, N5y, 0, N6y;
    N1y, N1x, N2y, N2x, N3y, N3x, N4y, N4x, N5y, N5x, N6y, N6x];
if p == 1
    D = (E/(1-NU*NU))*[1, NU, 0 ; NU, 1, 0 ; 0, 0, (1-NU)/2];
elseif p == 2
    D = (E/(1+NU)/(1-2*NU))*[1-NU, NU, 0 ; NU, 1-NU, 0 ; 0, 0,
        (1-2*NU)/2];
end
w = D*B*u
%
% We also calculate the stresses at the centroid of the element
%
xcent = (x1 + x2 + x3)/3;
ycent = (y1 + y2 + y3)/3;
wcent = subs(w, {x,y}, {xcent,ycent});
w = double(wcent);
```

```
function y = QuadTriangleElementPStresses(sigma)
%QuadTriangleElementPStresses          This function returns the element
%                                      principal stresses and their
%                                      angle given the element
%                                      stress vector.
R = (sigma(1) + sigma(2))/2;
Q = ((sigma(1) - sigma(2))/2)^2 + sigma(3)*sigma(3);
M = 2*sigma(3)/(sigma(1) - sigma(2));
s1 = R + sqrt(Q);
s2 = R - sqrt(Q);
theta = (atan(M)/2)*180/pi;
y = [s1 ; s2 ; theta];
```

Example 12.1:

Consider the thin plate subjected to a uniformly distributed load as shown in Fig. 12.2. This is the problem solved in Example 11.1 using linear triangles. It will be solved here using quadratic triangles. The plate is discretized using quadratic triangles as shown in Fig. 12.3. Given $E = 210\,\text{GPa}$, $\nu = 0.3$, $t = 0.025\,\text{m}$, and $w = 3000\,\text{kN/m}^2$, determine:

1. the global stiffness matrix for the structure.
2. the horizontal and vertical displacements at nodes 3, 6, and 9.
3. the reactions at nodes 1, 4, and 7.
4. the stresses in each element.
5. the principal stresses and principal angle for each element.

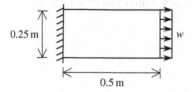

Fig. 12.2. Thin Plate for Example 12.1

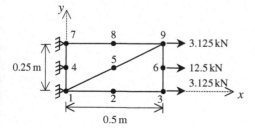

Fig. 12.3. Discretization of Thin Plate Using Two Quadratic Triangles

Solution:

Use the six steps outlined in Chap. 1 to solve this problem using the quadratic triangular element.

Step 1 – Discretizing the Domain:

We subdivide the plate into two elements only for illustration purposes. More elements must be used in order to obtain reliable results. Thus the domain is subdivided into two elements and nine nodes as shown in Fig. 12.3. The total force due to the distributed load is divided between nodes 3, 6, and 9 in the ratio 1/6 : 2/3 : 1/6. Since the plate is thin, a case of plane stress is assumed. The units used in the MATLAB calculations are kN and meter. Table 12.1 shows the element connectivity for this example.

Table 12.1. Element Connectivity for Example 12.1

Element Number	Node i	Node j	Node m	Node p	Node q	Node r
1	1	9	7	5	8	4
2	1	3	9	2	6	5

Step 2 – Writing the Element Stiffness Matrices:

The two element stiffness matrices k_1 and k_2 are obtained by making calls to the MATLAB function *QuadTriangleElementStiffness*. Each matrix has size 12×12.

```
» E=210e6

E =

    210000000

» NU=0.3

NU =

    0.3000

» t=0.025

t =

    0.0250
```

```
» k1=QuadTriangleElementStiffness(E,NU,t,0,0,0.5,0.25,0,
        0.25,1)

k1 =

   1.0e+007 *

   Columns 1 through 7
```

```
   0.2019        0        0   0.0337   0.0673  -0.0337        0
        0   0.5769   0.0288        0  -0.0288   0.1923  -0.1154
        0   0.0288   0.1442        0   0.0481  -0.0288        0
   0.0337        0        0   0.0505  -0.0337   0.0168  -0.1346
   0.0673  -0.0288   0.0481  -0.0337   0.3462  - 0.1875        0
  -0.0337   0.1923  -0.0288   0.0168  -0.1875   0.6274        0
        0  -0.1154        0  -0.1346        0        0   0.9231
  -0.1346        0  -0.1154        0        0        0  -0.2500
        0        0  -0.1923   0.1346  -0.1923   0.1154  -0.5385
        0        0   0.1154  -0.0673   0.1346  -0.0673   0.2500
  -0.2692   0.1154        0        0  -0.2692   0.1346  -0.3846
   0.1346  -0.7692        0        0   0.1154  -0.7692   0.2500
```

```
   Columns 8 through 12
```

```
  -0.1346        0        0  -0.2692   0.1346
        0        0        0   0.1154  -0.7692
  -0.1154  -0.1923   0.1154        0        0
        0   0.1346  -0.0673        0        0
        0  -0.1923   0.1346  -0.2692   0.1154
        0   0.1154  -0.0673   0.1346  -0.7692
  -0.2500  -0.5385   0.2500  -0.3846   0.2500
   1.6731   0.2500  -1.5385   0.2500  -0.1346
   0.2500   0.9231  -0.2500        0  -0.2500
  -1.5385  -0.2500   1.6731  -0.2500        0
   0.2500        0  -0.2500   0.9231  -0.2500
  -0.1346  -0.2500        0  -0.2500   1.6731
```

```
» k2=QuadTriangleElementStiffness(E,NU,t,0,0,0.5,0,0.5,
        0.25,1)

k2 =

   1.0e+007 *

   Columns 1 through 7
```

```
 0.1442        0   0.0481  -0.0288        0   0.0288  -0.1923
      0   0.0505  -0.0337   0.0168   0.0337        0   0.1346
 0.0481  -0.0337   0.3462  -0.1875   0.0673  -0.0288  -0.1923
-0.0288   0.0168  -0.1875   0.6274  -0.0337   0.1923   0.1154
      0   0.0337   0.0673  -0.0337   0.2019        0        0
 0.0288        0  -0.0288   0.1923        0   0.5769        0
-0.1923   0.1346  -0.1923   0.1154        0        0   0.9231
 0.1154  -0.0673   0.1346  -0.0673        0        0  -0.2500
      0        0  -0.2692   0.1346  -0.2692   0.1154        0
      0        0   0.1154  -0.7692   0.1346  -0.7692  -0.2500
      0  -0.1346        0        0        0  -0.1154  -0.5385
-0.1154        0        0        0  -0.1346        0   0.2500
```

Columns 8 through 12

```
 0.1154        0        0        0  -0.1154
-0.0673        0        0  -0.1346        0
 0.1346  -0.2692   0.1154        0        0
-0.0673   0.1346  -0.7692        0        0
      0  -0.2692   0.1346        0  -0.1346
      0   0.1154  -0.7692  -0.1154        0
-0.2500        0  -0.2500  -0.5385   0.2500
 1.6731  -0.2500        0   0.2500  -1.5385
-0.2500   0.9231  -0.2500  -0.3846   0.2500
      0  -0.2500   1.6731   0.2500  -0.1346
 0.2500  -0.3846   0.2500   0.9231  -0.2500
-1.5385   0.2500  -0.1346  -0.2500   1.6731
```

Step 3 – Assembling the Global Stiffness Matrix:

Since the structure has nine nodes, the size of the global stiffness matrix is 18×18. Therefore to obtain K we first set up a zero matrix of size 18×18 then make two calls to the MATLAB function *QuadTriangleAssemble* since we have two elements in the structure. Each call to the function will assemble one element. The following are the MATLAB commands:

```
» K=zeros(18,18);
```

```
» K=QuadTriangleAssemble(K,k1,1,9,7,5,8,4)
```

K =

    1.0e+007 *

Columns 1 through 7

```
  0.2019        0     0    0    0    0   -0.2692
       0   0.5769     0    0    0    0    0.1154
       0        0     0    0    0    0         0
       0        0     0    0    0    0         0
       0        0     0    0    0    0         0
       0        0     0    0    0    0         0
 -0.2692   0.1154     0    0    0    0    0.9231
  0.1346  -0.7692     0    0    0    0   -0.2500
       0  -0.1154     0    0    0    0   -0.3846
 -0.1346        0     0    0    0    0    0.2500
       0        0     0    0    0    0         0
       0        0     0    0    0    0         0
  0.0673  -0.0288     0    0    0    0   -0.2692
 -0.0337   0.1923     0    0    0    0    0.1346
       0        0     0    0    0    0         0
       0        0     0    0    0    0   -0.2500
       0   0.0288     0    0    0    0         0
  0.0337        0     0    0    0    0         0
```

Columns 8 through 14

```
  0.1346        0  -0.1346    0    0    0.0673  -0.0337
 -0.7692  -0.1154        0    0    0   -0.0288   0.1923
       0        0        0    0    0         0        0
       0        0        0    0    0         0        0
       0        0        0    0    0         0        0
       0        0        0    0    0         0        0
 -0.2500  -0.3846   0.2500    0    0   -0.2692   0.1346
  1.6731   0.2500  -0.1346    0    0    0.1154  -0.7692
  0.2500   0.9231  -0.2500    0    0         0        0
 -0.1346  -0.2500   1.6731    0    0         0        0
       0        0        0    0    0         0        0
       0        0        0    0    0         0        0
  0.1154        0        0    0    0    0.3462  -0.1875
 -0.7692        0        0    0    0   -0.1875   0.6274
 -0.2500  -0.5385   0.2500    0    0   -0.1923   0.1154
       0   0.2500  -1.5385    0    0    0.1346  -0.0673
       0        0  -0.1154    0    0    0.0481  -0.0288
       0  -0.1346        0    0    0   -0.0337   0.0168
```

Columns 15 through 18

```
      0          0          0     0.0337
      0          0     0.0288          0
      0          0          0          0
      0          0          0          0
      0          0          0          0
      0          0          0          0
      0    -0.2500          0          0
-0.2500          0          0          0
-0.5385     0.2500          0    -0.1346
 0.2500    -1.5385    -0.1154          0
      0          0          0          0
      0          0          0          0
-0.1923     0.1346     0.0481    -0.0337
 0.1154    -0.0673    -0.0288     0.0168
 0.9231    -0.2500    -0.1923     0.1346
-0.2500     1.6731     0.1154    -0.0673
-0.1923     0.1154     0.1442          0
 0.1346    -0.0673          0     0.0505
```

» K=QuadTriangleAssemble(K,k2,1,3,9,2,6,5)

K =

   1.0e+007 *

Columns 1 through 7

```
 0.3462          0    -0.1923     0.1154     0.0481    -0.0288    -0.2692
      0     0.6274     0.1346    -0.0673    -0.0337     0.0168     0.1154
-0.1923     0.1346     0.9231    -0.2500    -0.1923     0.1154          0
 0.1154    -0.0673    -0.2500     1.6731     0.1346    -0.0673          0
 0.0481    -0.0337    -0.1923     0.1346     0.3462    -0.1875          0
-0.0288     0.0168     0.1154    -0.0673    -0.1875     0.6274          0
-0.2692     0.1154          0          0          0          0     0.9231
 0.1346    -0.7692          0          0          0          0    -0.2500
      0    -0.2500    -0.5385     0.2500          0          0    -0.3846
-0.2500          0     0.2500    -1.5385          0          0     0.2500
      0          0          0    -0.2500    -0.2692     0.1346          0
      0          0    -0.2500          0     0.1154    -0.7692          0
 0.0673    -0.0288          0          0          0          0    -0.2692
-0.0337     0.1923          0          0          0          0     0.1346
      0          0          0          0          0          0          0
      0          0          0          0          0          0    -0.2500
      0     0.0625          0          0     0.0673    -0.0337          0
 0.0625          0          0          0    -0.0288     0.1923          0
```

```
Columns 8 through 14


   0.1346        0  -0.2500        0        0   0.0673  -0.0337
  -0.7692  -0.2500        0        0        0  -0.0288   0.1923
        0  -0.5385   0.2500        0  -0.2500        0        0
        0   0.2500  -1.5385  -0.2500        0        0        0
        0        0        0  -0.2692   0.1154        0        0
        0        0        0   0.1346  -0.7692        0        0
  -0.2500  -0.3846   0.2500        0        0  -0.2692   0.1346
   1.6731   0.2500  -0.1346        0        0   0.1154  -0.7692
   0.2500   1.8462  -0.5000  -0.3846   0.2500        0        0
  -0.1346  -0.5000   3.3462   0.2500  -0.1346        0        0
        0  -0.3846   0.2500   0.9231  -0.2500        0        0
        0   0.2500  -0.1346  -0.2500   1.6731        0        0
   0.1154        0        0        0        0   0.3462  -0.1875
  -0.7692        0        0        0        0  -0.1875   0.6274
  -0.2500  -0.5385   0.2500        0        0  -0.1923   0.1154
        0   0.2500  -1.5385        0        0   0.1346  -0.0673
        0        0  -0.2500  -0.2692   0.1346   0.0481  -0.0288
        0  -0.2500        0   0.1154  -0.7692  -0.0337   0.0168


Columns 15 through 18


        0        0        0   0.0625
        0        0   0.0625        0
        0        0        0        0
        0        0        0        0
        0        0   0.0673  -0.0288
        0        0  -0.0337   0.1923
        0  -0.2500        0        0
  -0.2500        0        0        0
  -0.5385   0.2500        0  -0.2500
   0.2500  -1.5385  -0.2500        0
        0        0  -0.2692   0.1154
        0        0   0.1346  -0.7692
  -0.1923   0.1346   0.0481  -0.0337
   0.1154  -0.0673  -0.0288   0.0168
   0.9231  -0.2500  -0.1923   0.1346
  -0.2500   1.6731   0.1154  -0.0673
  -0.1923   0.1154   0.3462        0
   0.1346  -0.0673        0   0.6274
```

Step 4 – Applying the Boundary Conditions:

The matrix (12.6) for this structure is obtained using the global stiffness matrix shown above. The equation is not written out explicitly below because it is too large. The boundary conditions for this problem are given as:

$$U_{1x} = U_{1y} = U_{4x} = U_{4y} = U_{7x} = U_{7y} = 0$$
$$F_{2x} = F_{2y} = F_{5x} = F_{5y} = F_{8x} = F_{8y} = 0$$
$$F_{3x} = F_{9x} = 3.125, \quad F_{6x} = 12.5, \quad F_{3y} = F_{6y} = F_{9y} = 0 \tag{12.8}$$

We next insert the above conditions into the equation (not shown) represented by (12.6) and we proceed to the solution step below.

Step 5 – Solving the Equations:

Solving the resulting system of equations will be performed by partitioning (manually) and Gaussian elimination (with MATLAB). First we partition the resulting equation by extracting the submatrix in rows 3 to 6, rows 9 to 12, rows 15 to 18, and columns 3 to 6, columns 9 to 12, columns 15 to 18. Therefore we obtain the following matrix equation showing the numbers to two decimal places only although the MATLAB calculations are carried out using at least four decimal places.

$$10^6 \begin{bmatrix}
9.2 & -2.5 & -1.9 & 1.2 & -5.4 & 2.5 & 0 & -2.5 & 0 & 0 & 0 & 0 \\
-2.5 & 16.7 & 1.3 & -0.7 & 2.5 & -15.4 & -2.5 & 0 & 0 & 0 & 0 & 0 \\
-1.9 & 1.3 & 3.5 & 1.9 & 0 & 0 & -2.7 & 1.2 & 0 & 0 & 0.7 & -0.3 \\
1.2 & -0.7 & -1.9 & 6.3 & 0 & 0 & 1.3 & -7.7 & 0 & 0 & -0.3 & 1.9 \\
-5.4 & 2.5 & 0 & 0 & 18.5 & -5.0 & -3.8 & 2.5 & -5.4 & 2.5 & 0 & -2.5 \\
2.5 & -15.4 & 0 & 0 & -5.0 & 33.5 & 2.5 & -1.3 & 2.5 & -15.4 & -2.5 & 0 \\
0 & -2.5 & -2.7 & 1.3 & -3.8 & 2.5 & 9.2 & -2.5 & 0 & 0 & -2.7 & 1.2 \\
-2.5 & 0 & 1.2 & -7.7 & 2.5 & -1.3 & -2.5 & 16.7 & 0 & 0 & 1.3 & -7.7 \\
0 & 0 & 0 & 0 & -5.4 & 2.5 & 0 & 0 & 9.2 & -2.5 & -1.9 & 1.3 \\
0 & 0 & 0 & 0 & 2.5 & -15.4 & 0 & 0 & -2.5 & 16.7 & 1.2 & -0.7 \\
0 & 0 & 0.7 & -0.3 & 0 & -2.5 & -2.7 & 1.3 & -1.9 & 1.2 & 3.5 & 0 \\
0 & 0 & -0.3 & 1.9 & -2.5 & 0 & 1.2 & -7.7 & 1.3 & -0.7 & 0 & 6.3
\end{bmatrix}$$

$$\begin{Bmatrix}
U_{2x} \\ U_{2y} \\ U_{3x} \\ U_{3y} \\ U_{5x} \\ U_{5y} \\ U_{6x} \\ U_{6y} \\ U_{8x} \\ U_{8y} \\ U_{9x} \\ U_{9y}
\end{Bmatrix} = \begin{Bmatrix}
F_{2x} \\ F_{2y} \\ F_{3x} \\ F_{3y} \\ F_{5x} \\ F_{5y} \\ F_{6x} \\ F_{6y} \\ F_{8x} \\ F_{8y} \\ F_{9x} \\ F_{9y}
\end{Bmatrix} \tag{12.9}$$

The solution of the above system is obtained using MATLAB as follows. Note that the backslash operator "\" is used for Gaussian elimination.

```
» k=[K(3:6,3:6)  K(3:6,9:12)  K(3:6,15:18)  ;  K(9:12,3:6)
      K(9:12,9:12)  K(9:12,15:18)  ;  K(15:18,3:6)
      K(15:18,9:12)  K(15:18,15:18)]

k =

  1.0e+007 *

  Columns 1 through 7


    0.9231  -0.2500  -0.1923   0.1154  -0.5385   0.2500        0
   -0.2500   1.6731   0.1346  -0.0673   0.2500  -1.5385  -0.2500
   -0.1923   0.1346   0.3462  -0.1875        0        0  -0.2692
    0.1154  -0.0673  -0.1875   0.6274        0        0   0.1346
   -0.5385   0.2500        0        0   1.8462  -0.5000  -0.3846
    0.2500  -1.5385        0        0  -0.5000   3.3462   0.2500
         0  -0.2500  -0.2692   0.1346  -0.3846   0.2500   0.9231
   -0.2500        0   0.1154  -0.7692   0.2500  -0.1346  -0.2500
         0        0        0        0  -0.5385   0.2500        0
         0        0        0        0   0.2500  -1.5385        0
         0        0   0.0673  -0.0337        0  -0.2500  -0.2692
         0        0  -0.0288   0.1923  -0.2500        0   0.1154


  Columns 8 through 12


   -0.2500        0        0        0        0
         0        0        0        0        0
    0.1154        0        0   0.0673  -0.0288
   -0.7692        0        0  -0.0337   0.1923
    0.2500  -0.5385   0.2500        0  -0.2500
   -0.1346   0.2500  -1.5385  -0.2500        0
   -0.2500        0        0  -0.2692   0.1154
    1.6731        0        0   0.1346  -0.7692
         0   0.9231  -0.2500  -0.1923   0.1346
         0  -0.2500   1.6731   0.1154  -0.0673
    0.1346  -0.1923   0.1154   0.3462        0
   -0.7692   0.1346  -0.0673        0   0.6274
```

```
» f=[0 ; 0 ; 3.125 ; 0 ; 0 ; 0 ; 12.5 ; 0 ; 0 ; 0 ;
        3.125 ; 0]

f =

        0
        0
   3.1250
        0
        0
        0
  12.5000
        0
        0
        0
   3.1250
        0

» u=k\f

u =

   1.0e-005 *

   0.3575
   0.0673
   0.7050
   0.0843
   0.3405
   0.0144
   0.7018
   0.0302
   0.3352
  -0.0372
   0.7045
  -0.0257
```

It is now clear that the horizontal and vertical displacements at node 3 are 0.7050 m (compared with 0.7111 m in Example 11.1) and 0.0843 m, respectively. The horizontal and vertical displacements at node 6 are 0.7018 m and 0.0302 m, respectively. The horizontal and vertical displacements at node 9 are 0.7045 m (compared with 0.6531 m in Example 11.1) and −0.0257 m, respectively. When a larger number of elements is used we expect to get the same result for the horizontal displacements at nodes 3, 6, and 9.

Step 6 – Post-processing:

In this step, we obtain the reactions at nodes 1, 4, and 7, and the stresses in each element using MATLAB as follows. First we set up the global nodal displacement vector U, then we calculate the global nodal force vector F.

```
» U=[0;0;u(1:4);0;0;u(5:8);0;0;u(9:12)]

U =

  1.0e-005 *

        0
        0
   0.3575
   0.0673
   0.7050
   0.0843
        0
        0
   0.3405
   0.0144
   0.7018
   0.0302
        0
        0
   0.3352
  -0.0372
   0.7045
  -0.0257

» F=K*U

F =

   -3.4721
   -1.9808
    0.0000
    0.0000
    3.1250
    0.0000
  -11.8059
   -0.0611
    0.0000
    0.0000
```

```
   12.5000
    0.0000
   -3.4721
    2.0419
    0.0000
    0.0000
    3.1250
    0.0000
```

Thus the horizontal and vertical reactions at node 1 are forces of 3.4721 kN (directed to the left) and 1.9808 kN (directed downwards). The horizontal and vertical reactions at node 4 are forces of 11.8059 kN (directed to the left) and 0.0611 kN (directed downwards). The horizontal and vertical reactions at node 7 are forces of 3.4721 kN (directed to the left) and 2.049 kN (directed upwards). Obviously force equilibrium is satisfied for this problem. It is noted that the results for the reactions are different than those obtained in Example 11.1 due to the small number of elements used. We need to use more elements to get reliable results for the reactions and stresses. Next we set up the element nodal displacement vectors u_1 and u_2 then we calculate the element stresses *sigma1* and *sigma2* by making calls to the MATLAB function *QuadTriangleElementStresses*. Closed-form solutions (linear functions in x and y) for the stresses in each element are obtained using the MATLAB Symbolic Math Toolbox as shown below. Then numerical values for the stresses are computed at the centroid of each element.

```
» u1=[U(1)  ;  U(2)  ;  U(17)  ;  U(18)  ;  U(13)  ;  U(14)  ;
       U(9)  ;  U(10);  U(15)  ;  U(16)  ;  U(7)    ;  U(8)]

u1 =

   1.0e-005  *

          0
          0
     0.7045
    -0.0257
          0
          0
     0.3405
     0.0144
     0.3352
     0.3352
          0
          0

» u2=[U(1)  ;  U(2)   ;  U(5)   ;  U(6)  ;  U(17)  ;  U(18)  ;
       U(3)  ;  U(4)  ;  U(11)  ;  U(12)  ;  U(9)  ;  U(10)]
```

```
u2 =

   1.0e-005 *

        0
        0
   0.7050
   0.0843
   0.7045
  -0.0257
   0.3575
   0.0673
   0.7018
   0.0302
   0.3405
   0.0144
```

» sigma1=QuadTriangleElementStresses(E,NU,0,0,0.5,0.25,
                0,0.25,1,u1)

```
w =

[150247392356873141099848423881109/49517601571415210995964968 96
+20716814651291511055787432217349/24758800785707605497982484 48*x
-48648038900486918478983768 3443/123794003928538027489912 4224*y]
[2253710885353096825445846737761/24758800785707605497982 48448
+29789159367528562094555864986393/123794003928538027489912 4224*x
-72972058350730368294616150047/618970019642690137449562 112*y]
[21334601311845219291187434866019/158456325028528675187087900672
-180871099018398204773158909579563/198070406285660843983859875 84*x
+3206927544451301888241358 35983/386856262276681335905976 32*y]

sigma1 =

   1.0e+003 *

    4.3633
    4.9012
   -0.0057
```

» sigma2=QuadTriangleElementStresses(E,NU,0,0,0.5,0,0.5,
                0.25,1,u2)

```
w =
```

```
[3792926942888974081216865984587/1237940039285380274899124224
+2323876295224990530218801701337/2475880078570760549798248448*y
-3884711216709091050973007090481/9903520314283042199192993792*x]
[1288199428919325167428812914795/1980704062856608439838598584
+1901660188245083100431630000563/4951760157141521099596496896*y
-3885899865930588250883343493417/1980704062856608439838598584*x]
[1283701383240506045827462488591/1584563250285286751870879006672
+5408416665748185847443906648743/1980704062856608439838598584*y
-5828255474284999942522475983389/1980704062856608439838598584*x]
```

```
sigma2 =
```

```
   1.0e+003 *
```

```
      3.0114
      0.0028
      0.0057
```

Thus it is clear that the stresses at the centroid of element 1 are $\sigma_x = 4.3633$ MPa (tensile), $\sigma_y = 4.9012$ MPa (tensile), and $\tau_{xy} = 0.0057$ MPa (negative). The stresses at the centroid of element 2 are $\sigma_x = 3.0114$ MPa (tensile), $\sigma_y = 0.0028$ MPa (tensile), and $\tau_{xy} = 0.0057$ MPa (positive). It is clear that the stresses in the x-direction approach closely the correct value of 3 MPa (tensile) at the centroid of element 2 because this element is located away from the supports at the left end of the plate. Next we calculate the principal stresses and principal angle for each element by making calls to the MATLAB function *QuadTriangleElementPStresses*.

```
» s1=QuadTriangleElementPStresses(sigma1)
```

```
s1 =
```

```
   1.0e+003 *
```

```
      4.9013
      4.3632
      0.0006
```

```
» s2=QuadTriangleElementPStresses(sigma2)
```

```
s2 =
```

```
   1.0e+003 *
```

```
      3.0114
      0.0028
      0.0001
```

Thus it is clear that the principal stresses at the centroid of element 1 are $\sigma_1 = 4.9013$ MPa (tensile), $\sigma_2 = 4.3632$ MPa (tensile), while the principal angle $\theta_p = 0.6°$. The principal stresses at the centroid of element 2 are $\sigma_1 = 3.0114$ MPa (tensile), $\sigma_2 = 0.0028$ MPa (tensile), while the principal angle $\theta_p = 0.1°$.

Problems:

Problem 12.1:

Consider the thin plate problem solved in Example 12.1. Solve the problem again using four quadratic triangular elements instead of two elements as shown in Fig. 12.4. Compare your answers for the displacements at nodes 3, 8, and 13 with the answers obtained in the example. Compare also the stresses obtained for the four elements with those obtained for the two elements in the example. Compare also your answers with those obtained in Example 11.1 and Problem 11.1.

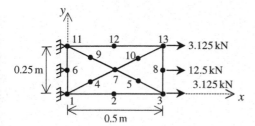

Fig. 12.4. Discretization of Thin Plate Using Four Quadratic Triangles

13 The Bilinear Quadrilateral Element

13.1
Basic Equations

The bilinear quadrilateral element is a two-dimensional finite element with both local and global coordinates. It is characterized by linear shape functions in each of the x and y directions. This element can be used for plane stress or plane strain problems in elasticity. It is a generalization of the 4-node rectangular element. This is the first isoparametric element we deal with in this book. The bilinear quadrilateral element has modulus of elasticity E, Poisson's ratio ν, and thickness t. Each bilinear quadrilateral element has four nodes with two in-plane degrees of freedom at each node as shown in Fig. 13.1. The global coordinates of the four nodes are denoted by (x_1, y_1), (x_2, y_2), (x_3, y_3), and (x_4, y_4). The order of the nodes for each element is important – they should be listed in a counterclockwise direction starting from any node. The area of each element should be positive – you can actually check this by using the MATLAB function *BilinearQuadElementArea* which is written specifically for this purpose. The element is mapped to a rectangle through the use of the natural coordinates ξ and η as shown in Fig. 13.2. In this case the element stiffness matrix is not written explicitly but calculated through symbolic integration with the aid of the MATLAB Symbolic Math Toolbox. The four shape functions for this element are listed explicitly as follows in terms of the natural coordinates ξ and η (see [1]).

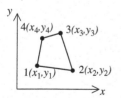

Fig. 13.1. The Bilinear Quadrilateral Element

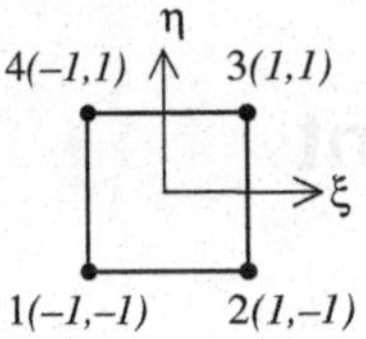

Fig. 13.2. The Bilinear Quadrilateral Element with Natural Coordinates

$$N_1 = \frac{1}{4}(1 - \xi)(1 - \eta)$$

$$N_2 = \frac{1}{4}(1 + \xi)(1 - \eta)$$

$$N_3 = \frac{1}{4}(1 + \xi)(1 + \eta)$$

$$N_4 = \frac{1}{4}(1 - \xi)(1 + \eta) \tag{13.1}$$

The $[B]$ matrix is given as follows for this element:

$$[B] = \frac{1}{|J|} [B_1 \ B_2 \ B_3 \ B_4] \tag{13.2}$$

where each $[B_i]$ is given by:

$$[B_i] = \begin{bmatrix} a\dfrac{\partial N_i}{\partial \xi} - b\dfrac{\partial N_i}{\partial \eta} & 0 \\ 0 & c\dfrac{\partial N_i}{\partial \eta} - d\dfrac{\partial N_i}{\partial \xi} \\ c\dfrac{\partial N_i}{\partial \eta} - d\dfrac{\partial N_i}{\partial \xi} & a\dfrac{\partial N_i}{\partial \xi} - b\dfrac{\partial N_i}{\partial \eta} \end{bmatrix} \tag{13.3}$$

and the parameters a, b, c, and d are given by:

$$a = \frac{1}{4}[y_1(\xi - 1) + y_2(-1 - \xi) + y_3(1 + \xi) + y_4(1 - \xi)]$$

$$b = \frac{1}{4}[y_1(\eta - 1) + y_2(1 - \eta) + y_3(1 + \eta) + y_4(-1 - \eta)]$$

$$c = \frac{1}{4}[x_1(\eta - 1) + x_2(1 - \eta) + x_3(1 + \eta) + x_4(-1 - \eta)]$$

$$d = \frac{1}{4}[x_1(\xi - 1) + x_2(-1 - \xi) + x_3(1 + \xi) + x_4(1 - \xi)] \tag{13.4}$$

In (13.2) the determinant $|J|$ is given by:

$$|J| = \frac{1}{8} \begin{bmatrix} x_1 & x_2 & x_3 & x_4 \end{bmatrix} \begin{bmatrix} 0 & 1-\eta & \eta-\xi & \xi-1 \\ \eta-1 & 0 & \xi+1 & -\xi-\eta \\ \xi-\eta & -\xi-1 & 0 & \eta+1 \\ 1-\xi & \xi+\eta & -\eta-1 & 0 \end{bmatrix} \begin{bmatrix} y_1 \\ y_2 \\ y_3 \\ y_4 \end{bmatrix} \tag{13.5}$$

For cases of plane stress the matrix $[D]$ is given by

$$[D] = \frac{E}{1-\nu^2} \begin{bmatrix} 1 & \nu & 0 \\ \nu & 1 & 0 \\ 0 & 0 & \dfrac{1-\nu}{2} \end{bmatrix} \tag{13.6}$$

For cases of plane strain the matrix $[D]$ is given by

$$[D] = \frac{E}{(1+\nu)(1-2\nu)} \begin{bmatrix} 1-\nu & \nu & 0 \\ \nu & 1-\nu & 0 \\ 0 & 0 & \dfrac{1-2\nu}{2} \end{bmatrix} \tag{13.7}$$

The element stiffness matrix for the bilinear quadrilateral element is written in terms of a double integral as follows:

$$[k] = t \int_{-1}^{1} \int_{-1}^{1} [B]^T [D][B] \, |J| \, d\xi \, d\eta \tag{13.8}$$

where t is the thickness of the element. The double integration of (13.8) is carried out symbolically with the aid of the MATLAB Symbolic Math Toolbox. See the details of the MATLAB code for the function *BilinearQuadElementStiffness* which calculates the element stiffness matrix for this element. The reader should note the calculation of this matrix will be somewhat slow due to the symbolic computations involved.

It is clear that the bilinear quadrilateral element has eight degrees of freedom – two at each node. Consequently for a structure with n nodes, the global stiffness matrix K will be of size $2n \times 2n$ (since we have two degrees of freedom at each node). The global stiffness matrix K is assembled by making calls to the MATLAB function *BilinearQuadAssemble* which is written specifically for this purpose. This process will be illustrated in detail in the examples.

Once the global stiffness matrix K is obtained we have the following structure equation:

$$[K]\{U\} = \{F\} \tag{13.9}$$

where U is the global nodal displacement vector and F is the global nodal force vector. At this step the boundary conditions are applied manually to the vectors U and F. Then the matrix (13.9) is solved by partitioning and Gaussian elimination. Finally once the unknown displacements and reactions are found, the stress vector is obtained for each element as follows:

$$\{\sigma\} = [D][B]\{u\} \tag{13.10}$$

where σ is the stress vector in the element (of size 3×1) and u is the 8×1 element displacement vector. The vector σ is written for each element as $\{\sigma\} = [\sigma_x \ \sigma_y \ \tau_{xy}]^{\mathrm{T}}$. It should be noted that this vector is a linear function of ξ and η. Usually numerical results are obtained at the centroid of the element where $\xi = \eta = 0$. The MATLAB function *BilinearQuadElementStresses* gives two results – the general linear stress functions in ξ and η, and the numerical values of the stresses at the centroid of the element.

13.2
MATLAB Functions Used

The six MATLAB functions used for the bilinear quadrilateral element are:

BilinearQuadElementArea(x_1, y_1, x_2, y_2, x_3, y_3, x_4, y_4) – This function returns the element area given the coordinates of the first node (x_1, y_1), the coordinates of the second node (x_2, y_2), the coordinates of the third node (x_3, y_3), and the coordinates of the fourth node (x_4, y_4).

BilinearQuadElementStiffness(E, *NU*, t, x_1, y_1, x_2, y_2, x_3, y_3, x_4, y_4, p) – This function calculates the element stiffness matrix for each bilinear quadrilateral element with modulus of elasticity E, Poisson's ratio *NU*, thickness t, and coordinates (x_1, y_1) for the first node, (x_2, y_2) for the second node, (x_3, y_3) for the third node, and (x_4, y_4) for the fourth node. Use $p = 1$ for cases of plane stress and $p = 2$ for cases of plane strain. It returns the 8×8 element stiffness matrix k.

BilinearQuadElementStiffness2(E, *NU*, t, x_1, y_1, x_2, y_2, x_3, y_3, x_4, y_4, p) – This function calculates the element stiffness matrix for each bilinear quadrilateral element with modulus of elasticity E, Poisson's ratio *NU*, thickness t, and coordinates (x_1, y_1)

for the first node, (x_2, y_2) for the second node, (x_3, y_3) for the third node, and (x_4, y_4) for the fourth node. Use $p = 1$ for cases of plane stress and $p = 2$ for cases of plane strain. It returns the 8×8 element stiffness matrix k. This function uses a different form of the element equations but produces exactly the same result as the function *BilinearQuadElementStiffness*. The MATALB code for this function is not shown here but is available on the accompanying CD-ROM.

BilinearQuadAssemble(K, k, i, j, m, n) – This function assembles the element stiffness matrix k of the bilinear quadrilateral element joining nodes i, j, m, and n into the global stiffness matrix K. It returns the $2n \times 2n$ global stiffness matrix K every time an element is assembled.

BilinearQuadElementStresses$(E, NU, x_1, y_1, x_2, y_2, x_3, y_3, x_4, y_4, p, u)$ – This function calculates the element stresses using the modulus of elasticity E, Poisson's ratio NU, the coordinates (x_1, y_1) for the first node, (x_2, y_2) for the second node, (x_3, y_3) for the third node, (x_4, y_4) for the fourth node, and the element displacement vector u. Use $p = 1$ for cases of plane stress and $p = 2$ for cases of plane strain. It returns the stress vector for the element.

BilinearQuadElementPStresses$(sigma)$ – This function calculates the element principal stresses using the element stress vector *sigma*. It returns a 3×1 vector in the form $[sigma1\ sigma2\ theta]^T$ where *sigma1* and *sigma2* are the principal stresses for the element and *theta* is the principal angle.

The following is a listing of the MATLAB source code for each function:

```
function y = BilinearQuadElementArea(x1,y1,x2,y2,x3,y3,x4,y4)
% BilinearQuadElementArea          This function returns the area
%                                  of the bilinear quadrilateral
%                                  element whose first node has
%                                  coordinates (x1,y1), second
%                                  node has coordinates (x2,y2),
%                                  third node has coordinates
%                                  (x3,y3), and fourth node has
%                                  coordinates (x4,y4).
yfirst = (x1*(y2-y3) + x2*(y3-y1) + x3*(y1-y2))/2;
ysecond = (x1*(y3-y4) + x3*(y4-y1) + x4*(y1-y3))/2;
y = yfirst + ysecond;
```

```
function w = BilinearQuadElementStiffness
             (E,NU,h,x1,y1,x2,y2,x3,y3,x4,y4,p)
%BilinearQuadElementStiffness      This function returns the element
%                                  stiffness matrix for a bilinear
%                                  quadrilateral element with modulus
%                                  of elasticity E, Poisson's ratio
%                                  NU, thickness h, coordinates of
%                                  node 1 (x1,y1), coordinates
```

```
%                                    node 3 (x3,y3), and coordinates of
%                                    node 4 (x4,y4). Use p = 1 for cases
%                                    of plane stress, and p = 2 for
%                                    cases of plane strain.
%                                    The size of the element
%                                    stiffness matrix is 8 x 8.
syms s t;
a = (y1*(s-1)+y2*(-1-s)+y3*(1+s)+y4*(1-s))/4;
b = (y1*(t-1)+y2*(1-t)+y3*(1+t)+y4*(-1-t))/4;
c = (x1*(t-1)+x2*(1-t)+x3*(1+t)+x4*(-1-t))/4;
d = (x1*(s-1)+x2*(-1-s)+x3*(1+s)+x4*(1-s))/4;
B1 = [a*(t-1)/4-b*(s-1)/4 0 ; 0 c*(s-1)/4-d*(t-1)/4 ;
   c*(s-1)/4-d*(t-1)/4 a*(t-1)/4-b*(s-1)/4];
B2 = [a*(1-t)/4-b*(-1-s)/4 0 ; 0 c*(-1-s)/4-d*(1-t)/4;
   c*(-1-s)/4-d*(1-t)/4 a*(1-t)/4-b*(-1-s)/4];
B3 = [a*(t+1)/4-b*(s+1)/4 0 ; 0 c*(s+1)/4-d*(t+1)/4 ;
   c*(s+1)/4-d*(t+1)/4 a*(t+1)/4-b*(s+1)/4];
B4 = [a*(-1-t)/4-b*(1-s)/4 0 ; 0 c*(1-s)/4-d*(-1-t)/4 ;
   c*(1-s)/4-d*(-1-t)/4 a*(-1-t)/4-b*(1-s)/4];
Bfirst = [B1 B2 B3 B4];
Jfirst = [0 1-t t-s s-1 ; t-1 0 s+1 -s-t ;
   s-t -s-1 0 t+1 ; 1-s s+t -t-1 0];
J = [x1 x2 x3 x4]*Jfirst*[y1 ; y2 ; y3 ; y4]/8;
B = Bfirst/J;
if p == 1
   D = (E/(1-NU*NU))*[1, NU, 0 ; NU, 1, 0 ; 0, 0, (1-NU)/2];
elseif p == 2
   D = (E/(1+NU))/(1-2*NU))*[1-NU, NU, 0 ; NU, 1-NU, 0 ; 0, 0,
      (1-2*NU)/2];
end
BD = J*transpose(B)*D*B;
r = int(int(BD, t, -1, 1), s, -1, 1);
z = h*r;
w = double(z);
```

```
function y = BilinearQuadAssemble(K,k,i,j,m,n)
%BilinearQuadAssemble       This function assembles the element
%                           stiffness matrix k of the bilinear
%                           quadrilateral element with nodes i, j,
%                           m, and n into the global stiffness
%                           matrix K.
%                           This function returns the global stiffness
%                           matrix K after the element stiffness matrix
%                           k is assembled.
K(2*i-1,2*i-1) = K(2*i-1,2*i-1) + k(1,1);
K(2*i-1,2*i) = K(2*i-1,2*i) + k(1,2);
K(2*i-1,2*j-1) = K(2*i-1,2*j-1) + k(1,3);
K(2*i-1,2*j) = K(2*i-1,2*j) + k(1,4);
K(2*i-1,2*m-1) = K(2*i-1,2*m-1) + k(1,5);
K(2*i-1,2*m) = K(2*i-1,2*m) + k(1,6);
K(2*i-1,2*n-1) = K(2*i-1,2*n-1) + k(1,7);
```

```
K(2*i-1,2*n) = K(2*i-1,2*n) + k(1,8);
K(2*i,2*i-1) = K(2*i,2*i-1) + k(2,1);
K(2*i,2*i) = K(2*i,2*i) + k(2,2);
K(2*i,2*j-1) = K(2*i,2*j-1) + k(2,3);
K(2*i,2*j) = K(2*i,2*j) + k(2,4);
K(2*i,2*m-1) = K(2*i,2*m-1) + k(2,5);
K(2*i,2*m) = K(2*i,2*m) + k(2,6);
K(2*i,2*n-1) = K(2*i,2*n-1) + k(2,7);
K(2*i,2*n) = K(2*i,2*n) + k(2,8);
K(2*j-1,2*i-1) = K(2*j-1,2*i-1) + k(3,1);
K(2*j-1,2*i) = K(2*j-1,2*i) + k(3,2);
K(2*j-1,2*j-1) = K(2*j-1,2*j-1) + k(3,3);
K(2*j-1,2*j) = K(2*j-1,2*j) + k(3,4);
K(2*j-1,2*m-1) = K(2*j-1,2*m-1) + k(3,5);
K(2*j-1,2*m) = K(2*j-1,2*m) + k(3,6);
K(2*j-1,2*n-1) = K(2*j-1,2*n-1) + k(3,7);
K(2*j-1,2*n) = K(2*j-1,2*n) + k(3,8);
K(2*j,2*i-1) = K(2*j,2*i-1) + k(4,1);
K(2*j,2*i) = K(2*j,2*i) + k(4,2);
K(2*j,2*j-1) = K(2*j,2*j-1) + k(4,3);
K(2*j,2*j) = K(2*j,2*j) + k(4,4);
K(2*j,2*m-1) = K(2*j,2*m-1) + k(4,5);
K(2*j,2*m) = K(2*j,2*m) + k(4,6);
K(2*j,2*n-1) = K(2*j,2*n-1) + k(4,7);
K(2*j,2*n) = K(2*j,2*n) + k(4,8);
K(2*m-1,2*i-1) = K(2*m-1,2*i-1) + k(5,1);
K(2*m-1,2*i) = K(2*m-1,2*i) + k(5,2);
K(2*m-1,2*j-1) = K(2*m-1,2*j-1) + k(5,3);
K(2*m-1,2*j) = K(2*m-1,2*j) + k(5,4);
K(2*m-1,2*m-1) = K(2*m-1,2*m-1) + k(5,5);
K(2*m-1,2*m) = K(2*m-1,2*m) + k(5,6);
K(2*m-1,2*n-1) = K(2*m-1,2*n-1) + k(5,7);
K(2*m-1,2*n) = K(2*m-1,2*n) + k(5,8);
K(2*m,2*i-y1) = K(2*m,2*i-1) + k(6,1);
K(2*m,2*i) = K(2*m,2*i) + k(6,2);
K(2*m,2*j-1) = K(2*m,2*j-1) + k(6,3);
K(2*m,2*j) = K(2*m,2*j) + k(6,4);
K(2*m,2*m-1) = K(2*m,2*m-1) + k(6,5);
K(2*m,2*m) = K(2*m,2*m) + k(6,6);
K(2*m,2*n-1) = K(2*m,2*n-1) + k(6,7);
K(2*m,2*n) = K(2*m,2*n) + k(6,8);
K(2*n-1,2*i-1) = K(2*n-1,2*i-1) + k(7,1);
K(2*n-1,2*i) = K(2*n-1,2*i) + k(7,2);
K(2*n-1,2*j-1) = K(2*n-1,2*j-1) + k(7,3);
K(2*n-1,2*j) = K(2*n-1,2*j) + k(7,4);
K(2*n-1,2*m-1) = K(2*n-1,2*m-1) + k(7,5);
K(2*n-1,2*m) = K(2*n-1,2*m) + k(7,6);
K(2*n-1,2*n-1) = K(2*n-1,2*n-1) + k(7,7);
K(2*n-1,2*n) = K(2*n-1,2*n) + k(7,8);
K(2*n,2*i-1) = K(2*n,2*i-1) + k(8,1);
K(2*n,2*i) = K(2*n,2*i) + k(8,2);
K(2*n,2*j-1) = K(2*n,2*j-1) + k(8,3);
K(2*n,2*j) = K(2*n,2*j) + k(8,4);
K(2*n,2*m-1) = K(2*n,2*m-1) + k(8,5);
K(2*n,2*m) = K(2*n,2*m) + k(8,6);
```

```
K(2*n,2*n-1) = K(2*n,2*n-1) + k(8,7);
K(2*n,2*n) = K(2*n,2*n) + k(8,8);
y = K;
```

```
function w = BilinearQuadElementStresses
            (E,NU,x1,y1,x2,y2,x3,y3,x4,y4,p,u)
%BilinearQuadElementStresses        This function returns the element
%                                   stress vector for a bilinear
%                                   quadrilateral element with modulus
%                                   of elasticity E, Poisson's ratio
%                                   NU, coordinates of
%                                   node 1 (x1,y1), coordinates
%                                   of node 2 (x2,y2), coordinates of
%                                   node 3 (x3,y3), and coordinates of
%                                   node 4 (x4,y4). Use p = 1 for cases
%                                   of plane stress, and p = 2 for
%                                   cases of plane strain.
syms s t;
a = (y1*(s-1)+y2*(-1-s)+y3*(1+s)+y4*(1-s))/4;
b = (y1*(t-1)+y2*(1-t)+y3*(1+t)+y4*(-1-t))/4;
c = (x1*(t-1)+x2*(1-t)+x3*(1+t)+x4*(-1-t))/4;
d = (x1*(s-1)+x2*(-1-s)+x3*(1+s)+x4*(1-s))/4;
B1 = [a*(t-1)/4-b*(s-1)/4 0 ; 0 c*(s-1)/4-d*(t-1)/4 ;
  c*(s-1)/4-d*(t-1)/4 a*(t-1)/4-b*(s-1)/4];
B2 = [a*(1-t)/4-b*(-1-s)/4 0 ; 0 c*(-1-s)/4-d*(1-t)/4 ;
  c*(-1-s)/4-d*(1-t)/4 a*(1-t)/4-b*(-1-s)/4];
B3 = [a*(t+1)/4-b*(s+1)/4 0 ; 0 c*(s+1)/4-d*(t+1)/4 ;
  c*(s+1)/4-d*(t+1)/4 a*(t+1)/4-b*(s+1)/4];
B4 = [a*(-1-t)/4-b*(1-s)/4 0 ; 0 c*(1-s)/4-d*(-1-t)/4 ;
  c*(1-s)/4-d*(-1-t)/4 a*(-1-t)/4-b*(1-s)/4];
Bfirst = [B1 B2 B3 B4];
Jfirst = [0 1-t t-s s-1 ; t-1 0 s+1 -s-t ;
  s-t -s-1 0 t+1 ; 1-s s+t -t-1 0];
J = [x1 x2 x3 x4]*Jfirst*[y1 ; y2 ; y3 ; y4]/8;
B = Bfirst/J;
if p == 1
  D = (E/(1-NU*NU))*[1, NU, 0 ; NU, 1, 0 ; 0, 0, (1-NU)/2];
elseif p == 2
  D = (E/(1+NU)/(1-2*NU))*[1-NU, NU, 0 ; NU, 1-NU, 0 ; 0, 0,
      (1- 2*NU)/2];
end
w = D*B*u
%
% We also calculate the stresses at the centroid of the element
%
  wcent = subs(w, s,t, 0,0);
  w = double(wcent);
```

```
function y = BilinearQuadElementPStresses(sigma)
%BilinearQuadElementPStresses       This function returns the element
%                                   principal stresses and their
%                                   angle given the element
%                                   stress vector.
```

```
R = (sigma(1) + sigma(2))/2;
Q = ((sigma(1) - sigma(2))/2)^2 + sigma(3)*sigma(3);
M = 2*sigma(3)/(sigma(1) - sigma(2));
s1 = R + sqrt(Q);
s2 = R - sqrt(Q);
theta = (atan(M)/2)*180/pi;
y = [s1 ; s2 ; theta];
```

Example 13.1:

Consider the thin plate subjected to a uniformly distributed load as shown in Fig. 13.3. This problem was solved in Example 11.1 using linear triangular elements. Solve this problem again using two bilinear quadrilateral elements as shown in Fig. 13.4. Given $E = 210\,\text{GPa}$, $\nu = 0.3$, $t = 0.025\,\text{m}$, and $w = 3000\,\text{kN/m}^2$, determine:

1. the global stiffness matrix for the structure.
2. the horizontal and vertical displacements at nodes 3 and 6.
3. the reactions at nodes 1 and 4.
4. the stresses in each element.
5. the principal stresses and principal angle for each element.

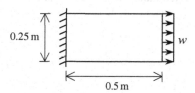

Fig. 13.3. Thin Plate for Example 13.1

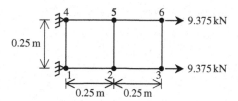

Fig. 13.4. Discretization of Thin Plate Using Two Bilinear Quadrilaterals

Solution:

Use the six steps outlined in Chap. 1 to solve this problem using the bilinear quadrilateral element.

Step 1 – Discretizing the Domain:

We subdivide the plate into two elements only for illustration purposes. More elements must be used in order to obtain reliable results. Thus the domain is subdivided into two elements and six nodes as shown in Fig. 13.4. The total force due to the distributed load is divided equally between nodes 3 and 6. Since the plate is thin, a case of plane stress is assumed. The units used in the MATLAB calculations are kN and meter. Table 13.1 shows the element connectivity for this example.

Table 13.1. Element Connectivity for Example 13.1

Element Number	Node i	Node j	Node m	Node n
1	1	2	5	4
2	2	3	6	5

Step 2 – Writing the Element Stiffness Matrices:

The two element stiffness matrices k_1 and k_2 are obtained by making calls to the MATLAB function *BilinearQuadElementStiffness*. Each matrix has size 8×8.

```
» E=210e6

E =

   210000000

» NU=0.3

NU =

   0.3000

» h=0.025

h =

   0.0250

» k1=BilinearQuadElementStiffness
        (E,NU,h,0,0,0.25,0,0.25,0.25,0,0.25,1)
```

```
k1 =

   1.0e+006 *

   Columns 1 through 7

     2.5962     0.9375    -1.5865    -0.0721    -1.2981    -0.9375     0.2885
     0.9375     2.5962     0.0721     0.2885    -0.9375    -1.2981    -0.0721
    -1.5865     0.0721     2.5962    -0.9375     0.2885    -0.0721    -1.2981
    -0.0721     0.2885    -0.9375     2.5962     0.0721    -1.5865     0.9375
    -1.2981    -0.9375     0.2885     0.0721     2.5962     0.9375    -1.5865
    -0.9375    -1.2981    -0.0721    -1.5865     0.9375     2.5962     0.0721
     0.2885    -0.0721    -1.2981     0.9375    -1.5865     0.0721     2.5962
     0.0721    -1.5865     0.9375    -1.2981    -0.0721     0.2885    -0.9375


   Column 8

     0.0721
    -1.5865
     0.9375
    -1.2981
    -0.0721
     0.2885
    -0.9375
     2.5962


» k2=BilinearQuadElementStiffness
       (E,NU,h,0.25,0,0.5,0,0.5,0.25,0.25,0.25,1)

k2 =

   1.0e+006 *

   Columns 1 through 7

     2.5962     0.9375    -1.5865    -0.0721    -1.2981    -0.9375     0.2885
     0.9375     2.5962     0.0721     0.2885    -0.9375    -1.2981    -0.0721
    -1.5865     0.0721     2.5962    -0.9375     0.2885    -0.0721    -1.2981
    -0.0721     0.2885    -0.9375     2.5962     0.0721    -1.5865     0.9375
    -1.2981    -0.9375     0.2885     0.0721     2.5962     0.9375    -1.5865
    -0.9375    -1.2981    -0.0721    -1.5865     0.9375     2.5962     0.0721
     0.2885    -0.0721    -1.2981     0.9375    -1.5865     0.0721     2.5962
     0.0721    -1.5865     0.9375    -1.2981    -0.0721     0.2885    -0.9375
```

```
Column 8

   0.0721
  -1.5865
   0.9375
  -1.2981
  -0.0721
   0.2885
  -0.9375
   2.5962
```

Step 3 – Assembling the Global Stiffness Matrix:

Since the structure has six nodes, the size of the global stiffness matrix is 12×12. Therefore to obtain K we first set up a zero matrix of size 12×12 then make two calls to the MATLAB function *BilinearQuadAssemble* since we have two elements in the structure. Each call to the function will assemble one element. The following are the MATLAB commands:

» K=zeros(12,12)

K =

```
  0   0   0   0   0   0   0   0   0   0   0   0
  0   0   0   0   0   0   0   0   0   0   0   0
  0   0   0   0   0   0   0   0   0   0   0   0
  0   0   0   0   0   0   0   0   0   0   0   0
  0   0   0   0   0   0   0   0   0   0   0   0
  0   0   0   0   0   0   0   0   0   0   0   0
  0   0   0   0   0   0   0   0   0   0   0   0
  0   0   0   0   0   0   0   0   0   0   0   0
  0   0   0   0   0   0   0   0   0   0   0   0
  0   0   0   0   0   0   0   0   0   0   0   0
  0   0   0   0   0   0   0   0   0   0   0   0
  0   0   0   0   0   0   0   0   0   0   0   0
```

» K=BilinearQuadAssemble(K,k1,1,2,5,4)

K =

     1.0e+006 *

Columns 1 through 7

```
   2.5962      0.9375     -1.5865     -0.0721      0    0     0.2885
   0.9375      2.5962      0.0721      0.2885      0    0    -0.0721
  -1.5865      0.0721      2.5962     -0.9375      0    0    -1.2981
  -0.0721      0.2885     -0.9375      2.5962      0    0     0.9375
        0           0           0           0      0    0          0
        0           0           0           0      0    0          0
   0.2885     -0.0721     -1.2981      0.9375      0    0     2.5962
   0.0721     -1.5865      0.9375     -1.2981      0    0    -0.9375
  -1.2981     -0.9375      0.2885      0.0721      0    0    -1.5865
  -0.9375     -1.2981     -0.0721     -1.5865      0    0     0.0721
        0           0           0           0      0    0          0
        0           0           0           0      0    0          0
```

Columns 8 through 12

```
   0.0721     -1.2981     -0.9375      0    0
  -1.5865     -0.9375     -1.2981      0    0
   0.9375      0.2885     -0.0721      0    0
  -1.2981      0.0721     -1.5865      0    0
        0           0           0      0    0
        0           0           0      0    0
  -0.9375     -1.5865      0.0721      0    0
   2.5962     -0.0721      0.2885      0    0
  -0.0721      2.5962      0.9375      0    0
   0.2885      0.9375      2.5962      0    0
        0           0           0      0    0
        0           0           0      0    0
```

» K=BilinearQuadAssemble(K,k2,2,3,6,5)

K =

1.0e+006 *

Columns 1 through 7

```
   2.5962   0.9375  -1.5865  -0.0721          0          0   0.2885
   0.9375   2.5962   0.0721   0.2885          0          0  -0.0721
  -1.5865   0.0721   5.1923        0    -1.5865    -0.0721  -1.2981
  -0.0721   0.2885        0   5.1923     0.0721     0.2885   0.9375
        0        0  -1.5865   0.0721     2.5962    -0.9375        0
```

```
      0        0   -0.0721    0.2885   -0.9375    2.5962        0
  0.2885  -0.0721   -1.2981    0.9375        0         0    2.5962
  0.0721  -1.5865    0.9375   -1.2981        0         0   -0.9375
 -1.2981  -0.9375    0.5769        0   -1.2981    0.9375   -1.5865
 -0.9375  -1.2981        0   -3.1731    0.9375   -1.2981    0.0721
      0        0   -1.2981   -0.9375    0.2885    0.0721        0
      0        0   -0.9375   -1.2981   -0.0721   -1.5865        0
```

Columns 8 through 12

```
  0.0721   -1.2981   -0.9375        0         0
 -1.5865   -0.9375   -1.2981        0         0
  0.9375    0.5769        0   -1.2981   -0.9375
 -1.2981        0   -3.1731   -0.9375   -1.2981
      0   -1.2981    0.9375    0.2885   -0.0721
      0    0.9375   -1.2981    0.0721   -1.5865
 -0.9375   -1.5865    0.0721        0         0
  2.5962   -0.0721    0.2885        0         0
 -0.0721    5.1923        0   -1.5865    0.0721
  0.2885        0    5.1923   -0.0721    0.2885
      0   -1.5865   -0.0721    2.5962    0.9375
      0    0.0721    0.2885    0.9375    2.5962
```

Step 4 – Applying the Boundary Conditions:

The matrix (13.9) for this structure is obtained as follows using the global stiffness matrix obtained in the previous step:

$$
10^6
\begin{bmatrix}
2.60 & 0.94 & -1.59 & -0.07 & 0 & 0 & 0.29 & 0.07 & -1.30 & -0.94 & 0 & 0 \\
0.94 & 2.60 & 0.07 & 0.29 & 0 & 0 & -0.07 & -1.59 & -0.94 & -1.30 & 0 & 0 \\
-1.59 & 0.07 & 5.19 & 0 & -1.59 & -0.07 & -1.30 & 0.94 & 0.58 & 0 & -1.30 & -0.94 \\
-0.07 & 0.29 & 0 & 5.19 & 0.07 & 0.29 & 0.94 & -1.30 & 0 & -3.17 & -0.94 & -1.30 \\
0 & 0 & -1.59 & 0.07 & 2.60 & -0.94 & 0 & 0 & -1.30 & 0.94 & 0.29 & -0.07 \\
0 & 0 & -0.07 & 0.29 & -0.94 & 2.60 & 0 & 0 & 0.94 & -1.30 & 0.07 & -1.59 \\
0.29 & -0.07 & -1.30 & 0.94 & 0 & 0 & 2.60 & -0.94 & -1.59 & 0.07 & 0 & 0 \\
0.07 & -1.59 & 0.94 & -1.30 & 0 & 0 & -0.94 & 2.60 & -0.07 & 0.29 & 0 & 0 \\
-1.30 & -0.94 & 0.58 & 0 & -1.30 & 0.94 & -1.59 & -0.07 & 5.19 & 0 & -1.59 & 0.07 \\
-0.94 & -1.30 & 0 & -3.17 & 0.94 & -1.30 & 0.07 & 0.29 & 0 & 5.19 & -0.07 & 0.29 \\
0 & 0 & -1.30 & -0.94 & 0.29 & 0.07 & 0 & 0 & -1.59 & -0.07 & 2.60 & 0.94 \\
0 & 0 & -0.94 & -1.30 & -0.07 & -1.59 & 0 & 0 & 0.07 & 0.29 & 0.94 & 2.60
\end{bmatrix}
$$

$$
\begin{Bmatrix}
U_{1x} \\ U_{1y} \\ U_{2x} \\ U_{2y} \\ U_{3x} \\ U_{3y} \\ U_{4x} \\ U_{4y} \\ U_{5x} \\ U_{5y} \\ U_{6x} \\ U_{6y}
\end{Bmatrix}
=
\begin{Bmatrix}
F_{1x} \\ F_{1y} \\ F_{2x} \\ F_{2y} \\ F_{3x} \\ F_{3y} \\ F_{4x} \\ F_{4y} \\ F_{5x} \\ F_{5y} \\ F_{6x} \\ F_{6y}
\end{Bmatrix}
\tag{13.11}
$$

The boundary conditions for this problem are given as:

$$U_{1x} = U_{1y} = U_{4x} = U_{4y} = 0$$
$$F_{2x} = F_{2y} = F_{5x} = F_{5y} = 0$$
$$F_{3x} = 9.375, F_{3y} = 0, F_{6x} = 9.375, F_{6y} = 0 \tag{13.12}$$

Inserting the above conditions into (13.11) we obtain:

$$
10^6
\begin{bmatrix}
2.60 & 0.94 & -1.59 & -0.07 & 0 & 0 & 0.29 & 0.07 & -1.30 & -0.94 & 0 & 0 \\
0.94 & 2.60 & 0.07 & 0.29 & 0 & 0 & -0.07 & -0.59 & -0.94 & -1.30 & 0 & 0 \\
-1.59 & 0.07 & 5.19 & 0 & -1.59 & -0.07 & -1.30 & 0.94 & 0.58 & 0 & -1.30 & -0.94 \\
-0.07 & 0.29 & 0 & 5.19 & 0.07 & 0.29 & 0.94 & -1.30 & 0 & -3.17 & -0.94 & -1.30 \\
0 & 0 & -1.59 & 0.07 & 2.60 & -0.94 & 0 & 0 & -1.30 & 0.94 & 0.29 & -0.07 \\
0 & 0 & -0.07 & 0.29 & -0.94 & 2.60 & 0 & 0 & 0.94 & -1.30 & 0.07 & -1.59 \\
0.29 & -0.07 & -1.30 & 0.94 & 0 & 0 & 2.60 & -0.94 & -1.59 & 0.07 & 0 & 0 \\
0.07 & -1.59 & 0.94 & -1.30 & 0 & 0 & -0.94 & 2.60 & -0.07 & 0.29 & 0 & 0 \\
-1.30 & -0.94 & 0.58 & 0 & -1.30 & 0.94 & -1.59 & -0.07 & 5.19 & 0 & -1.59 & 0.07 \\
-0.94 & -1.30 & 0 & -3.17 & 0.94 & -1.30 & 0.07 & 0.29 & 0 & 5.19 & -0.07 & 0.29 \\
0 & 0 & -1.30 & -0.94 & 0.29 & 0.07 & 0 & 0 & -1.59 & -0.07 & 2.60 & 0.94 \\
0 & 0 & -0.94 & -1.30 & -0.07 & -1.59 & 0 & 0 & 0.07 & 0.29 & 0.94 & 2.60
\end{bmatrix}
\begin{Bmatrix}
0 \\ 0 \\ U_{2x} \\ U_{2y} \\ U_{3x} \\ U_{3y} \\ 0 \\ 0 \\ U_{5x} \\ U_{5y} \\ U_{6x} \\ U_{6y}
\end{Bmatrix}
=
\begin{Bmatrix}
F_{1x} \\ F_{1y} \\ 0 \\ 0 \\ 9.375 \\ 0 \\ F_{4x} \\ F_{4y} \\ 0 \\ 0 \\ 9.375 \\ 0
\end{Bmatrix}
\tag{13.13}
$$

Step 5 – Solving the Equations:

Solving the system of equations in (13.13) will be performed by partitioning (manually) and Gaussian elimination (with MATLAB). First we partition (13.13) by extracting the submatrices in rows 3 to 6, rows 9 to 12, and columns 3 to 6, columns 9 to 12. Therefore we obtain:

$$
10^6
\begin{bmatrix}
5.19 & 0 & -1.59 & -0.07 & 0.58 & 0 & -1.30 & -0.94 \\
0 & 5.19 & 0.07 & 0.29 & 0 & -3.17 & -0.94 & -1.30 \\
-1.59 & 0.07 & 2.60 & -0.94 & -1.30 & 0.94 & 0.29 & -0.07 \\
-0.07 & 0.29 & -0.94 & 2.60 & 0.94 & -1.30 & 0.07 & -1.59 \\
0.58 & 0 & -1.30 & 0.94 & 5.19 & 0 & -1.59 & 0.07 \\
0 & -3.17 & 0.94 & -1.30 & 0 & 5.19 & -0.07 & 0.29 \\
-1.30 & -0.94 & 0.29 & 0.07 & -1.59 & -0.07 & 2.60 & 0.94 \\
-0.94 & -1.30 & -0.07 & -1.59 & 0.07 & 0.29 & 0.94 & 2.60
\end{bmatrix}
$$

$$\begin{Bmatrix} U_{2x} \\ U_{2y} \\ U_{3x} \\ U_{3y} \\ U_{5x} \\ U_{5y} \\ U_{6x} \\ U_{6y} \end{Bmatrix} = \begin{Bmatrix} 0 \\ 0 \\ 9.375 \\ 0 \\ 0 \\ 0 \\ 9.375 \\ 0 \end{Bmatrix} \qquad (13.14)$$

The solution of the above system is obtained using MATLAB as follows. Note that the backslash operator "\" is used for Gaussian elimination.

```
» k=[K(3:6,3:6)  K(3:6,9:12)  ;  K(9:12,3:6)  K(9:12,9:12)]

k =

   1.0e+006 *

   Columns 1 through 7


    5.1923         0   -1.5865   -0.0721    0.5769         0   -1.2981
         0    5.1923    0.0721    0.2885         0   -3.1731   -0.9375
   -1.5865    0.0721    2.5962   -0.9375   -1.2981    0.9375    0.2885
   -0.0721    0.2885   -0.9375    2.5962    0.9375   -1.2981    0.0721
    0.5769         0   -1.2981    0.9375    5.1923         0   -1.5865
         0   -3.1731    0.9375   -1.2981         0    5.1923   -0.0721
   -1.2981   -0.9375    0.2885    0.0721   -1.5865   -0.0721    2.5962
   -0.9375   -1.2981   -0.0721   -1.5865    0.0721    0.2885    0.9375


   Column 8

   -0.9375
   -1.2981
   -0.0721
   -1.5865
    0.0721
    0.2885
    0.9375
    2.5962


» f=[0 ; 0 ; 9.375 ; 0 ; 0 ; 0 ; 9.375 ; 0]
```

```
f =

            0
            0
       9.3750
            0
            0
            0
       9.3750
            0
```

» u=k\f

```
u =

   1.0e-005  *

       0.3440
       0.0632
       0.7030
       0.0503
       0.3440
      -0.0632
       0.7030
      -0.0503
```

It is now clear that the horizontal and vertical displacements at node 3 are 0.7030 m (compared with 0.7111 m in Example 11.1) and 0.0503 m, respectively, and the horizontal and vertical displacements at node 6 are 0.7030 m (compared with 0.6531 m in Example 11.1) and −0.0503 m, respectively.

Step 6 – Post-processing:

In this step, we obtain the reactions at nodes 1 and 4, and the stresses in each element using MATLAB as follows. First we set up the global nodal displacement vector U, then we calculate the global nodal force vector F.

» U=[0;0;u(1:4);0;0;u(5:8)]

```
U =

   1.0e-005  *

            0
            0
       0.3440
```

```
  0.0632
  0.7030
  0.0503
       0
       0
  0.3440
 -0.0632
  0.7030
 -0.0503
```

» F=K*U

F =

```
 -9.3750
 -1.9741
  0.0000
       0
  9.3750
  0.0000
 -9.3750
  1.9741
  0.0000
  0.0000
  9.3750
  0.0000
```

Thus the horizontal and vertical reactions at node 1 are forces of 9.375 kN (directed to the left) and 1.9741 kN (directed downwards). The horizontal and vertical reactions at node 4 are forces of 9.375 kN (directed to the left) and 1.9741 kN (directed upwards). Obviously force equilibrium is satisfied for this problem. Next we set up the element nodal displacement vectors u_1 and u_2 then we calculate the element stresses *sigma1* and *sigma2* by making calls to the MATLAB function *BilinearQuadElementStresses*. Closed-form solutions (linear functions in ξ and η, denoted by s and t in the results below) for the stresses in each element are obtained using the MATLAB Symbolic Math Toolbox as shown below. Then numerical values for the stresses are computed at the centroid of each element.

» u1=[U(1) ; U(2) ; U(3) ; U(4) ; U(9) ; U(10) ; U(7) ;
     U(8)]

u1 =

  1.0e-005 *

```
           0
           0
      0.3440
      0.0632
      0.3440
     -0.0632
           0
           0

» u2=[U(3) ; U(4) ; U(5) ; U(6) ; U(11) ; U(12) ; U(9) ;
     U(10)]

u2 =

   1.0e-005 *

      0.3440
      0.0632
      0.7030
      0.0503
      0.7030
     -0.0503
      0.3440
     -0.0632

» sigma1=BilinearQuadElementStresses(E,NU,0,0,0.25,0,
         0.25,0.25,0,0.25,1,u1)

w =
[23768448754279299619168990954126S/79228162514264337593S
43950336-69689974153846149/39614081257132168796771975168
*t-138619752168264953076103248987B1/79228162514264337593
543950336*s]
[58514709543595028089622024752447/15845632502852867518708
7900672-10453496123076921/1980704062856608439838598758S4
*t-9241316811217664731861929143O489/15845632502852867518
7087900672*s]
[8130496984615383/15845632502852867518708790O672-2439149
O953846149/39614081257132168796771975168*s-3234460883926
1822384424091430489/15845632502852867518708790O672*t]

sigma1 =

   1.0e+003 *
```

```
    3.0000
    0.3693
    0.0000
```

» sigma2=BilinearQuadElementStresses(E,NU,0.25,0,0.5,0,
        0.5,0.25,0.25,0.25,1,u2)

```
w =
[-7743330461538461/3961408125713216879677197516 8*t+23768
4487542792979807838973608303/7922816251426433759354395 03
36+2821463982185924070363071984163/79228162514264337593
543950336*s]
[-1161499569230769/198070406285660843983859875 84*t-8464
39194655779094086818388929 9/15845632502852867518708790 0
672+1880975988123949623157676796304 7/15845632502852867 5
187087900672*s]
[-2710165661538461/396140812571321687967719751 68*s+40652
484923076915/15845632502852867518708790067 2+658341595843
3822830847167963047/15845632502852867518708790067 2*t]
```

sigma2 =

    1.0e+003 *

```
    3.0000
   -0.0534
    0.0000
```

Thus it is clear that the stresses at the centroid of element 1 are $\sigma_x = 3.0000$ MPa (tensile), $\sigma_y = 0.3693$ MPa (tensile), and $\tau_{xy} = 0.0000$ MPa. The stresses at the centroid of element 2 are $\sigma_x = 3.0000$ MPa (tensile), $\sigma_y = 0.0534$ MPa (compressive), and $\tau_{xy} = 0.0000$ MPa. It is clear that the stresses in the x-direction approach closely the correct value of 3 MPa (tensile) at the centroids of both elements. Next we calculate the principal stresses and principal angle for each element by making calls to the MATLAB function *BilinearQuadElementPStresses*.

» s1=BilinearQuadElementPStresses(sigma1)

s1 =

    1.0e+003 *

```
  3.0000
  0.3693
  0.0000
```

» s2=BilinearQuadElementPStresses(sigma2)

s2 =

  1.0e+003 *

    3.0000
   -0.0534
    0.0000

Thus it is clear that the principal stresses at the centroid of element 1 are $\sigma_1 = 3.0000$ MPa (tensile), $\sigma_2 = 0.3693$ MPa (tensile), while the principal angle $\theta_p = 0°$. The principal stresses at the centroid of element 2 are $\sigma_1 = 3.0000$ MPa (tensile), $\sigma_2 = 0.0534$ MPa (compressive), while the principal angle $\theta_p = 0°$.

Example 13.2:

Consider the thin plate subjected to both a uniformly distributed load and a concentrated load as shown in Fig. 13.5. This problem was solved previously in Example 11.2 using linear triangular elements. Solve this problem again using three bilinear quadrilateral elements as shown in Fig. 13.6. Given $E = 210$ GPa, $\nu = 0.3$, $t = 0.025$ m, and $w = 100$ kN/m, determine:

1. the global stiffness matrix for the structure.
2. the horizontal and vertical displacements at each node.
3. the reactions at nodes 1, 4, and 7.
4. the stresses in each element.
5. the principal stresses in each element.

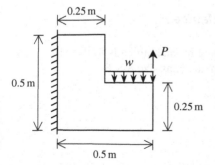

Fig. 13.5. Thin Plate with a Distributed Load and a Concentrated Load for Example 13.2

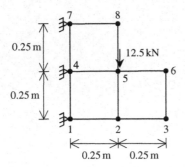

Fig. 13.6. Discretization of Thin Plate Using Three Bilinear Quadrilaterals

Solution:

Use the six steps outlined in Chap. 1 to solve this problem using the bilinear quadrilateral element.

Step 1 – Discretizing the Domain:

We subdivide the plate into three elements and eight nodes as shown in Fig. 13.6. The total force due to the distributed load is divided equally between nodes 5 and 6. However, the resultant applied force at node 6 cancels out and we are left with a concentrated force of 12.5 kN applied vertically downwards at node 5. Since the plate is thin, a case of plane stress is assumed. The units used in the MATLAB calculations are kN and meter. Table 13.2 shows the element connectivity for this example.

Table 13.2. Element Connectivity for Example 13.2

Element Number	Node i	Node j	Node m	Node n
1	1	2	5	4
2	2	3	6	5
3	4	5	8	7

Step 2 – Writing the Element Stiffness Matrices:

The three element stiffness matrices are obtained by making calls to the MATLAB function *BilinearQuadElementStiffness*. Each matrix has size 8×8.

```
» E=210e6

E =

   210000000
```

```
» NU=0.3

NU =

   0.3000

» h=0.025

h =

   0.0250


» k1=BilinearQuadElementStiffness
     (E,NU,h,0,0,0.25,0,0.25,0.25,0,0.25,1)

k1 =

   1.0e+006 *

   Columns 1 through 7

   2.5962    0.9375   -1.5865   -0.0721   -1.2981   -0.9375    0.2885
   0.9375    2.5962    0.0721    0.2885   -0.9375   -1.2981   -0.0721
  -1.5865    0.0721    2.5962   -0.9375    0.2885   -0.0721   -1.2981
  -0.0721    0.2885   -0.9375    2.5962    0.0721   -1.5865    0.9375
  -1.2981   -0.9375    0.2885    0.0721    2.5962    0.9375   -1.5865
  -0.9375   -1.2981   -0.0721   -1.5865    0.9375    2.5962    0.0721
   0.2885   -0.0721   -1.2981    0.9375   -1.5865    0.0721    2.5962
   0.0721   -1.5865    0.9375   -1.2981   -0.0721    0.2885   -0.9375

   Column 8

    0.0721
   -1.5865
    0.9375
   -1.2981
   -0.0721
    0.2885
   -0.9375
    2.5962

» k2=BilinearQuadElementStiffness
     (E,NU,h,0.25,0,0.5,0,0.5,0.25,0.25,0.25,1)
```

k2 =

   1.0e+006 *

   Columns 1 through 7

```
     2.5962    0.9375   -1.5865   -0.0721   -1.2981   -0.9375    0.2885
     0.9375    2.5962    0.0721    0.2885   -0.9375   -1.2981   -0.0721
    -1.5865    0.0721    2.5962   -0.9375    0.2885   -0.0721   -1.2981
    -0.0721    0.2885   -0.9375    2.5962    0.0721   -1.5865    0.9375
    -1.2981   -0.9375    0.2885    0.0721    2.5962    0.9375   -1.5865
    -0.9375   -1.2981   -0.0721   -1.5865    0.9375    2.5962    0.0721
     0.2885   -0.0721   -1.2981    0.9375   -1.5865    0.0721    2.5962
     0.0721   -1.5865    0.9375   -1.2981   -0.0721    0.2885   -0.9375
```

   Column 8

```
     0.0721
    -1.5865
     0.9375
    -1.2981
    -0.0721
     0.2885
    -0.9375
     2.5962
```

» k3=BilinearQuadElementStiffness
     (E,NU,h,0,0.25,0.25,0.25,0.25,0.5,0,0.5,1)

k3 =

   1.0e+006 *

   Columns 1 through 7

```
     2.5962    0.9375   -1.5865   -0.0721   -1.2981   -0.9375    0.2885
     0.9375    2.5962    0.0721    0.2885   -0.9375   -1.2981   -0.0721
    -1.5865    0.0721    2.5962   -0.9375    0.2885   -0.0721   -1.2981
    -0.0721    0.2885   -0.9375    2.5962    0.0721   -1.5865    0.9375
    -1.2981   -0.9375    0.2885    0.0721    2.5962    0.9375   -1.5865
    -0.9375   -1.2981   -0.0721   -1.5865    0.9375    2.5962    0.0721
     0.2885   -0.0721   -1.2981    0.9375   -1.5865    0.0721    2.5962
     0.0721   -1.5865    0.9375   -1.2981   -0.0721    0.2885   -0.9375
```

```
Column 8

    0.0721
   -1.5865
    0.9375
   -1.2981
   -0.0721
    0.2885
   -0.9375
    2.5962
```

Step 3 – Assembling the Global Stiffness Matrix:

Since the structure has eight nodes, the size of the global stiffness matrix is 16×16. Therefore to obtain K we first set up a zero matrix of size 16×16 then make three calls to the MATLAB function *BilinearQuadAssemble* since we have three elements in the structure. Each call to the function will assemble one element. The following are the MATLAB commands The result is shown only after the last element is assembled.

» K=zeros(16,16);

» K=BilinearQuadAssemble(K,k1,1,2,5,4);

» K=BilinearQuadAssemble(K,k2,2,3,6,5);

» K=BilinearQuadAssemble(K,k3,4,5,8,7)

K =

```
   1.0e+006 *

   Columns 1 through 7

    2.5962    0.9375   -1.5865   -0.0721         0         0    0.2885
    0.9375    2.5962    0.0721    0.2885         0         0   -0.0721
   -1.5865    0.0721    5.1923         0   -1.5865   -0.0721   -1.2981
   -0.0721    0.2885         0    5.1923    0.0721    0.2885    0.9375
         0         0   -1.5865    0.0721    2.5962   -0.9375         0
         0         0   -0.0721    0.2885   -0.9375    2.5962         0
    0.2885   -0.0721   -1.2981    0.9375         0         0    5.1923
    0.0721   -1.5865    0.9375   -1.2981         0         0         0
   -1.2981   -0.9375    0.5769         0   -1.2981    0.9375   -3.1731
   -0.9375   -1.2981         0   -3.1731    0.9375   -1.2981         0
         0         0   -1.2981   -0.9375    0.2885    0.0721         0
```

0	0	-0.9375	-1.2981	-0.0721	-1.5865	0
0	0	0	0	0	0	0.2885
0	0	0	0	0	0	0.0721
0	0	0	0	0	0	-1.2981
0	0	0	0	0	0	-0.9375

Columns 8 through 14

0.0721	-1.2981	-0.9375	0	0	0	0
-1.5865	-0.9375	-1.2981	0	0	0	0
0.9375	0.5769	0	-1.2981	-0.9375	0	0
-1.2981	0	-3.1731	-0.9375	-1.2981	0	0
0	-1.2981	0.9375	0.2885	-0.0721	0	0
0	0.9375	-1.2981	0.0721	-1.5865	0	0
0	-3.1731	0	0	0	0.2885	0.0721
5.1923	0	0.5769	0	0	-0.0721	-1.5865
0	7.7885	-0.9375	-1.5865	0.0721	-1.2981	0.9375
0.5769	-0.9375	7.7885	-0.0721	0.2885	0.9375	-1.2981
0	-1.5865	-0.0721	2.5962	0.9375	0	0
0	0.0721	0.2885	0.9375	2.5962	0	0
-0.0721	-1.2981	0.9375	0	0	2.5962	-0.9375
-1.5865	0.9375	-1.2981	0	0	-0.9375	2.5962
-0.9375	0.2885	0.0721	0	0	-1.5865	-0.0721
-1.2981	-0.0721	-1.5865	0	0	0.0721	0.2885

Columns 15 through 16

0	0
0	0
0	0
0	0
0	0
0	0
-1.2981	-0.9375
-0.9375	-1.2981
0.2885	-0.0721
0.0721	-1.5865
0	0
0	0
-1.5865	0.0721
-0.0721	0.2885
2.5962	0.9375
0.9375	2.5962

Step 4 – Applying the Boundary Conditions:

The matrix (13.9) for this structure is obtained using the global stiffness matrix shown above. The boundary conditions for this problem are given as:

$$U_{1x} = U_{1y} = U_{4x} = U_{4y} = U_{7x} = U_{7y} = 0$$
$$F_{2x} = F_{2y} = F_{3x} = F_{3y} = F_{6x} = F_{6y} = F_{8x} = F_{8y} = 0$$
$$F_{5x} = 0 \quad , \quad F_{5y} = -12.5 \tag{13.15}$$

Step 5 – Solving the Equations:

Solving the resulting system of equations will be performed by partitioning (manually) and Gaussian elimination (with MATLAB). First we partition the matrix equation by extracting the submatrices in rows 3 to 6, rows 9 to 12, rows 15 to 16, and columns 3 to 6, columns 9 to 12, columns 15 to 16. Therefore we obtain:

$$10^6
\begin{bmatrix}
5.19 & 0 & -1.59 & -0.07 & 0.58 & 0 & -1.30 & -0.94 & 0 & 0 \\
0 & 5.19 & 0.07 & 0.29 & 0 & -3.17 & 0.94 & -1.30 & 0 & 0 \\
-1.59 & 0.07 & 2.60 & -0.94 & -1.30 & 0.94 & 0.29 & -0.07 & 0 & 0 \\
-0.07 & 0.29 & -0.94 & 2.60 & 0.94 & -1.30 & 0.07 & -1.59 & 0 & 0 \\
0.58 & 0 & -1.30 & 0.94 & 7.79 & -0.94 & -1.59 & 0.07 & 0.29 & -0.07 \\
0 & -3.17 & 0.94 & -1.30 & -0.94 & 7.79 & -0.07 & 0.29 & 0.07 & -1.59 \\
-1.30 & -0.94 & 0.29 & 0.07 & -1.59 & -0.07 & 2.60 & 0.94 & 0 & 0 \\
-0.94 & -1.30 & -0.07 & -1.59 & 0.07 & 0.29 & 0.94 & 2.60 & 0 & 0 \\
0 & 0 & 0 & 0 & 0.29 & 0.07 & 0 & 0 & 2.60 & 0.94 \\
0 & 0 & 0 & 0 & -0.07 & -1.59 & 0 & 0 & 0.94 & 2.60
\end{bmatrix}$$

$$\begin{Bmatrix} U_{2x} \\ U_{2y} \\ U_{3x} \\ U_{3y} \\ U_{5x} \\ U_{5y} \\ U_{6x} \\ U_{6y} \\ U_{8x} \\ U_{8y} \end{Bmatrix} = \begin{Bmatrix} F_{2x} \\ F_{2y} \\ F_{3x} \\ F_{3y} \\ F_{5x} \\ F_{5y} \\ F_{6x} \\ F_{6y} \\ F_{8x} \\ F_{8y} \end{Bmatrix} \tag{13.16}$$

The solution of the above system is obtained using MATLAB as follows. Note that the backslash operator "\" is used for Gaussian elimination.

```
» k= [K(3:6,3:6)  K(3:6,9:12)  K(3:6,15:16)  ;  K(9:12,3:6)
K(9:12,9:12)  K(9:12,15:16)  ;  K(15:16,3:6)  K(15:16,9:12)
K(15:16,15:16)]
```

k =

1.0e+006 *

Columns 1 through 7

```
    5.1923         0  -1.5865  -0.0721   0.5769         0  -1.2981
         0    5.1923   0.0721   0.2885        0  -3.1731  -0.9375
   -1.5865    0.0721   2.5962  -0.9375  -1.2981   0.9375   0.2885
   -0.0721    0.2885  -0.9375   2.5962   0.9375  -1.2981   0.0721
    0.5769         0  -1.2981   0.9375   7.7885  -0.9375  -1.5865
         0   -3.1731   0.9375  -1.2981  -0.9375   7.7885  -0.0721
   -1.2981   -0.9375   0.2885   0.0721  -1.5865  -0.0721   2.5962
   -0.9375   -1.2981  -0.0721  -1.5865   0.0721   0.2885   0.9375
         0         0        0        0   0.2885   0.0721        0
         0         0        0        0  -0.0721  -1.5865        0
```

Columns 8 through 10

```
   -0.9375         0        0
   -1.2981         0        0
   -0.0721         0        0
   -1.5865         0        0
    0.0721    0.2885  -0.0721
    0.2885    0.0721  -1.5865
    0.9375         0        0
    2.5962         0        0
         0    2.5962   0.9375
         0    0.9375   2.5962
```

» f=[0 ; 0 ; 0 ; 0 ; 0 ; -12.5 ; 0 ; 0 ; 0 ; 0]

f =

```
         0
         0
         0
         0
         0
  -12.5000
         0
         0
         0
         0
```

```
» u=k\f

u =

   1.0e-005 *

    -0.1396
    -0.3536
    -0.1329
    -0.5406
     0.0018
    -0.4216
     0.0086
    -0.5176
     0.1202
    -0.3010
```

The horizontal and vertical displacements are shown clearly above for the five free nodes in the structure. These results can be compared with those obtained in Example 11.2 using linear triangular elements. We note that the vertical displacement at node 5 is obtained here as –0.4216 m while the result obtained in Example 11.2 was –0.4170 m. It is clear that these results are very close to each other indicating that the two types of elements and meshes used give similar results in this case.

Step 6 – Post-processing:

In this step, we obtain the reactions at nodes 1, 4, and 7 using MATLAB as follows. We set up the global nodal displacement vector U, then we calculate the global nodal force vector F.

```
» U=[0;0;u(1:4);0;0;u(5:8);0;0;u(9:10)]

U =

   1.0e-005 *

          0
          0
    -0.1396
    -0.3536
    -0.1329
    -0.5406
          0
          0
     0.0018
```

```
        -0.4216
         0.0086
        -0.5176
              0
              0
         0.1202
        -0.3010
```

» F=K*U

F =

```
         6.3994
         4.3354
         0.0000
         0.0000
              0
         0.0000
        -0.2988
         3.6296
         0.0000
       -12.5000
         0.0000
         0.0000
        -6.1006
         4.5350
         0.0000
         0.0000
```

The reactions are shown clearly above. These results can also be compared to those obtained in Example 11.2. We note that the horizontal and vertical reactions at node 1 are 6.3394 kN and 4.3354 kN, respectively, while those obtained in Example 11.2 were 6.4061 kN and 4.2228 kN, respectively. It is also seen that the results obtained using the two types of elements and meshes give very similar results. The stresses in each element are obtained by making successive calls to the MATLAB function *BilinearQuadElementStresses* as follows:

» u1=[U(1) ; U(2) ; U(3) ; U(4) ; U(9) ; U(10) ;
              U(7) ; U(8)]

u1 =

    1.0e-005 *

```
                0
                0
          -0.1396
          -0.3536
           0.0018
          -0.4216
                0
                0

» u2=[U(3)  ;  U(4)  ;  U(5)  ;  U(6)  ;  U(11)  ;  U(12)  ;
        U(9)  ;  U(10)]

u2 =

    1.0e-005 *

      -0.1396
      -0.3536
      -0.1329
      -0.5406
       0.0086
      -0.5176
       0.0018
      -0.4216

» u3=[U(7)  ;  U(8)  ;  U(9)  ;  U(10)  ;  U(15)  ;  U(16)  ;
        U(13)  ;  U(14)]

u3 =

    1.0e-005 *

             0
             0
        0.0018
       -0.4216
        0.1202
       -0.3010
             0
             0


» sigma1=BilinearQuadElementStresses
        (E,NU,0,0,0.25,0,0.25,0.25,0,0.25,1,u1)
```

```
w =
[-3702445501098258470367893952681855/507060240091291760
5986812821504+33096598543339198882197421667956449/507060
2400912917605986812821504*t-1864946507323907280831865840
0023/198070406285660843983859875844*s]
[-12794633890750345444134379977636111/253530120045645880
2993406410752+49644897815008791911985384291242121/2535301
200456458802993406410752*t-124329767154927168111826202929
3387/396140812571321687967719751688*s]
[-519179243554852828621045311548902323/507060240091291760
5986812821504+11583809490168718112796589667956449/507060
2400912917605986812821504*s-4351541850422450321941020229
3387/396140812571321687967719751688*t]

sigma1 =

   1.0e+003 *

     -0.7302
     -0.5047
     -1.0239
```

» sigma2=BilinearQuadElementStresses
            (E,NU,0.25,0,0.5,0,0.5,0.25,0.25,0.25,1,u2)

```
w =
[-5799754515692307289/5070602400912917605986812821504*t
+88810714646684071/5070602400912917605986812821504+2495
7372377422904005019308696233/19807040628566084398385987
584*s]
[-869963177353845981/253530120045645880299340641075*t
-4791815496465182417759008406721571/2535301200456458802
993406410752+16638248251615271485399438695787/39614081
2571321687967719751688*s]
[-2029914080492307289/5070602400912917605986812821504
*s+2767079140430768681/5070602400912917605986812821504
+5823386888065344267837838695787/396140812571321687967
6771975168*t]

sigma2 =

       0.0000
    -189.0038
       0.0000
```

» sigma3=BilinearQuadElementStresses
            (E,NU,0,0.25,0.25,0.25,0.25,0.5,0,0.5,1,u3)

```
w =

[3702445501098258450545924735720001/5070602400912917605
986812821504+27710374781756561856735113630533775/5070602
4009129176059868128215044*t+16538425074453527014211084405
287/99035203142830421991929937922*s]
[18396306602796367789306535146660837/2535301200456458802
993406410752+41565562172634837417184772702287755/25353012
004564588029934064107552*t+11025616716302352766692986279
003/198070406285660843983859877584*s]
[-4949412366277303518648709229360065/507060240091291760
5986812821504+96986311736147953973431136305533755/5070602
4009129176059868128215044*s+38589658507058229699825862779
003/198070406285660843983859877584*t]

sigma3 =

    730.1786
    725.6064
   -976.0995
```

Thus it is clear that the stresses at the centroid of element 1 are $\sigma_x = 0.7302$ MPa (compressive), $\sigma_y = 0.5047$ MPa (compressive), and $\tau_{xy} = 1.0239$ MPa (negative). The stresses at the centroid of element 2 are $\sigma_x = 0.0000$ MPa, $\sigma_y = 0.1890$ MPa (compressive), and $\tau_{xy} = 0.0000$ MPa. The stresses at the centroid of element 3 are $\sigma_x = 0.7302$ MPa (tensile), $\sigma_y = 0.7256$ MPa (tensile), and $\tau_{xy} = 0.9761$ MPa (negative). We cannot compare these results with Example 11.2 because the stresses were not calculated in that example. However, accurate results for the stresses can be obtained by refining the mesh and using more elements. Next we calculate the principal stresses and principal angle for each element by making calls to the MATLAB function *BilinearQuadElementPStresses*.

```
» s1=BilinearQuadElementPStresses(sigma1)

s1 =

   1.0e+003 *

    0.4127
   -1.6475
    0.0419

» s2=BilinearQuadElementPStresses(sigma2)

s2 =

    0.0000
 -189.0038
    0.0000
```

```
» s3=BilinearQuadElementPStresses(sigma3)

s3 =

    1.0e+003 *

      1.7040
     -0.2482
     -0.0449
```

Thus it is clear that the principal stresses at the centroid of element 1 are $\sigma_1 = 0.4127$ MPa (tensile), $\sigma_2 = 1.6475$ MPa (compressive), while the principal angle $\theta_p = 41.9°$. The principal stresses at the centroid of element 2 are $\sigma_1 = 0.0000$ MPa, $\sigma_2 = 0.1890$ MPa (compressive), while the principal angle $\theta_p = 0°$. The principal stresses at the centroid of element 3 are $\sigma_1 = 1.7040$ MPa (tensile), $\sigma_2 = 0.2482$ MPa (compressive), while the principal angle $\theta_p = -44.9°$.

Problems:

Problem 13.1:

Consider the thin plate subjected to a uniformly distributed load as shown in Fig. 13.3. This problem was solved in Example 13.1 using two bilinear quadrilateral elements. Solve this problem again using eight bilinear quadrilateral elements as shown in Fig. 13.7. Given $E = 210$ GPa, $\nu = 0.3$, $t = 0.025$ m, and $w = 3000$ kN/m^2, determine:

1. the global stiffness matrix for the structure.
2. the horizontal and vertical displacements at nodes 5, 10, and 15.
3. the reactions at nodes 1, 6, and 11.
4. the stresses in each element.
5. the principal stresses and principal angle for each element.

Compare your answers with those obtained in Example 13.1, Example 11.1, Problem 11.1, Example 12.1, and Problem 12.1.

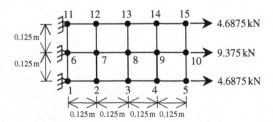

Fig. 13.7. Discretization of Thin Plate Using Eight Bilinear Quadrilaterals

Problem 13.2:

Consider the thin plate with a hole in the middle that is subjected to a concentrated load at the corner as shown in Fig. 13.8. Given $E = 70\,\text{GPa}$, $\nu = 0.25$, $t = 0.02\,\text{m}$, and $P = 20\,\text{kN}$, use the finite element mesh (comprised of bilinear quadrilaterals) shown in Fig. 13.9 to determine:

1. the global stiffness matrix for the structure.
2. the horizontal and vertical displacements at nodes 4, 6, 7, 10, 11 and 16.
3. the reactions at nodes 1, 5, 9, and 13.
4. the stresses in each element.
5. the principal stresses and principal angle for each element.

Compare the results obtained for the displacements and reactions with those obtained in Problem 11.2.

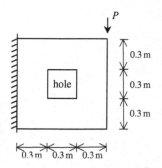

Fig. 13.8. Thin Plate with a Hole for Problem 13.2

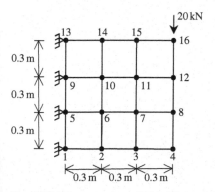

Fig. 13.9. Discretization of Thin Plate with a Hole Using Eight Bilinear Quadrilaterals

Problem 13.3:

Consider the thin plate supported on three springs and subjected to a uniformly distributed load as shown in Fig. 13.10. Given $E = 200\,\text{GPa}$, $\nu = 0.3$, $t = 0.01\,\text{m}$, $k = 4000\,\text{kN/m}$, and $w = 5000\,\text{kN/m}^2$, determine:

1. the global stiffness matrix for the structure.
2. the horizontal and vertical displacements at nodes 1, 2, 3, 4, 5, and 6.
3. the reactions at nodes 7, 8, and 9.
4. the stresses in each element.
5. the principal stresses and principal angle for each element.
6. the force in each spring.

Use two bilinear quadrilaterals to solve this problem. Compare your answers with those obtained for Problem 11.3.

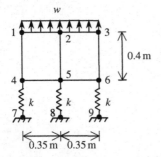

Fig. 13.10. Thin Plate Supported on Three Springs for Problem 13.3

14 The Quadratic Quadrilateral Element

14.1
Basic Equations

The quadratic quadrilateral element is a two-dimensional finite element with both local and global coordinates. It is characterized by quadratic shape functions in each of the x and y directions. This element can be used for plane stress or plane strain problems in elasticity. This is the second isoparametric element we deal with in this book. The quadratic quadrilateral element has modulus of elasticity E, Poisson's ratio ν, and thickness t. Each quadratic quadrilateral element has eight nodes with two in-plane degrees of freedom at each node as shown in Fig. 14.1. The global coordinates of the eight nodes are denoted by (x_1, y_1), (x_2, y_2), (x_3, y_3), (x_4, y_4), (x_5, y_5), (x_6, y_6), (x_7, y_7), and (x_8, y_8). The order of the nodes for each element is important – they should be listed in a counterclockwise direction starting from the corner nodes followed by the midside nodes. The area of each element should be positive – you can actually check this by using the MATLAB function *QuadraticQuadElementArea* which is written specifically for this purpose. The element is mapped to a rectangle through the use of the natural coordinates ξ and η as shown in Fig. 14.2. In this case the element stiffness matrix is not written explicitly but calculated through symbolic integration with the aid of the MATLAB Symbolic Math Toolbox. The eight shape functions for this element are listed explicitly as follows in terms of the natural coordinates ξ and η (see [1]).

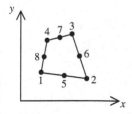

Fig. 14.1. The Quadratic Quadrilateral Element

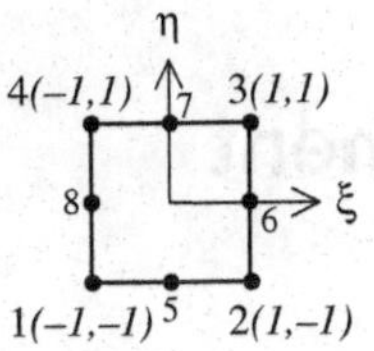

Fig. 14.2. The Quadratic Quadrilateral Element with Natural Coordinates

$$N_1 = \frac{1}{4}(1-\xi)(1-\eta)(-\xi-\eta-1)$$

$$N_2 = \frac{1}{4}(1+\xi)(1-\eta)(\xi-\eta-1)$$

$$N_3 = \frac{1}{4}(1+\xi)(1+\eta)(\xi+\eta-1)$$

$$N_4 = \frac{1}{4}(1-\xi)(1+\eta)(-\xi+\eta-1)$$

$$N_5 = \frac{1}{2}(1-\eta)(1+\xi)(1-\xi)$$

$$N_6 = \frac{1}{2}(1+\xi)(1+\eta)(1-\eta)$$

$$N_7 = \frac{1}{2}(1+\eta)(1+\xi)(1-\xi)$$

$$N_8 = \frac{1}{2}(1-\xi)(1+\eta)(1-\eta) \tag{14.1}$$

The Jacobian matrix for this element is given by

$$[J] = \begin{bmatrix} \dfrac{\partial x}{\partial \xi} & \dfrac{\partial y}{\partial \xi} \\[2mm] \dfrac{\partial x}{\partial \eta} & \dfrac{\partial y}{\partial \eta} \end{bmatrix} \tag{14.2}$$

where x and y are given by

$$x = N_1 x_1 + N_2 x_2 + N_3 x_3 + N_4 x_4 + N_5 x_5 + N_6 x_6 + N_7 x_7 + N_8 x_8$$
$$y = N_1 y_1 + N_2 y_2 + N_3 y_3 + N_4 y_4 + N_5 y_5 + N_6 y_6 + N_7 y_7 + N_8 y_8 \tag{14.3}$$

The $[B]$ matrix is given as follows for this element:

$$[B] = [D'][N] \tag{14.4}$$

where $[D']$ and $[N]$ are given by:

$$[D'] = \frac{1}{|J|} \begin{bmatrix} \dfrac{\partial y}{\partial \eta}\dfrac{\partial(\ )}{\partial \xi} - \dfrac{\partial y}{\partial \xi}\dfrac{\partial(\ )}{\partial \eta} & 0 \\[3mm] 0 & \dfrac{\partial x}{\partial \xi}\dfrac{\partial(\ )}{\partial \eta} - \dfrac{\partial x}{\partial \eta}\dfrac{\partial(\ )}{\partial \xi} \\[3mm] \dfrac{\partial x}{\partial \xi}\dfrac{\partial(\ )}{\partial \eta} - \dfrac{\partial x}{\partial \eta}\dfrac{\partial(\ )}{\partial \xi} & \dfrac{\partial y}{\partial \eta}\dfrac{\partial(\ )}{\partial \xi} - \dfrac{\partial y}{\partial \xi}\dfrac{\partial(\ )}{\partial \eta} \end{bmatrix} \tag{14.5}$$

$$[N] = \begin{bmatrix} N_1 & 0 & N_2 & 0 & N_3 & 0 & N_4 & 0 & N_5 & 0 & N_6 & 0 & N_7 & 0 & N_8 & 0 \\ 0 & N_1 & 0 & N_2 & 0 & N_3 & 0 & N_4 & 0 & N_5 & 0 & N_6 & 0 & N_7 & 0 & N_8 \end{bmatrix} \tag{14.6}$$

For cases of plane stress the matrix $[D]$ is given by

$$[D] = \frac{E}{1-\nu^2} \begin{bmatrix} 1 & \nu & 0 \\ \nu & 1 & 0 \\ 0 & 0 & \dfrac{1-\nu}{2} \end{bmatrix} \tag{14.7}$$

For cases of plane strain the matrix $[D]$ is given by

$$[D] = \frac{E}{(1+\nu)(1-2\nu)} \begin{bmatrix} 1-\nu & \nu & 0 \\ \nu & 1-\nu & 0 \\ 0 & 0 & \dfrac{1-2\nu}{2} \end{bmatrix} \tag{14.8}$$

The element stiffness matrix for the quadratic quadrilateral element is written in terms of a double integral as follows:

$$[k] = t \int_{-1}^{1} \int_{-1}^{1} [B]^T[D][B]\,|J|\,d\xi d\eta \tag{14.9}$$

where t is the thickness of the element. The partial differentiation of (14.5) and the double integration of (14.9) are carried out symbolically with the aid of the MATLAB Symbolic Math Toolbox. See the details of the MATLAB code for the function *QuadraticQuadElementStiffness* which calculates the element stiffness matrix for this element. The reader should note the calculation of this matrix will be somewhat slow due to the symbolic computations involved.

It is clear that the quadratic quadrilateral element has sixteen degrees of freedom – two at each node. Consequently for a structure with n nodes, the global stiffness matrix

K will be of size $2n \times 2n$ (since we have two degrees of freedom at each node). The global stiffness matrix K is assembled by making calls to the MATLAB function *QuadraticQuadAssemble* which is written specifically for this purpose. This process will be illustrated in detail in the examples.

Once the global stiffness matrix K is obtained we have the following structure equation:

$$[K]\{U\} = \{F\} \tag{14.10}$$

where U is the global nodal displacement vector and F is the global nodal force vector. At this step the boundary conditions are applied manually to the vectors U and F. Then the matrix (14.10) is solved by partitioning and Gaussian elimination. Finally once the unknown displacements and reactions are found, the stress vector is obtained for each element as follows:

$$\{\sigma\} = [D][B]\{u\} \tag{14.11}$$

where σ is the stress vector in the element (of size 3×1) and u is the 16×1 element displacement vector. The vector σ is written for each element as $\{\sigma\} = [\sigma_x \sigma_y \tau_{xy}]^\mathrm{T}$. It should be noted that in this case this vector is a quadratic function of ξ and η. Usually numerical results are obtained at the centroid of the element where $\xi = \eta = 0$. The MATLAB function *QuadraticQuadElementStresses* gives two results – the general quadratic stress functions in ξ and η, and the numerical values of the stresses at the centroid of the element.

14.2
MATLAB Functions Used

The five MATLAB functions used for the quadratic quadrilateral element are:

QuadraticQuadElementArea(x_1, y_1, x_2, y_2, x_3, y_3, x_4, y_4) – This function returns the element area given the coordinates of the first node (x_1, y_1), the coordinates of the second node (x_2, y_2), the coordinates of the third node (x_3, y_3), and the coordinates of the fourth node (x_4, y_4).

QuadraticQuadElementStiffness(E, NU, t, x_1, y_1, x_2, y_2, x_3, y_3, x_4, y_4, p) – This function calculates the element stiffness matrix for each quadratic quadrilateral element with modulus of elasticity E, Poisson's ratio NU, thickness t, and coordinates (x_1, y_1) for the first node, (x_2, y_2) for the second node, (x_3, y_3) for the third node, and (x_4, y_4) for the fourth node. Use $p = 1$ for cases of plane stress and $p = 2$ for cases of plane strain. It returns the 16×16 element stiffness matrix k.

QuadraticQuadAssemble(K, k, i, j, m, p, q, r, s, t) – This function assembles the element stiffness matrix k of the quadratic quadrilateral element joining nodes i, j, m, p, q, r, s, and t into the global stiffness matrix K. It returns the $2n \times 2n$ global stiffness matrix K every time an element is assembled.

QuadraticQuadElementStresses(E, *NU*, x_1, y_1, x_2, y_2, x_3, y_3, x_4, y_4, p, u) – This function calculates the element stresses using the modulus of elasticity E, Poisson's ratio *NU*, the coordinates (x_1, y_1) for the first node, (x_2, y_2) for the second node, (x_3, y_3) for the third node, (x_4, y_4) for the fourth node, and the element displacement vector u. Use $p = 1$ for cases of plane stress and $p = 2$ for cases of plane strain. It returns the stress vector for the element.

QuadraticQuadElementPStresses(*sigma*) – This function calculates the element principal stresses using the element stress vector *sigma*. It returns a 3×1 vector in the form [*sigma1 sigma2 theta*]^T where *sigma1* and *sigma2* are the principal stresses for the element and *theta* is the principal angle.

The following is a listing of the MATLAB source code for each function:

```
function y = QuadraticQuadElementArea(x1,y1,x2,y2,x3,y3,x4,y4)
% QuadraticQuadElementArea          This function returns the area
%                                   of the quadratic quadrilateral
%                                   element whose first node has
%                                   coordinates (x1,y1), second
%                                   node has coordinates (x2,y2),
%                                   third node has coordinates
%                                   (x3,y3), and fourth node has
%                                   coordinates (x4,y4).
yfirst = (x1*(y2-y3) + x2*(y3-y1) + x3*(y1-y2))/2;
ysecond = (x1*(y3-y4) + x3*(y4-y1) + x4*(y1-y3))/2;
y = yfirst + ysecond;
```

```
function w = QuadraticQuadElementStiffness
             (E,NU,h,x1,y1,x2,y2,x3,y3,x4,y4,p)
%QuadraticQuadElementStiffness      This function returns the element
%                                   stiffness matrix for a quadratic
%                                   quadrilateral element with modulus
%                                   of elasticity E, Poisson's ratio
%                                   NU, thickness h, coordinates of
%                                   node 1 (x1,y1), coordinates
%                                   of node 2 (x2,y2), coordinates of
%                                   node 3 (x3,y3), and coordinates of
%                                   node 4 (x4,y4). Use p = 1 for cases
%                                   of plane stress, and p = 2 for
%                                   cases of plane strain.
%                                   The size of the element
%                                   stiffness matrix is 16 x 16.
```

```
syms s t;
x5 = (x1 + x2)/2;
x6 = (x2 + x3)/2;
x7 = (x3 + x4)/2;
x8 = (x4 + x1)/2;
y5 = (y1 + y2)/2;
y6 = (y2 + y3)/2;
y7 = (y3 + y4)/2;
y8 = (y4 + y1)/2;
N1 = (1-s)*(1-t)*(-s-t-1)/4;
N2 = (1+s)*(1-t)*(s-t-1)/4;
N3 = (1+s)*(1+t)*(s+t-1)/4;
N4 = (1-s)*(1+t)*(-s+t-1)/4;
N5 = (1-t)*(1+s)*(1-s)/2;
N6 = (1+s)*(1+t)*(1-t)/2;
N7 = (1+t)*(1+s)*(1-s)/2;
N8 = (1-s)*(1+t)*(1-t)/2;
x = N1*x1 + N2*x2 + N3*x3 + N4*x4 + N5*x5 + N6*x6 + N7*x7 + N8*x8;
y = N1*y1 + N2*y2 + N3*y3 + N4*y4 + N5*y5 + N6*y6 + N7*y7 + N8*y8;
xs = diff(x,s);
xt = diff(x,t);
ys = diff(y,s);
yt = diff(y,t);
J = xs*yt - ys*xt;
N1s = diff(N1,s);
N2s = diff(N2,s);
N3s = diff(N3,s);
N4s = diff(N4,s);
N5s = diff(N5,s);
N6s = diff(N6,s);
N7s = diff(N7,s);
N8s = diff(N8,s);
N1t = diff(N1,t);
N2t = diff(N2,t);
N3t = diff(N3,t);
N4t = diff(N4,t);
N5t = diff(N5,t);
N6t = diff(N6,t);
N7t = diff(N7,t);
N8t = diff(N8,t);
B11 = yt*N1s - ys*N1t;
B12 = 0;
B13 = yt*N2s - ys*N2t;
B14 = 0;
B15 = yt*N3s - ys*N3t;
B16 = 0;
B17 = yt*N4s - ys*N4t;
B18 = 0;
B19 = yt*N5s - ys*N5t;
B110 = 0;
B111 = yt*N6s - ys*N6t;
B112 = 0;
B113 = yt*N7s - ys*N7t;
```

```
B114 = 0;
B115 = yt*N8s - ys*N8t;
B116 = 0;
B21 = 0;
B22 = xs*N1t - xt*N1s;
B23 = 0;
B24 = xs*N2t - xt*N2s;
B25 = 0;
B26 = xs*N3t - xt*N3s;
B27 = 0;
B28 = xs*N4t - xt*N4s;
B29 = 0;
B210 = xs*N5t - xt*N5s;
B211 = 0;
B212 = xs*N6t - xt*N6s;
B213 = 0;
B214 = xs*N7t - xt*N7s;
B215 = 0;
B216 = xs*N8t - xt*N8s;
B31 = xs*N1t - xt*N1s;
B32 = yt*N1s - ys*N1t;
B33 = xs*N2t - xt*N2s;
B34 = yt*N2s - ys*N2t;
B35 = xs*N3t - xt*N3s;
B36 = yt*N3s - ys*N3t;
B37 = xs*N4t - xt*N4s;
B38 = yt*N4s - ys*N4t;
B39 = xs*N5t - xt*N5s;
B310 = yt*N5s - ys*N5t;
B311 = xs*N6t - xt*N6s;
B312 = yt*N6s - ys*N6t;
B313 = xs*N7t - xt*N7s;
B314 = yt*N7s - ys*N7t;
B315 = xs*N8t - xt*N8s;
B316 = yt*N8s - ys*N8t;
B = [B11 B12 B13 B14 B15 B16 B17 B18 B19 B110 B111 B112 B113 B114
     B115 B116;
  B21 B22 B23 B24 B25 B26 B27 B28 B29 B210 B211 B212 B213 B214
     B215 B216;
  B31 B32 B33 B34 B35 B36 B37 B38 B39 B310 B311 B312 B313 B314
     B315 B316];
if p == 1
   D = (E/(1-NU*NU))*[1, NU, 0 ; NU, 1, 0 ; 0, 0, (1-NU)/2];
elseif p == 2
   D = (E/(1+NU)/(1-2*NU))
        *[1-NU, NU, 0 ;NU, 1-NU, 0 ; 0, 0, (1-2*NU)/2];
end
Bnew = simplify(B);
Jnew = simplify(J);
BD = transpose(Bnew)*D*Bnew/Jnew;
r = int(int(BD, t, -1, 1), s, -1, 1);
z = h*r;
w = double(z);
```

```
function y = QuadraticQuadAssemble(K,k,i,j,m,p,q,r,s,t)
%QuadraticQuadAssemble            This function assembles the element
%                                 stiffness matrix k of the quadratic
%                                 quadrilateral element with nodes i, j,
%                                 m, p, q, r, s, and t into the global
%                                 stiffness matrix K.
%                                 This function returns the global
%                                 stiffness matrix K after the element
%                                 stiffness matrix k is assembled.
K(2*i-1,2*i-1) = K(2*i-1,2*i-1) + k(1,1);
K(2*i-1,2*i) = K(2*i-1,2*i) + k(1,2);
K(2*i-1,2*j-1) = K(2*i-1,2*j-1) + k(1,3);
K(2*i-1,2*j) = K(2*i-1,2*j) + k(1,4);
K(2*i-1,2*m-1) = K(2*i-1,2*m-1) + k(1,5);
K(2*i-1,2*m) = K(2*i-1,2*m) + k(1,6);
K(2*i-1,2*p-1) = K(2*i-1,2*p-1) + k(1,7);
K(2*i-1,2*p) = K(2*i-1,2*p) + k(1,8);
K(2*i-1,2*q-1) = K(2*i-1,2*q-1) + k(1,9);
K(2*i-1,2*q) = K(2*i-1,2*q) + k(1,10);
K(2*i-1,2*r-1) = K(2*i-1,2*r-1) + k(1,11);
K(2*i-1,2*r) = K(2*i-1,2*r) + k(1,12);
K(2*i-1,2*s-1) = K(2*i-1,2*s-1) + k(1,13);
K(2*i-1,2*s) = K(2*i-1,2*s) + k(1,14);
K(2*i-1,2*t-1) = K(2*i-1,2*t-1) + k(1,15);
K(2*i-1,2*t) = K(2*i-1,2*t) + k(1,16);
K(2*i,2*i-1) = K(2*i,2*i-1) + k(2,1);
K(2*i,2*i) = K(2*i,2*i) + k(2,2);
K(2*i,2*j-1) = K(2*i,2*j-1) + k(2,3);
K(2*i,2*j) = K(2*i,2*j) + k(2,4);
K(2*i,2*m-1) = K(2*i,2*m-1) + k(2,5);
K(2*i,2*m) = K(2*i,2*m) + k(2,6);
K(2*i,2*p-1) = K(2*i,2*p-1) + k(2,7);
K(2*i,2*p) = K(2*i,2*p) + k(2,8);
K(2*i,2*q-1) = K(2*i,2*q-1) + k(2,9);
K(2*i,2*q) = K(2*i,2*q) + k(2,10);
K(2*i,2*r-1) = K(2*i,2*r-1) + k(2,11);
K(2*i,2*r) = K(2*i,2*r) + k(2,12);
K(2*i,2*s-1) = K(2*i,2*s-1) + k(2,13);
K(2*i,2*s) = K(2*i,2*s) + k(2,14);
K(2*i,2*t-1) = K(2*i,2*t-1) + k(2,15);
K(2*i,2*t) = K(2*i,2*t) + k(2,16);
K(2*j-1,2*i-1) = K(2*j-1,2*i-1) + k(3,1);
K(2*j-1,2*i) = K(2*j-1,2*i) + k(3,2);
K(2*j-1,2*j-1) = K(2*j-1,2*j-1) + k(3,3);
K(2*j-1,2*j) = K(2*j-1,2*j) + k(3,4);
K(2*j-1,2*m-1) = K(2*j-1,2*m-1) + k(3,5);
K(2*j-1,2*m) = K(2*j-1,2*m) + k(3,6);
K(2*j-1,2*p-1) = K(2*j-1,2*p-1) + k(3,7);
K(2*j-1,2*p) = K(2*j-1,2*p) + k(3,8);
K(2*j-1,2*q-1) = K(2*j-1,2*q-1) + k(3,9);
K(2*j-1,2*q) = K(2*j-1,2*q) + k(3,10);
K(2*j-1,2*r-1) = K(2*j-1,2*r-1) + k(3,11);
K(2*j-1,2*r) = K(2*j-1,2*r) + k(3,12);
```

```
K(2*j-1,2*s-1) = K(2*j-1,2*s-1) + k(3,13);
K(2*j-1,2*s) = K(2*j-1,2*s) + k(3,14);
K(2*j-1,2*t-1) = K(2*j-1,2*t-1) + k(3,15);
K(2*j-1,2*t) = K(2*j-1,2*t) + k(3,16);
K(2*j,2*i-1) = K(2*j,2*i-1) + k(4,1);
K(2*j,2*i) = K(2*j,2*i) + k(4,2);
K(2*j,2*j-1) = K(2*j,2*j-1) + k(4,3);
K(2*j,2*j) = K(2*j,2*j) + k(4,4);
K(2*j,2*m-1) = K(2*j,2*m-1) + k(4,5);
K(2*j,2*m) = K(2*j,2*m) + k(4,6);
K(2*j,2*p-1) = K(2*j,2*p-1) + k(4,7);
K(2*j,2*p) = K(2*j,2*p) + k(4,8);
K(2*j,2*q-1) = K(2*j,2*q-1) + k(4,9);
K(2*j,2*q) = K(2*j,2*q) + k(4,10);
K(2*j,2*r-1) = K(2*j,2*r-1) + k(4,11);
K(2*j,2*r) = K(2*j,2*r) + k(4,12);
K(2*j,2*s-1) = K(2*j,2*s-1) + k(4,13);
K(2*j,2*s) = K(2*j,2*s) + k(4,14);
K(2*j,2*t-1) = K(2*j,2*t-1) + k(4,15);
K(2*j,2*t) = K(2*j,2*t) + k(4,16);
K(2*m-1,2*i-1) = K(2*m-1,2*i-1) + k(5,1);
K(2*m-1,2*i) = K(2*m-1,2*i) + k(5,2);
K(2*m-1,2*j-1) = K(2*m-1,2*j-1) + k(5,3);
K(2*m-1,2*j) = K(2*m-1,2*j) + k(5,4);
K(2*m-1,2*m-1) = K(2*m-1,2*m-1) + k(5,5);
K(2*m-1,2*m) = K(2*m-1,2*m) + k(5,6);
K(2*m-1,2*p-1) = K(2*m-1,2*p-1) + k(5,7);
K(2*m-1,2*p) = K(2*m-1,2*p) + k(5,8);
K(2*m-1,2*q-1) = K(2*m-1,2*q-1) + k(5,9);
K(2*m-1,2*q) = K(2*m-1,2*q) + k(5,10);
K(2*m-1,2*r-1) = K(2*m-1,2*r-1) + k(5,11);
K(2*m-1,2*r) = K(2*m-1,2*r) + k(5,12);
K(2*m-1,2*s-1) = K(2*m-1,2*s-1) + k(5,13);
K(2*m-1,2*s) = K(2*m-1,2*s) + k(5,14);
K(2*m-1,2*t-1) = K(2*m-1,2*t-1) + k(5,15);
K(2*m-1,2*t) = K(2*m-1,2*t) + k(5,16);
K(2*m,2*i-1) = K(2*m,2*i-1) + k(6,1);
K(2*m,2*i) = K(2*m,2*i) + k(6,2);
K(2*m,2*j-1) = K(2*m,2*j-1) + k(6,3);
K(2*m,2*j) = K(2*m,2*j) + k(6,4);
K(2*m,2*m-1) = K(2*m,2*m-1) + k(6,5);
K(2*m,2*m) = K(2*m,2*m) + k(6,6);
K(2*m,2*p-1) = K(2*m,2*p-1) + k(6,7);
K(2*m,2*p) = K(2*m,2*p) + k(6,8);
K(2*m,2*q-1) = K(2*m,2*q-1) + k(6,9);
K(2*m,2*q) = K(2*m,2*q) + k(6,10);
K(2*m,2*r-1) = K(2*m,2*r-1) + k(6,11);
K(2*m,2*r) = K(2*m,2*r) + k(6,12);
K(2*m,2*s-1) = K(2*m,2*s-1) + k(6,13);
K(2*m,2*s) = K(2*m,2*s) + k(6,14);
K(2*m,2*t-1) = K(2*m,2*t-1) + k(6,15);
K(2*m,2*t) = K(2*m,2*t) + k(6,16);
K(2*p-1,2*i-1) = K(2*p-1,2*i-1) + k(7,1);
```

```
K(2*p-1,2*i) = K(2*p-1,2*i) + k(7,2);
K(2*p-1,2*j-1) = K(2*p-1,2*j-1) + k(7,3);
K(2*p-1,2*j) = K(2*p-1,2*j) + k(7,4);
K(2*p-1,2*m-1) = K(2*p-1,2*m-1) + k(7,5);
K(2*p-1,2*m) = K(2*p-1,2*m) + k(7,6);
K(2*p-1,2*p-1) = K(2*p-1,2*p-1) + k(7,7);
K(2*p-1,2*p) = K(2*p-1,2*p) + k(7,8);
K(2*p-1,2*q-1) = K(2*p-1,2*q-1) + k(7,9);
K(2*p-1,2*q) = K(2*p-1,2*q) + k(7,10);
K(2*p-1,2*r-1) = K(2*p-1,2*r-1) + k(7,11);
K(2*p-1,2*r) = K(2*p-1,2*r) + k(7,12);
K(2*p-1,2*s-1) = K(2*p-1,2*s-1) + k(7,13);
K(2*p-1,2*s) = K(2*p-1,2*s) + k(7,14);
K(2*p-1,2*t-1) = K(2*p-1,2*t-1) + k(7,15);
K(2*p-1,2*t) = K(2*p-1,2*t) + k(7,16);
K(2*p,2*i-1) = K(2*p,2*i-1) + k(8,1);
K(2*p,2*i) = K(2*p,2*i) + k(8,2);
K(2*p,2*j-1) = K(2*p,2*j-1) + k(8,3);
K(2*p,2*j) = K(2*p,2*j) + k(8,4);
K(2*p,2*m-1) = K(2*p,2*m-1) + k(8,5);
K(2*p,2*m) = K(2*p,2*m) + k(8,6);
K(2*p,2*p-1) = K(2*p,2*p-1) + k(8,7);
K(2*p,2*p) = K(2*p,2*p) + k(8,8);
K(2*p,2*q-1) = K(2*p,2*q-1) + k(8,9);
K(2*p,2*q) = K(2*p,2*q) + k(8,10);
K(2*p,2*r-1) = K(2*p,2*r-1) + k(8,11);
K(2*p,2*r) = K(2*p,2*r) + k(8,12);
K(2*p,2*s-1) = K(2*p,2*s-1) + k(8,13);
K(2*p,2*s) = K(2*p,2*s) + k(8,14);
K(2*p,2*t-1) = K(2*p,2*t-1) + k(8,15);
K(2*p,2*t) = K(2*p,2*t) + k(8,16);
K(2*q-1,2*i-1) = K(2*q-1,2*i-1) + k(9,1);
K(2*q-1,2*i) = K(2*q-1,2*i) + k(9,2);
K(2*q-1,2*j-1) = K(2*q-1,2*j-1) + k(9,3);
K(2*q-1,2*j) = K(2*q-1,2*j) + k(9,4);
K(2*q-1,2*m-1) = K(2*q-1,2*m-1) + k(9,5);
K(2*q-1,2*m) = K(2*q-1,2*m) + k(9,6);
K(2*q-1,2*p-1) = K(2*q-1,2*p-1) + k(9,7);
K(2*q-1,2*p) = K(2*q-1,2*p) + k(9,8);
K(2*q-1,2*q-1) = K(2*q-1,2*q-1) + k(9,9);
K(2*q-1,2*q) = K(2*q-1,2*q) + k(9,10);
K(2*q-1,2*r-1) = K(2*q-1,2*r-1) + k(9,11);
K(2*q-1,2*r) = K(2*q-1,2*r) + k(9,12);
K(2*q-1,2*s-1) = K(2*q-1,2*s-1) + k(9,13);
K(2*q-1,2*s) = K(2*q-1,2*s) + k(9,14);
K(2*q-1,2*t-1) = K(2*q-1,2*t-1) + k(9,15);
K(2*q-1,2*t) = K(2*q-1,2*t) + k(9,16);
K(2*q,2*i-1) = K(2*q,2*i-1) + k(10,1);
K(2*q,2*i) = K(2*q,2*i) + k(10,2);
K(2*q,2*j-1) = K(2*q,2*j-1) + k(10,3);
K(2*q,2*j) = K(2*q,2*j) + k(10,4);
K(2*q,2*m-1) = K(2*q,2*m-1) + k(10,5);
K(2*q,2*m) = K(2*q,2*m) + k(10,6);
```

```
K(2*q,2*p-1) = K(2*q,2*p-1) + k(10,7);
K(2*q,2*p) = K(2*q,2*p) + k(10,8);
K(2*q,2*q-1) = K(2*q,2*q-1) + k(10,9);
K(2*q,2*q) = K(2*q,2*q) + k(10,10);
K(2*q,2*r-1) = K(2*q,2*r-1) + k(10,11);
K(2*q,2*r) = K(2*q,2*r) + k(10,12);
K(2*q,2*s-1) = K(2*q,2*s-1) + k(10,13);
K(2*q,2*s) = K(2*q,2*s) + k(10,14);
K(2*q,2*t-1) = K(2*q,2*t-1) + k(10,15);
K(2*q,2*t) = K(2*q,2*t) + k(10,16);
K(2*r-1,2*i-1) = K(2*r-1,2*i-1) + k(11,1);
K(2*r-1,2*i) = K(2*r-1,2*i) + k(11,2);
K(2*r-1,2*j-1) = K(2*r-1,2*j-1) + k(11,3);
K(2*r-1,2*j) = K(2*r-1,2*j) + k(11,4);
K(2*r-1,2*m-1) = K(2*r-1,2*m-1) + k(11,5);
K(2*r-1,2*m) = K(2*r-1,2*m) + k(11,6);
K(2*r-1,2*p-1) = K(2*r-1,2*p-1) + k(11,7);
K(2*r-1,2*p) = K(2*r-1,2*p) + k(11,8);
K(2*r-1,2*q-1) = K(2*r-1,2*q-1) + k(11,9);
K(2*r-1,2*q) = K(2*r-1,2*q) + k(11,10);
K(2*r-1,2*r-1) = K(2*r-1,2*r-1) + k(11,11);
K(2*r-1,2*r) = K(2*r-1,2*r) + k(11,12);
K(2*r-1,2*s-1) = K(2*r-1,2*s-1) + k(11,13);
K(2*r-1,2*s) = K(2*r-1,2*s) + k(11,14);
K(2*r-1,2*t-1) = K(2*r-1,2*t-1) + k(11,15);
K(2*r-1,2*t) = K(2*r-1,2*t) + k(11,16);
K(2*r,2*i-1) = K(2*r,2*i-1) + k(12,1);
K(2*r,2*i) = K(2*r,2*i) + k(12,2);
K(2*r,2*j-1) = K(2*r,2*j-1) + k(12,3);
K(2*r,2*j) = K(2*r,2*j) + k(12,4);
K(2*r,2*m-1) = K(2*r,2*m-1) + k(12,5);
K(2*r,2*m) = K(2*r,2*m) + k(12,6);
K(2*r,2*p-1) = K(2*r,2*p-1) + k(12,7);
K(2*r,2*p) = K(2*r,2*p) + k(12,8);
K(2*r,2*q-1) = K(2*r,2*q-1) + k(12,9);
K(2*r,2*q) = K(2*r,2*q) + k(12,10);
K(2*r,2*r-1) = K(2*r,2*r-1) + k(12,11);
K(2*r,2*r) = K(2*r,2*r) + k(12,12);
K(2*r,2*s-1) = K(2*r,2*s-1) + k(12,13);
K(2*r,2*s) = K(2*r,2*s) + k(12,14);
K(2*r,2*t-1) = K(2*r,2*t-1) + k(12,15);
K(2*r,2*t) = K(2*r,2*t) + k(12,16);
K(2*s-1,2*i-1) = K(2*s-1,2*i-1) + k(13,1);
K(2*s-1,2*i) = K(2*s-1,2*i) + k(13,2);
K(2*s-1,2*j-1) = K(2*s-1,2*j-1) + k(13,3);
K(2*s-1,2*j) = K(2*s-1,2*j) + k(13,4);
K(2*s-1,2*m-1) = K(2*s-1,2*m-1) + k(13,5);
K(2*s-1,2*m) = K(2*s-1,2*m) + k(13,6);
K(2*s-1,2*p-1) = K(2*s-1,2*p-1) + k(13,7);
K(2*s-1,2*p) = K(2*s-1,2*p) + k(13,8);
K(2*s-1,2*q-1) = K(2*s-1,2*q-1) + k(13,9);
K(2*s-1,2*q) = K(2*s-1,2*q) + k(13,10);
K(2*s-1,2*r-1) = K(2*s-1,2*r-1) + k(13,11);
```

```
K(2*s-1,2*r) = K(2*s-1,2*r) + k(13,12);
K(2*s-1,2*s-1) = K(2*s-1,2*s-1) + k(13,13);
K(2*s-1,2*s) = K(2*s-1,2*s) + k(13,14);
K(2*s-1,2*t-1) = K(2*s-1,2*t-1) + k(13,15);
K(2*s-1,2*t) = K(2*s-1,2*t) + k(13,16);
K(2*s,2*i-1) = K(2*s,2*i-1) + k(14,1);
K(2*s,2*i) = K(2*s,2*i) + k(14,2);
K(2*s,2*j-1) = K(2*s,2*j-1) + k(14,3);
K(2*s,2*j) = K(2*s,2*j) + k(14,4);
K(2*s,2*m-1) = K(2*s,2*m-1) + k(14,5);
K(2*s,2*m) = K(2*s,2*m) + k(14,6);
K(2*s,2*p-1) = K(2*s,2*p-1) + k(14,7);
K(2*s,2*p) = K(2*s,2*p) + k(14,8);
K(2*s,2*q-1) = K(2*s,2*q-1) + k(14,9);
K(2*s,2*q) = K(2*s,2*q) + k(14,10);
K(2*s,2*r-1) = K(2*s,2*r-1) + k(14,11);
K(2*s,2*r) = K(2*s,2*r) + k(14,12);
K(2*s,2*s-1) = K(2*s,2*s-1) + k(14,13);
K(2*s,2*s) = K(2*s,2*s) + k(14,14);
K(2*s,2*t-1) = K(2*s,2*t-1) + k(14,15);
K(2*s,2*t) = K(2*s,2*t) + k(14,16);
K(2*t-1,2*i-1) = K(2*t-1,2*i-1) + k(15,1);
K(2*t-1,2*i) = K(2*t-1,2*i) + k(15,2);
K(2*t-1,2*j-1) = K(2*t-1,2*j-1) + k(15,3);
K(2*t-1,2*j) = K(2*t-1,2*j) + k(15,4);
K(2*t-1,2*m-1) = K(2*t-1,2*m-1) + k(15,5);
K(2*t-1,2*m) = K(2*t-1,2*m) + k(15,6);
K(2*t-1,2*p-1) = K(2*t-1,2*p-1) + k(15,7);
K(2*t-1,2*p) = K(2*t-1,2*p) + k(15,8);
K(2*t-1,2*q-1) = K(2*t-1,2*q-1) + k(15,9);
K(2*t-1,2*q) = K(2*t-1,2*q) + k(15,10);
K(2*t-1,2*r-1) = K(2*t-1,2*r-1) + k(15,11);
K(2*t-1,2*r) = K(2*t-1,2*r) + k(15,12);
K(2*t-1,2*s-1) = K(2*t-1,2*s-1) + k(15,13);
K(2*t-1,2*s) = K(2*t-1,2*s) + k(15,14);
K(2*t-1,2*t-1) = K(2*t-1,2*t-1) + k(15,15);
K(2*t-1,2*t) = K(2*t-1,2*t) + k(15,16);
K(2*t,2*i-1) = K(2*t,2*i-1) + k(16,1);
K(2*t,2*i) = K(2*t,2*i) + k(16,2);
K(2*t,2*j-1) = K(2*t,2*j-1) + k(16,3);
K(2*t,2*j) = K(2*t,2*j) + k(16,4);
K(2*t,2*m-1) = K(2*t,2*m-1) + k(16,5);
K(2*t,2*m) = K(2*t,2*m) + k(16,6);
K(2*t,2*p-1) = K(2*t,2*p-1) + k(16,7);
K(2*t,2*p) = K(2*t,2*p) + k(16,8);
K(2*t,2*q-1) = K(2*t,2*q-1) + k(16,9);
K(2*t,2*q) = K(2*t,2*q) + k(16,10);
K(2*t,2*r-1) = K(2*t,2*r-1) + k(16,11);
K(2*t,2*r) = K(2*t,2*r) + k(16,12);
K(2*t,2*s-1) = K(2*t,2*s-1) + k(16,13);
K(2*t,2*s) = K(2*t,2*s) + k(16,14);
K(2*t,2*t-1) = K(2*t,2*t-1) + k(16,15);
K(2*t,2*t) = K(2*t,2*t) + k(16,16);
y = K;
```

```
function w =QuadraticQuadElementSresses
          (E,NU,x1,y1,x2,y2,x3,y3,x4,y4,p,u)
%QuadraticQuadElementStresses    This function returns the element
%                                stress vector for a quadratic
%                                quadrilateral element with modulus
%                                of elasticity E, Poisson's ratio
%                                NU, coordinates of
%                                node 1 (x1,y1), coordinates
%                                of node 2 (x2,y2), coordinates of
%                                node 3 (x3,y3), coordinates of
%                                node 4 (x4,y4), and element
%                                displacement vector u.
%                                Use p = 1 for cases
%                                of plane stress, and p = 2 for
%                                cases of plane strain.
syms s t;
x5 = (x1 + x2)/2;
x6 = (x2 + x3)/2;
x7 = (x3 + x4)/2;
x8 = (x4 + x1)/2;
y5 = (y1 + y2)/2;
y6 = (y2 + y3)/2;
y7 = (y3 + y4)/2;
y8 = (y4 + y1)/2;
N1 = (1-s)*(1-t)*(-s-t-1)/4;
N2 = (1+s)*(1-t)*(s-t-1)/4;
N3 = (1+s)*(1+t)*(s+t-1)/4;
N4 = (1-s)*(1+t)*(-s+t-1)/4;
N5 = (1-t)*(1+s)*(1-s)/2;
N6 = (1+s)*(1+t)*(1-t)/2;
N7 = (1+t)*(1+s)*(1-s)/2;
N8 = (1-s)*(1+t)*(1-t)/2;
x = N1*x1 + N2*x2 + N3*x3 + N4*x4 + N5*x5 + N6*x6 + N7*x7 + N8*x8;
y = N1*y1 + N2*y2 + N3*y3 + N4*y4 + N5*y5 + N6*y6 + N7*y7 + N8*y8;
xs = diff(x,s);
xt = diff(x,t);
ys = diff(y,s);
yt = diff(y,t);
J = xs*yt - ys*xt;
N1s = diff(N1,s);
N2s = diff(N2,s);
N3s = diff(N3,s);
N4s = diff(N4,s);
N5s = diff(N5,s);
N6s = diff(N6,s);
N7s = diff(N7,s);
N8s = diff(N8,s);
N1t = diff(N1,t);
N2t = diff(N2,t);
N3t = diff(N3,t);
N4t = diff(N4,t);
N5t = diff(N5,t);
N6t = diff(N6,t);
```

```
N7t = diff(N7,t);
N8t = diff(N8,t);
B11 = yt*N1s - ys*N1t;
B12 = 0;
B13 = yt*N2s - ys*N2t;
B14 = 0;
B15 = yt*N3s - ys*N3t;
B16 = 0;
B17 = yt*N4s - ys*N4t;
B18 = 0;
B19 = yt*N5s - ys*N5t;
B110 = 0;
B111 = yt*N6s - ys*N6t;
B112 = 0;
B113 = yt*N7s - ys*N7t;
B114 = 0;
B115 = yt*N8s - ys*N8t;
B116 = 0;
B21 = 0;
B22 = xs*N1t - xt*N1s;
B23 = 0;
B24 = xs*N2t - xt*N2s;
B25 = 0;
B26 = xs*N3t - xt*N3s;
B27 = 0;
B28 = xs*N4t - xt*N4s;
B29 = 0;
B210 = xs*N5t - xt*N5s;
B211 = 0;
B212 = xs*N6t - xt*N6s;
B213 = 0;
B214 = xs*N7t - xt*N7s;
B215 = 0;
B216 = xs*N8t - xt*N8s;
B31 = xs*N1t - xt*N1s;
B32 = yt*N1s - ys*N1t;
B33 = xs*N2t - xt*N2s;
B34 = yt*N2s - ys*N2t;
B35 = xs*N3t - xt*N3s;
B36 = yt*N3s - ys*N3t;
B37 = xs*N4t - xt*N4s;
B38 = yt*N4s - ys*N4t;
B39 = xs*N5t - xt*N5s;
B310 = yt*N5s - ys*N5t;
B311 = xs*N6t - xt*N6s;
B312 = yt*N6s - ys*N6t;
B313 = xs*N7t - xt*N7s;
B314 = yt*N7s - ys*N7t;
B315 = xs*N8t - xt*N8s;
B316 = yt*N8s - ys*N8t;
Jnew = simplify(J);
```

```
B = [B11 B12 B13 B14 B15 B16 B17 B18 B19 B110 B111 B112 B113 B114
     B115 B116;
   B21 B22 B23 B24 B25 B26 B27 B28 B29 B210 B211 B212 B213 B214
     B215 B216;
   B31 B32 B33 B34 B35 B36 B37 B38 B39 B310 B311 B312 B313 B314
     B315 B316]/Jnew;
if p == 1
   D = (E/(1-NU*NU))*[1, NU, 0 ; NU, 1, 0 ; 0, 0, (1-NU)/2];
elseif p == 2
   D = (E/(1+NU)/(1-2*NU))
         *[1-NU, NU, 0 ;NU, 1-NU, 0 ; 0, 0, (1-2*NU)/2];
end
Bnew = simplify(B);
w = D*Bnew*u
%
% We also calculate the stresses at the centroid of the element
%
wcent = subs(w, {s,t}, {0,0});
w = double(wcent);
```

```
function y = QuadraticQuadElementPStresses(sigma)
%QuadraticQuadElementPStresses            This function returns the element
%                                         principal stresses and their
%                                         angle given the element
%                                         stress vector.
R = (sigma(1) + sigma(2))/2;
Q = ((sigma(1) - sigma(2))/2)^2 + sigma(3)*sigma(3);
M = 2*sigma(3)/(sigma(1) - sigma(2));
s1 = R + sqrt(Q);
s2 = R - sqrt(Q);
theta = (atan(M)/2)*180/pi;
y = [s1 ; s2 ; theta];
```

Example 14.1:

Consider the thin plate subjected to a uniformly distributed load as shown in Fig. 14.3. This problem was solved in Example 13.1 using bilinear quadrilateral elements. Solve this problem again using one quadratic quadrilateral element as shown in Fig. 14.4. Given $E = 210\,\text{GPa}$, $\nu = 0.3$, $t = 0.025\,\text{m}$, and $w = 3000\,\text{kN/m}^2$, determine:

1. the global stiffness matrix for the structure.
2. the horizontal and vertical displacements at nodes 3, 5, and 8.
3. the reactions at nodes 1, 4, and 6.
4. the stresses in the element.
5. the principal stresses and principal angle for the element.

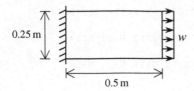

Fig. 14.3. Thin Plate for Example 14.1

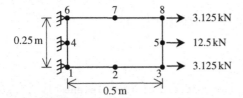

Fig. 14.4. Discretization of Thin Plate Using One Quadratic Quadrilateral

Solution:

Use the six steps outlined in Chap. 1 to solve this problem using the quadratic quadrilateral element.

Step 1 – Discretizing the Domain:

We use one quadratic quadrilateral element to model the plate only for illustration purposes. More elements must be used in order to obtain reliable results. Thus the domain is subdivided into one element and eight nodes as shown in Fig. 14.4. The total force due to the distributed load is divided between nodes 3, 5, and 8 in the ratio 1/6 : 2/3 : 1/6, respectively. Since the plate is thin, a case of plane stress is assumed. The units used in the MATLAB calculations are kN and meter. Table 14.1 shows the element connectivity for this example.

Table 14.1. Element Connectivity for Example 14.1

Element Number	Node i	Node j	Node m	Node p	Node q	Node r	Node s	Node t
1	1	3	8	6	2	5	7	4

Step 2 – Writing the Element Stiffness Matrices:

The element stiffness matrix k_1 is obtained by making calls to the MATLAB function *QuadraticQuadElementStiffness*. The matrix has size 16×16.

```
E=210e6

E =

   210000000

» NU=0.3

NU =

   0.3000

» h=0.025

h =

   0.0250

» k1=QuadraticQuadElementStiffness
   (E,NU,h,0,0,0.5,0,0.5,0.25,0,0.25,1)

k1 =

 1.0e+007 *

 Columns 1 through 7


   0.4000    0.1771    0.1660    0.0024    0.1769    0.0729    0.1801
   0.1771    0.7250   -0.0024    0.2494    0.0729    0.3207    0.0024
   0.1660   -0.0024    0.4000   -0.1771    0.1801    0.0024    0.1769
   0.0024    0.2494   -0.1771    0.7250   -0.0024    0.3780   -0.0729
   0.1769    0.0729    0.1801   -0.0024    0.4000    0.1771    0.1660
   0.0729    0.3207    0.0024    0.3780    0.1771    0.7250   -0.0024
   0.1801    0.0024    0.1769   -0.0729    0.1660   -0.0024    0.4000
  -0.0024    0.3780   -0.0729    0.3207    0.0024    0.2494   -0.1771
  -0.2295   -0.0737   -0.2295    0.0737   -0.1551   -0.0417   -0.1551
  -0.0929   -0.0128    0.0929   -0.0128   -0.0417   -0.1218    0.0417
  -0.1987   -0.0417   -0.3397    0.0929   -0.3397   -0.0929   -0.1987
  -0.0417   -0.5196    0.0737   -1.0189   -0.0737   -1.0189    0.0417
  -0.1551   -0.0417   -0.1551    0.0417   -0.2295   -0.0737   -0.2295
  -0.0417   -0.1218    0.0417   -0.1218   -0.0929   -0.0128    0.0929
  -0.3397   -0.0929   -0.1987    0.0417   -0.1987   -0.0417   -0.3397
  -0.0737   -1.0189    0.0417   -0.5196   -0.0417   -0.5196    0.0737
```

```
Columns 8 through 14

-0.0024  -0.2295  -0.0929  -0.1987  -0.0417  -0.1551  -0.0417
 0.3780  -0.0737  -0.0128  -0.0417  -0.5196  -0.0417  -0.1218
-0.0729  -0.2295   0.0929  -0.3397   0.0737  -0.1551   0.0417
 0.3207   0.0737  -0.0128   0.0929  -1.0189   0.0417  -0.1218
 0.0024  -0.1551  -0.0417  -0.3397  -0.0737  -0.2295  -0.0929
 0.2494  -0.0417  -0.1218  -0.0929  -1.0189  -0.0737  -0.0128
-0.1771  -0.1551   0.0417  -0.1987   0.0417  -0.2295   0.0929
 0.7250   0.0417  -0.1218   0.0417  -0.5196   0.0737  -0.0128
 0.0417   0.7282        0        0  -0.1667   0.0410        0
-0.1218        0   0.7949  -0.1667        0        0  -0.5256
 0.0417        0  -0.1667   0.8718        0        0   0.1667
-0.5196  -0.1667        0        0   2.1051   0.1667        0
 0.0737   0.0410        0        0   0.1667   0.7282        0
-0.0128        0  -0.5256   0.1667        0        0   0.7949
 0.0929        0   0.1667   0.2051        0        0  -0.1667
-1.0189   0.1667        0        0   0.9718  -0.1667        0

Columns 15 through 16

-0.3397  -0.0737
-0.0929  -1.0189
-0.1987   0.0417
 0.0417  -0.5196
-0.1987  -0.0417
-0.0417  -0.5196
-0.3397   0.0737
 0.0929  -1.0189
      0   0.1667
 0.1667        0
 0.2051        0
      0   0.9718
      0  -0.1667
-0.1667        0
 0.8718        0
      0   2.1051
```

Step 3 – Assembling the Global Stiffness Matrix:

Since the structure has eight nodes, the size of the global stiffness matrix is 16×16. Therefore to obtain K we first set up a zero matrix of size 16×16 then make one call to the MATLAB function *QuadraticQuadAssemble* since we have one element in the structure. Each call to the function will assemble one element. The following are the MATLAB commands. Only the final result is shown after the element is assembled.

```
» K=zeros(16,16);

» K=QuadraticQuadAssemble(K,k1,1,3,8,6,2,5,7,4)

K =

   1.0e+007 *

   Columns 1 through 7

    0.4000    0.1771   -0.2295   -0.0929    0.1660    0.0024   -0.3397
    0.1771    0.7250   -0.0737   -0.0128   -0.0024    0.2494   -0.0929
   -0.2295   -0.0737    0.7282         0   -0.2295    0.0737         0
   -0.0929   -0.0128         0    0.7949    0.0929   -0.0128    0.1667
    0.1660   -0.0024   -0.2295    0.0929    0.4000   -0.1771   -0.1987
    0.0024    0.2494    0.0737   -0.0128   -0.1771    0.7250    0.0417
   -0.3397   -0.0929         0    0.1667   -0.1987    0.0417    0.8718
   -0.0737   -1.0189    0.1667         0    0.0417   -0.5196         0
   -0.1987   -0.0417         0   -0.1667   -0.3397    0.0929    0.2051
   -0.0417   -0.5196   -0.1667         0    0.0737   -1.0189         0
    0.1801    0.0024   -0.1551    0.0417    0.1769   -0.0729   -0.3397
   -0.0024    0.3780    0.0417   -0.1218   -0.0729    0.3207    0.0929
   -0.1551   -0.0417    0.0410         0   -0.1551    0.0417         0
   -0.0417   -0.1218         0   -0.5256    0.0417   -0.1218   -0.1667
    0.1769    0.0729   -0.1551   -0.0417    0.1801   -0.0024   -0.1987
    0.0729    0.3207   -0.0417   -0.1218    0.0024    0.3780   -0.0417

   Columns 8 through 14

   -0.0737   -0.1987   -0.0417    0.1801   -0.0024   -0.1551   -0.0417
   -1.0189   -0.0417   -0.5196    0.0024    0.3780   -0.0417   -0.1218
    0.1667         0   -0.1667   -0.1551    0.0417    0.0410         0
         0   -0.1667         0    0.0417   -0.1218         0   -0.5256
    0.0417   -0.3397    0.0737    0.1769   -0.0729   -0.1551    0.0417
   -0.5196    0.0929   -1.0189   -0.0729    0.3207    0.0417   -0.1218
         0    0.2051         0   -0.3397    0.0929         0   -0.1667
    2.1051         0    0.9718    0.0737   -1.0189   -0.1667         0
         0    0.8718         0   -0.1987    0.0417         0    0.1667
    0.9718         0    2.1051    0.0417   -0.5196    0.1667         0
    0.0737   -0.1987    0.0417    0.4000   -0.1771   -0.2295    0.0929
   -1.0189    0.0417   -0.5196   -0.1771    0.7250    0.0737   -0.0128
```

```
-0.1667           0   0.1667 -0.2295   0.0737   0.7282            0
       0   0.1667           0   0.0929 -0.0128            0   0.7949
-0.0417 -0.3397 -0.0737   0.1660   0.0024 -0.2295   -0.0929
-0.5196 -0.0929 -1.0189 -0.0024   0.2494 -0.0737 -   0.0128
```

Columns 15 through 16

```
  0.1769        0.0729
  0.0729        0.3207
 -0.1551       -0.0417
 -0.0417       -0.1218
  0.1801        0.0024
 -0.0024        0.3780
 -0.1987       -0.0417
 -0.0417       -0.5196
 -0.3397       -0.0929
 -0.0737       -1.0189
  0.1660       -0.0024
  0.0024        0.2494
 -0.2295       -0.0737
 -0.0929       -0.0128
  0.4000        0.1771
  0.1771        0.7250
```

Step 4 – Applying the Boundary Conditions:

The matrix (14.10) for this structure is obtained directly using the global stiffness matrix obtained in the previous step; however, the full matrix equation is not shown here explicitly because it is very large. The boundary conditions for this problem are given as:

$$U_{1x} = U_{1y} = U_{4x} = U_{4y} = U_{6x} = U_{6y} = 0$$
$$F_{2x} = F_{2y} = F_{7x} = F_{7y} = F_{3y} = F_{5y} = F_{8y} = 0$$
$$F_{3x} = 3.125, \ F_{5x} = 12.5, \ F_{8x} = 3.125 \tag{14.12}$$

We next insert the above conditions into the matrix equation for this structure and proceed to the solution step below.

Step 5 – Solving the Equations:

Solving the resulting system of equations will be performed by partitioning (manually) and Gaussian elimination (with MATLAB). First we partition the matrix equation by extracting the submatrices in rows 3 to 6, rows 9 to 10, rows 13 to 16, and columns 3 to 6, columns 9 to 10, columns 13 to 16. Therefore we obtain:

$$
10^6
\begin{bmatrix}
7.28 & 0 & -2.30 & 0.74 & 0 & -1.67 & 0.41 & 0 & -1.55 & -0.42 \\
0 & 7.95 & 0.93 & -0.13 & -1.67 & 0 & 0 & -5.26 & -0.42 & -1.22 \\
-2.30 & 0.93 & 4.00 & -1.77 & -3.40 & 0.74 & -1.55 & 0.42 & 1.80 & 0.02 \\
0.74 & -0.13 & -1.77 & 7.25 & 0.93 & -10.19 & 0.42 & -1.22 & -0.02 & 3.78 \\
0 & -1.67 & -3.40 & 0.93 & 8.72 & 0 & 0 & 1.67 & -3.40 & -0.93 \\
-1.67 & 0 & 0.74 & -10.19 & 0 & 21.05 & 1.67 & 0 & -0.74 & -10.19 \\
0.41 & 0 & -1.55 & 0.42 & 0 & 1.67 & 7.28 & 0 & -2.30 & -0.74 \\
0 & -5.26 & 0.42 & -1.22 & 1.67 & 0 & 0 & 7.95 & -0.93 & -0.13 \\
-1.55 & -0.42 & 1.80 & -0.02 & -3.40 & -0.74 & -2.30 & -0.93 & 4.00 & 1.77 \\
-0.42 & -1.22 & 0.02 & 3.78 & -0.93 & -10.19 & -0.74 & -0.13 & 1.77 & 7.25
\end{bmatrix}
$$

$$
\begin{Bmatrix}
U_{2x} \\ U_{2y} \\ U_{3x} \\ U_{3y} \\ U_{5x} \\ U_{5y} \\ U_{7x} \\ U_{7y} \\ U_{8x} \\ U_{8y}
\end{Bmatrix}
=
\begin{Bmatrix}
0 \\ 0 \\ 3.125 \\ 0 \\ 12.5 \\ 0 \\ 0 \\ 0 \\ 3.125 \\ 0
\end{Bmatrix}
\tag{14.13}
$$

The solution of the above system is obtained using MATLAB as follows. Note that the backslash operator "\" is used for Gaussian elimination.

```
» k=[K(3:6,3:6)  K(3:6,9:10)  K(3:6,13:16) ; K(9:10,3:6)
   K(9:10,9:10)  K(9:10,13:16) ;K(13:16,3:6)  K(13:16,9:10)
   K(13:16,13:16)]

k =

   1.0e+007 *

   Columns 1 through 7


 0.7282           0 -0.2295   0.0737          0 -0.1667   0.0410
      0   0.7949   0.0929  -0.0128  -0.1667         0        0
-0.2295   0.0929   0.4000  -0.1771  -0.3397   0.0737  -0.1551
 0.0737  -0.0128  -0.1771   0.7250   0.0929  -1.0189   0.0417
      0  -0.1667  -0.3397   0.0929   0.8718        0        0
-0.1667        0   0.0737  -1.0189        0   2.1051   0.1667
 0.0410        0  -0.1551   0.0417        0   0.1667   0.7282
      0  -0.5256   0.0417  -0.1218   0.1667        0        0
-0.1551  -0.0417   0.1801  -0.0024  -0.3397  -0.0737  -0.2295
-0.0417  -0.1218   0.0024   0.3780  -0.0929  -1.0189  -0.0737
```

Columns 8 through 10

```
      0     -0.1551    -0.0417
-0.5256    -0.0417    -0.1218
 0.0417     0.1801     0.0024
-0.1218    -0.0024     0.3780
 0.1667    -0.3397    -0.0929
      0     -0.0737    -1.0189
      0     -0.2295    -0.0737
 0.7949    -0.0929    -0.0128
-0.0929     0.4000     0.1771
-0.0128     0.1771     0.7250
```

» f=[0 ; 0 ; 3.125 ; 0 ; 12.5 ; 0 ; 0 ; 0 ; 3.125 ; 0]

f =

```
      0
      0
 3.1250
      0
12.5000
      0
      0
      0
 3.1250
      0
```

» u=k\f

u =

   1.0e-005 *

```
 0.3457
 0.0582
 0.7040
 0.0420
 0.7054
 0.0000
 0.3457
-0.0582
 0.7040
-0.0420
```

It is now clear that the horizontal and vertical displacements at node 3 are 0.7040 m (compared with 0.7030 m in Example 13.1) and 0.0420 m, respectively. The horizontal and vertical displacements at node 8 are 0.7040 m (compared with 0.7030 m in Example 13.1) and –0.0420 m, respectively. The horizontal and vertical displacements at node 5 are 0.7054 m and 0.0000 m, respectively.

Step 6 – Post-processing:

In this step, we obtain the reactions at nodes 1, 4, and 6, and the stresses in the element using MATLAB as follows. First we set up the global nodal displacement vector U, then we calculate the global nodal force vector F.

```
» U=[0;0;u(1:4);0;0;u(5:6);0;0;u(7:10)]

U =

     1.0e-005  *

           0
           0
      0.3457
      0.0582
      0.7040
      0.0420
           0
           0
      0.7054
      0.0000
           0
           0
      0.3457
     -0.0582
      0.7040
     -0.0420

» F=K*U

F =

     -3.7650
     -1.6288
      0.0000
      0.0000
      3.1250
      0.0000
```

```
     -11.2200
       0.0000
      12.5000
       0.0000
      -3.7650
       1.6288
       0.0000
       0.0000
       3.1250
       0.0000
```

Thus the horizontal and vertical reactions at node 1 are forces of 3.7650 kN (directed to the left) and 1.6288 kN (directed downwards). The horizontal and vertical reactions at node 4 are forces of 11.2200 kN (directed to the left) and 0.0000 kN. Thus the horizontal and vertical reactions at node 6 are forces of 3.7650 kN (directed to the left) and 1.6288 kN (directed upwards). Obviously force equilibrium is satisfied for this problem. Next we set up the element nodal displacement vector u_1 then we calculate the element stresses *sigma1* by making calls to the MATLAB function *QuadraticQuadElementStresses*. Closed-form solutions (quadratic functions in ξ and η, denoted by s and t in the results below) for the stresses in the element are obtained using the MATLAB Symbolic Math Toolbox as shown below. Then numerical values for the stresses are computed at the centroid of the element.

```
» u1=[U(1); U(2); U(5); U(6); U(15); U(16); U(11); U(12);
       U(3); U(4); U(9); U(10); U(13); U(14); U(7); U(8)]

u1 =

   1.0e-005 *

            0
            0
       0.7040
       0.0420
       0.7040
      -0.0420
            0
            0
       0.3457
       0.0582
       0.7054
       0.0000
       0.3457
      -0.0582
            0
            0
```

```
» sigma1=QuadraticQuadElementStresses(E,NU,0,0,0.5,0,0.5,
                0.25,0,0.25,1,u1)

w =

[
21707267240782058883129160650 7/34844914 3727040986586495 5
98010130648530944*t-96764073971696635/79228162514264337 5
93543950336*s-46403787800607050 0395899812373/34844914372
70409865864955980101306485309 44*t*s-12742907563074467741
5664158725/198070406285660843983 85987584*t^2+16344902767
57110059183885037387 9/79228162514264337593543950336*s^2+
11620304602722202291145284 3874333/3961408125713216879677
1975168]
[
-20567711961890518643246222187 3/6968982874540819731729 91
196020261297061888*t-558573554038 9175408229514503718 7/15
84563250285286751870879006 72*s-6143434498672397871922189
39953/696898287454081973172991196 020261297061888*t*s-191
143613446116991438560680 25/990352031428304219919299379 2
*t^2+108966018450474018017 834117539051/15845632502852867
5187087900672*s^2-77863884805 7832273378362287167 9/792281
6251426433759354395033 6]
[
-24338009526394714014472697258771/31691265005705735037 41
75801344*t-1372648485561246015119741029839/27875931498 16
3278926919647840810451882475 52+62333810215384603/158456 3
2502852867518708790067 2*s+36710900810601561178216064459 8
51/158456325028528675187087900672*t*s-12920027810138364 2
672364909105/2787593149816327892691964784081045188247552
*t^2-8130496984615383/198070406285660843983 85987584*s^2]

sigma1 =

  1.0e+003 *

    2.9334
   -0.0983
    0.0000
```

Thus it is clear that the stresses at the centroid of the element are $\sigma_x = 2.9334\,\text{MPa}$ (tensile), $\sigma_y = 0.0983\,\text{MPa}$ (compressive), and $\tau_{xy} = 0.0000\,\text{MPa}$. It is clear that the stress in the x- direction approaches closely the correct value of $3\,\text{MPa}$ (tensile) at the centroid of the element. It is seen that only one quadratic quadrilateral element gives accurate results in this problem. Next we calculate the principal stresses and principal angle for the element by making calls to the MATLAB function *QuadraticQuadElementPStresses*.

```
» s1=QuadraticQuadElementPStresses(sigma1)

s1 =

    1.0e+003 *

      2.9334
     -0.0983
      0.0000
```

Thus it is clear that the principal stresses at the centroid of the element are $\sigma_1 = 2.9334$ MPa (tensile), $\sigma_2 = 0.0983$ MPa (compressive), while the principal angle $\theta_p = 0°$.

Problems:

Problem 14.1:

Consider the thin plate supported on three springs and subjected to a uniformly distributed load as shown in Fig. 14.5. Solve this problem using one quadratic quadrilateral element as shown in the figure. Given $E = 200$ GPa, $\nu = 0.3$, $t = 0.01$ m, $k = 4000$ kN/m, and $w = 5000$ kN/m^2, determine:

1. the global stiffness matrix for the structure.
2. the horizontal and vertical displacements at nodes 1, 2, 3, 6, 7, and 8.
3. the reactions at nodes 9, 10, and 11.
4. the stresses in the element.
5. the principal stresses and principal angle for the element.
6. the force in each spring.

Compare your answers with those obtained for Problem 11.3 and Problem 13.3.

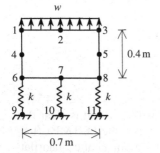

Fig. 14.5. Thin Plate Supported on Three Springs

15 The Linear Tetrahedral (Solid) Element

15.1
Basic Equations

The linear tetrahedral (solid) element is a three-dimensional finite element with both local and global coordinates. It is characterized by linear shape functions. It is also called the constant strain tetrahedron. The linear tetrahedral element has modulus of elasticity E and Poisson's ratio ν. Each linear tetrahedron has four nodes with three degrees of freedom at each node as shown in Fig. 15.1. The global coordinates of the four nodes are denoted by (x_1, y_1, z_1), (x_2, y_2, z_2), (x_3, y_3, z_3), and (x_4, y_4, z_4). The numbering of the nodes for each element is very important – you should number the nodes such that the volume of the element is positive. It is advised to actually check this by using the MATLAB function *TetrahedronElementVolume* which is written specifically for this purpose. In this case the element stiffness matrix is given by (see [1] and [8]).

$$[k] = V[B]^T[D][B] \tag{15.1}$$

where V is the volume of the element given by

$$6V = \begin{vmatrix} 1 & x_1 & y_1 & z_1 \\ 1 & x_2 & y_2 & z_2 \\ 1 & x_3 & y_3 & z_3 \\ 1 & x_4 & y_4 & z_4 \end{vmatrix} \tag{15.2}$$

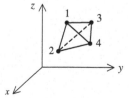

Fig. 15.1. The Linear Tetrahedral (Solid) Element

and the matrix $[B]$ is given by

$$[B] = \begin{bmatrix} \dfrac{\partial N_1}{\partial x} & 0 & 0 & \dfrac{\partial N_2}{\partial x} & 0 & 0 & \dfrac{\partial N_3}{\partial x} & 0 & 0 & \dfrac{\partial N_4}{\partial x} & 0 & 0 \\[2mm] 0 & \dfrac{\partial N_1}{\partial y} & 0 & 0 & \dfrac{\partial N_2}{\partial y} & 0 & 0 & \dfrac{\partial N_3}{\partial y} & 0 & 0 & \dfrac{\partial N_4}{\partial y} & 0 \\[2mm] 0 & 0 & \dfrac{\partial N_1}{\partial z} & 0 & 0 & \dfrac{\partial N_2}{\partial z} & 0 & 0 & \dfrac{\partial N_3}{\partial z} & 0 & 0 & \dfrac{\partial N_4}{\partial z} \\[2mm] \dfrac{\partial N_1}{\partial y} & \dfrac{\partial N_1}{\partial x} & 0 & \dfrac{\partial N_2}{\partial y} & \dfrac{\partial N_2}{\partial x} & 0 & \dfrac{\partial N_3}{\partial y} & \dfrac{\partial N_3}{\partial x} & 0 & \dfrac{\partial N_4}{\partial y} & \dfrac{\partial N_4}{\partial x} & 0 \\[2mm] 0 & \dfrac{\partial N_1}{\partial z} & \dfrac{\partial N_1}{\partial y} & 0 & \dfrac{\partial N_2}{\partial z} & \dfrac{\partial N_2}{\partial y} & 0 & \dfrac{\partial N_3}{\partial z} & \dfrac{\partial N_3}{\partial y} & 0 & \dfrac{\partial N_4}{\partial z} & \dfrac{\partial N_4}{\partial y} \\[2mm] \dfrac{\partial N_1}{\partial z} & 0 & \dfrac{\partial N_1}{\partial x} & \dfrac{\partial N_2}{\partial z} & 0 & \dfrac{\partial N_2}{\partial x} & \dfrac{\partial N_3}{\partial z} & 0 & \dfrac{\partial N_3}{\partial x} & \dfrac{\partial N_4}{\partial z} & 0 & \dfrac{\partial N_4}{\partial x} \end{bmatrix} \tag{15.3}$$

In (15.3), the shape functions N_1, N_2, N_3, and N_4 are given by:

$$N_1 = \frac{1}{6V}(\alpha_1 + \beta_1 x + \gamma_1 y + \delta_1 z)$$

$$N_2 = \frac{1}{6V}(\alpha_2 + \beta_2 x + \gamma_2 y + \delta_2 z)$$

$$N_3 = \frac{1}{6V}(\alpha_3 + \beta_3 x + \gamma_3 y + \delta_3 z)$$

$$N_4 = \frac{1}{6V}(\alpha_4 + \beta_4 x + \gamma_4 y + \delta_4 z) \tag{15.4}$$

where $\alpha_1, \alpha_2, \alpha_3, \alpha_4, \beta_1, \beta_2, \beta_3, \beta_4, \gamma_1, \gamma_2, \gamma_3, \gamma_4, \delta_1, \delta_2, \delta_3,$ and δ_4 are given by:

$$\alpha_1 = \begin{vmatrix} x_2 & y_2 & z_2 \\ x_3 & y_3 & z_3 \\ x_4 & y_4 & z_4 \end{vmatrix} \qquad \alpha_2 = - \begin{vmatrix} x_1 & y_1 & z_1 \\ x_3 & y_3 & z_3 \\ x_4 & y_4 & z_4 \end{vmatrix}$$

$$\alpha_3 = \begin{vmatrix} x_1 & y_1 & z_1 \\ x_2 & y_2 & z_2 \\ x_4 & y_4 & z_4 \end{vmatrix} \qquad \alpha_4 = - \begin{vmatrix} x_1 & y_1 & z_1 \\ x_2 & y_2 & z_2 \\ x_3 & y_3 & z_3 \end{vmatrix} \tag{15.5}$$

$$\beta_1 = - \begin{vmatrix} 1 & y_2 & z_2 \\ 1 & y_3 & z_3 \\ 1 & y_4 & z_4 \end{vmatrix} \qquad \beta_2 = \begin{vmatrix} 1 & y_1 & z_1 \\ 1 & y_3 & z_3 \\ 1 & y_4 & z_4 \end{vmatrix}$$

$$\beta_3 = - \begin{vmatrix} 1 & y_1 & z_1 \\ 1 & y_2 & z_2 \\ 1 & y_4 & z_4 \end{vmatrix} \qquad \beta_4 = \begin{vmatrix} 1 & y_1 & z_1 \\ 1 & y_2 & z_2 \\ 1 & y_3 & z_3 \end{vmatrix} \tag{15.6}$$

$$\gamma_1 = \begin{vmatrix} 1 & x_2 & z_2 \\ 1 & x_3 & z_3 \\ 1 & x_4 & z_4 \end{vmatrix} \qquad \gamma_2 = - \begin{vmatrix} 1 & x_1 & z_1 \\ 1 & x_3 & z_3 \\ 1 & x_4 & z_4 \end{vmatrix}$$

$$\gamma_3 = \begin{vmatrix} 1 & x_1 & z_1 \\ 1 & x_2 & z_2 \\ 1 & x_4 & z_4 \end{vmatrix} \qquad \gamma_4 = - \begin{vmatrix} 1 & x_1 & z_1 \\ 1 & x_2 & z_2 \\ 1 & x_3 & z_3 \end{vmatrix} \tag{15.7}$$

$$\delta_1 = - \begin{vmatrix} 1 & x_2 & y_2 \\ 1 & x_3 & y_3 \\ 1 & x_4 & y_4 \end{vmatrix} \qquad \delta_2 = \begin{vmatrix} 1 & x_1 & y_1 \\ 1 & x_3 & y_3 \\ 1 & x_4 & y_4 \end{vmatrix}$$

$$\delta_3 = - \begin{vmatrix} 1 & x_1 & y_1 \\ 1 & x_2 & y_2 \\ 1 & x_4 & y_4 \end{vmatrix} \qquad \delta_4 = \begin{vmatrix} 1 & x_1 & y_1 \\ 1 & x_2 & y_2 \\ 1 & x_3 & y_3 \end{vmatrix} \tag{15.8}$$

In (15.1), the matrix $[D]$ is given by

$$[D] = \frac{E}{(1+\nu)(1-2\nu)} \begin{bmatrix} 1-\nu & \nu & \nu & 0 & 0 & 0 \\ \nu & 1-\nu & \nu & 0 & 0 & 0 \\ \nu & \nu & 1-\nu & 0 & 0 & 0 \\ 0 & 0 & 0 & \frac{1-2\nu}{2} & 0 & 0 \\ 0 & 0 & 0 & 0 & \frac{1-2\nu}{2} & 0 \\ 0 & 0 & 0 & 0 & 0 & \frac{1-2\nu}{2} \end{bmatrix} \tag{15.9}$$

It is clear that the linear tetrahedral element has twelve degrees of freedom – three at each node. Consequently for a structure with n nodes, the global stiffness matrix K will be of size $3n \times 3n$ (since we have three degrees of freedom at each node). The global stiffness matrix K is assembled by making calls to the MATLAB function *TetrahedronAssemble* which is written specifically for this purpose. This process will be illustrated in detail in the examples.

Once the global stiffness matrix K is obtained we have the following structure equation:

$$[K]\{U\} = \{F\} \tag{15.10}$$

where U is the global nodal displacement vector and F is the global nodal force vector. At this step the boundary conditions are applied manually to the vectors U and F. Then the matrix (15.10) is solved by partitioning and Gaussian elimination.

Finally once the unknown displacements and reactions are found, the stress vector is obtained for each element as follows:

$$\{\sigma\} = [D][B]\{u\}$$ (15.11)

where σ is the stress vector in the element (of size 6×1) and u is the 12×1 element displacement vector. The vector σ is written for each element as $\{\sigma\} = [\sigma_x \; \sigma_y \; \sigma_z \; \tau_{xy} \; \tau_{yz} \; \tau_{zx}]^T$.

15.2
MATLAB Functions Used

The five MATLAB functions used for the linear tetrahedral (solid) element are:

TetrahedronElementVolume$(x_1, y_1, z_1, x_2, y_2, z_2, x_3, y_3, z_3, x_4, y_4, z_4)$ – This function returns the element volume given the coordinates of the first node (x_1, y_1, z_1), the coordinates of the second node (x_2, y_2, z_2), the coordinates of the third node (x_3, y_3, z_3), and the coordinates of the fourth node (x_4, y_4, z_4),.

TetrahedronElementStiffness$(E, NU, x_1, y_1, z_1, x_2, y_2, z_2, x_3, y_3, z_3, x_4, y_4, z_4)$ – This function calculates the element stiffness matrix for each linear tetrahedron with modulus of elasticity E, Poisson's ratio NU, and coordinates (x_1, y_1, z_1) for the first node, (x_2, y_2, z_2) for the second node, (x_3, y_3, z_3) for the third node, and (x_4, y_4, z_4) for the fourth node. It returns the 12×12 element stiffness matrix k.

TetrahedronAssemble(K, k, i, j, m, n) – This function assembles the element stiffness matrix k of the linear tetrahedron joining nodes i, j, m, and n into the global stiffness matrix K. It returns the $3n \times 3n$ global stiffness matrix K every time an element is assembled.

TetrahedronElementStresses$(E, NU, x_1, y_1, z_1, x_2, y_2, z_2, x_3, y_3, z_3, x_4, y_4, z_4, u)$ – This function calculates the element stresses using the modulus of elasticity E, Poisson's ratio NU, the coordinates (x_1, y_1, z_1) for the first node, (x_2, y_2, z_2) for the second node, (x_3, y_3, z_3) for the third node, and (x_4, y_4, z_4) for the fourth node, and the element displacement vector u. It returns the stress vector for the element.

TetrahedronElementPStresses$(sigma)$ – This function calculates the three principal stresses for the element using the element stress vector *sigma*. It returns a 3×1 vector in the form $[sigma1 \; sigma2 \; sigma3]^T$ where *sigma1*, *sigma2*, and *sigma3* are the principal stresses for the element. This function does not return the principal angles.

The following is a listing of the MATLAB source code for each function:

```
function y = TetrahedronElementVolume
              (x1,y1,z1,x2,y2,z2,x3,y3,z3,x4,y4,z4)
%TetrahedronElementVolume          This function returns the volume
%                                  of the linear tetrahedral element
%                                  whose first node has coordinates
%                                  (x1,y1,z1), second node has
%                                  coordinates (x2,y2,z2), third node
%                                  has coordinates (x3,y3,z3), and
%                                  fourth node has coordiantes
%                                  (x4,y4,z4).
xyz = [1 x1 y1 z1 ; 1 x2 y2 z2 ; 1 x3 y3 z3 ; 1 x4 y4 z4];
y = det(xyz)/6;
```

```
function y = TetrahedronElementStiffness
              (E,NU,x1,y1,z1,x2,y2,z2,x3,y3,z3,x4,y4,z4)
%TetrahedronElementStiffness       This function returns the element
%                                  stiffness matrix for a linear
%                                  tetrahedral (solid) element with
%                                  modulus of elasticity E,
%                                  Poisson's ratio NU, coordinates
%                                  of the first node (x1,y1,z1),
%                                  coordinates of the second node
%                                  (x2,y2,z2), coordinates of the
%                                  third node (x3,y3,z3), and
%                                  coordinates of the fourth node
%                                  (x4,y4,z4).
%                                  The size of the element stiffness
%                                  matrix is 12 x 12.
xyz = [1 x1 y1 z1 ; 1 x2 y2 z2 ; 1 x3 y3 z3 ; 1 x4 y4 z4];
V = det(xyz)/6;
mbeta1 = [1 y2 z2 ; 1 y3 z3 ; 1 y4 z4];
mbeta2 = [1 y1 z1 ; 1 y3 z3 ; 1 y4 z4];
mbeta3 = [1 y1 z1 ; 1 y2 z2 ; 1 y4 z4];
mbeta4 = [1 y1 z1 ; 1 y2 z2 ; 1 y3 z3];
mgamma1 = [1 x2 z2 ; 1 x3 z3 ; 1 x4 z4];
mgamma2 = [1 x1 z1 ; 1 x3 z3 ; 1 x4 z4];
mgamma3 = [1 x1 z1 ; 1 x2 z2 ; 1 x4 z4];
mgamma4 = [1 x1 z1 ; 1 x2 z2 ; 1 x3 z3];
mdelta1 = [1 x2 y2 ; 1 x3 y3 ; 1 x4 y4];
mdelta2 = [1 x1 y1 ; 1 x3 y3 ; 1 x4 y4];
mdelta3 = [1 x1 y1 ; 1 x2 y2 ; 1 x4 y4];
mdelta4 = [1 x1 y1 ; 1 x2 y2 ; 1 x3 y3];
beta1 = -1*det(mbeta1);
beta2 = det(mbeta2);
beta3 = -1*det(mbeta3);
beta4 = det(mbeta4);
gamma1 = det(mgamma1);
gamma2 = -1*det(mgamma2);
gamma3 = det(mgamma3);
gamma4 = -1*det(mgamma4);
delta1 = -1*det(mdelta1);
delta2 = det(mdelta2);
```

```
delta3 = -1*det(mdelta3);
delta4 = det(mdelta4);
B1 = [beta1 0 0 ; 0 gamma1 0 ; 0 0 delta1 ;
   gamma1 beta1 0 ; 0 delta1 gamma1 ; delta1 0 beta1];
B2 = [beta2 0 0 ; 0 gamma2 0 ; 0 0 delta2 ;
   gamma2 beta2 0 ; 0 delta2 gamma2 ; delta2 0 beta2];
B3 = [beta3 0 0 ; 0 gamma3 0 ; 0 0 delta3 ;
   gamma3 beta3 0 ; 0 delta3 gamma3 ; delta3 0 beta3];
B4 = [beta4 0 0 ; 0 gamma4 0 ; 0 0 delta4 ;
   gamma4 beta4 0 ; 0 delta4 gamma4 ; delta4 0 beta4];
B = [B1 B2 B3 B4]/(6*V);
D = (E/((1+NU)*(1-2*NU)))*[1-NU NU NU 0 0 0 ;
    NU 1-NU NU 0 0 0 ;
    NU NU 1-NU 0 0 0 ;
    0 0 0 (1-2*NU)/2 0 0 ; 0 0 0 0 (1- 2*NU)/2 0 ; 0 0 0 0 0 (1-
    2*NU)/2];
y = V*B'*D*B;
```

```
function y = TetrahedronAssemble(K,k,i,j,m,n)
%TetrahedronAssemble    This function assembles the element stiffness
%                       matrix k of the linear tetrahedral (solid)
%                       element with nodes i, j, m, and n into the
%                       global stiffness matrix K.
%                       This function returns the global stiffness
%                       matrix K after the element stiffness matrix
%                       k is assembled.
K(3*i-2,3*i-2) = K(3*i-2,3*i-2) + k(1,1);
K(3*i-2,3*i-1) = K(3*i-2,3*i-1) + k(1,2);
K(3*i-2,3*i) = K(3*i-2,3*i) + k(1,3);
K(3*i-2,3*j-2) = K(3*i-2,3*j-2) + k(1,4);
K(3*i-2,3*j-1) = K(3*i-2,3*j-1) + k(1,5);
K(3*i-2,3*j) = K(3*i-2,3*j) + k(1,6);
K(3*i-2,3*m-2) = K(3*i-2,3*m-2) + k(1,7);
K(3*i-2,3*m-1) = K(3*i-2,3*m-1) + k(1,8);
K(3*i-2,3*m) = K(3*i-2,3*m) + k(1,9);
K(3*i-2,3*n-2) = K(3*i-2,3*n-2) + k(1,10);
K(3*i-2,3*n-1) = K(3*i-2,3*n-1) + k(1,11);
K(3*i-2,3*n) = K(3*i-2,3*n) + k(1,12);
K(3*i-1,3*i-2) = K(3*i-1,3*i-2) + k(2,1);
K(3*i-1,3*i-1) = K(3*i-1,3*i-1) + k(2,2);
K(3*i-1,3*i) = K(3*i-1,3*i) + k(2,3);
K(3*i-1,3*j-2) = K(3*i-1,3*j-2) + k(2,4);
K(3*i-1,3*j-1) = K(3*i-1,3*j-1) + k(2,5);
K(3*i-1,3*j) = K(3*i-1,3*j) + k(2,6);
K(3*i-1,3*m-2) = K(3*i-1,3*m-2) + k(2,7);
K(3*i-1,3*m-1) = K(3*i-1,3*m-1) + k(2,8);
K(3*i-1,3*m) = K(3*i-1,3*m) + k(2,9);
K(3*i-1,3*n-2) = K(3*i-1,3*n-2) + k(2,10);
K(3*i-1,3*n-1) = K(3*i-1,3*n-1) + k(2,11);
K(3*i-1,3*n) = K(3*i-1,3*n) + k(2,12);
K(3*i,3*i-2) = K(3*i,3*i-2) + k(3,1);
K(3*i,3*i-1) = K(3*i,3*i-1) + k(3,2);
K(3*i,3*i) = K(3*i,3*i) + k(3,3);
K(3*i,3*j-2) = K(3*i,3*j-2) + k(3,4);
K(3*i,3*j-1) = K(3*i,3*j-1) + k(3,5);
K(3*i,3*j) = K(3*i,3*j) + k(3,6);
```

```
K(3*i,3*m-2) = K(3*i,3*m-2) + k(3,7);
K(3*i,3*m-1) = K(3*i,3*m-1) + k(3,8);
K(3*i,3*m) = K(3*i,3*m) + k(3,9);
K(3*i,3*n-2) = K(3*i,3*n-2) + k(3,10);
K(3*i,3*n-1) = K(3*i,3*n-1) + k(3,11);
K(3*i,3*n) = K(3*i,3*n) + k(3,12);
K(3*j-2,3*i-2) = K(3*j-2,3*i-2) + k(4,1);
K(3*j-2,3*i-1) = K(3*j-2,3*i-1) + k(4,2);
K(3*j-2,3*i) = K(3*j-2,3*i) + k(4,3);
K(3*j-2,3*j-2) = K(3*j-2,3*j-2) + k(4,4);
K(3*j-2,3*j-1) = K(3*j-2,3*j-1) + k(4,5);
K(3*j-2,3*j) = K(3*j-2,3*j) + k(4,6);
K(3*j-2,3*m-2) = K(3*j-2,3*m-2) + k(4,7);
K(3*j-2,3*m-1) = K(3*j-2,3*m-1) + k(4,8);
K(3*j-2,3*m) = K(3*j-2,3*m) + k(4,9);
K(3*j-2,3*n-2) = K(3*j-2,3*n-2) + k(4,10);
K(3*j-2,3*n-1) = K(3*j-2,3*n-1) + k(4,11);
K(3*j-2,3*n) = K(3*j-2,3*n) + k(4,12);
K(3*j-1,3*i-2) = K(3*j-1,3*i-2) + k(5,1);
K(3*j-1,3*i-1) = K(3*j-1,3*i-1) + k(5,2);
K(3*j-1,3*i) = K(3*j-1,3*i) + k(5,3);
K(3*j-1,3*j-2) = K(3*j-1,3*j-2) + k(5,4);
K(3*j-1,3*j-1) = K(3*j-1,3*j-1) + k(5,5);
K(3*j-1,3*j) = K(3*j-1,3*j) + k(5,6);
K(3*j-1,3*m-2) = K(3*j-1,3*m-2) + k(5,7);
K(3*j-1,3*m-1) = K(3*j-1,3*m-1) + k(5,8);
K(3*j-1,3*m) = K(3*j-1,3*m) + k(5,9);
K(3*j-1,3*n-2) = K(3*j-1,3*n-2) + k(5,10);
K(3*j-1,3*n-1) = K(3*j-1,3*n-1) + k(5,11);
K(3*j-1,3*n) = K(3*j-1,3*n) + k(5,12);
K(3*j,3*i-2) = K(3*j,3*i-2) + k(6,1);
K(3*j,3*i-1) = K(3*j,3*i-1) + k(6,2);
K(3*j,3*i) = K(3*j,3*i) + k(6,3);
K(3*j,3*j-2) = K(3*j,3*j-2) + k(6,4);
K(3*j,3*j-1) = K(3*j,3*j-1) + k(6,5);
K(3*j,3*j) = K(3*j,3*j) + k(6,6);
K(3*j,3*m-2) = K(3*j,3*m-2) + k(6,7);
K(3*j,3*m-1) = K(3*j,3*m-1) + k(6,8);
K(3*j,3*m) = K(3*j,3*m) + k(6,9);
K(3*j,3*n-2) = K(3*j,3*n-2) + k(6,10);
K(3*j,3*n-1) = K(3*j,3*n-1) + k(6,11);
K(3*j,3*n) = K(3*j,3*n) + k(6,12);
K(3*m-2,3*i-2) = K(3*m-2,3*i-2) + k(7,1);
K(3*m-2,3*i-1) = K(3*m-2,3*i-1) + k(7,2);
K(3*m-2,3*i) = K(3*m-2,3*i) + k(7,3);
K(3*m-2,3*j-2) = K(3*m-2,3*j-2) + k(7,4);
K(3*m-2,3*j-1) = K(3*m-2,3*j-1) + k(7,5);
K(3*m-2,3*j) = K(3*m-2,3*j) + k(7,6);
K(3*m-2,3*m-2) = K(3*m-2,3*m-2) + k(7,7);
K(3*m-2,3*m-1) = K(3*m-2,3*m-1) + k(7,8);
K(3*m-2,3*m) = K(3*m-2,3*m) + k(7,9);
K(3*m-2,3*n-2) = K(3*m-2,3*n-2) + k(7,10);
K(3*m-2,3*n-1) = K(3*m-2,3*n-1) + k(7,11);
K(3*m-2,3*n) = K(3*m-2,3*n) + k(7,12);
K(3*m-1,3*i-2) = K(3*m-1,3*i-2) + k(8,1);
K(3*m-1,3*i-1) = K(3*m-1,3*i-1) + k(8,2);
```

```
K(3*m-1,3*i)   = K(3*m-1,3*i)   + k(8,3);
K(3*m-1,3*j-2) = K(3*m-1,3*j-2) + k(8,4);
K(3*m-1,3*j-1) = K(3*m-1,3*j-1) + k(8,5);
K(3*m-1,3*j)   = K(3*m-1,3*j)   + k(8,6);
K(3*m-1,3*m-2) = K(3*m-1,3*m-2) + k(8,7);
K(3*m-1,3*m-1) = K(3*m-1,3*m-1) + k(8,8);
K(3*m-1,3*m)   = K(3*m-1,3*m)   + k(8,9);
K(3*m-1,3*n-2) = K(3*m-1,3*n-2) + k(8,10);
K(3*m-1,3*n-1) = K(3*m-1,3*n-1) + k(8,11);
K(3*m-1,3*n)   = K(3*m-1,3*n)   + k(8,12);
K(3*m,3*i-2)   = K(3*m,3*i-2)   + k(9,1);
K(3*m,3*i-1)   = K(3*m,3*i-1)   + k(9,2);
K(3*m,3*i)     = K(3*m,3*i)     + k(9,3);
K(3*m,3*j-2)   = K(3*m,3*j-2)   + k(9,4);
K(3*m,3*j-1)   = K(3*m,3*j-1)   + k(9,5);
K(3*m,3*j)     = K(3*m,3*j)     + k(9,6);
K(3*m,3*m-2)   = K(3*m,3*m-2)   + k(9,7);
K(3*m,3*m-1)   = K(3*m,3*m-1)   + k(9,8);
K(3*m,3*m)     = K(3*m,3*m)     + k(9,9);
K(3*m,3*n-2)   = K(3*m,3*n-2)   + k(9,10);
K(3*m,3*n-1)   = K(3*m,3*n-1)   + k(9,11);
K(3*m,3*n)     = K(3*m,3*n)     + k(9,12);
K(3*n-2,3*i-2) = K(3*n-2,3*i-2) + k(10,1);
K(3*n-2,3*i-1) = K(3*n-2,3*i-1) + k(10,2);
K(3*n-2,3*i)   = K(3*n-2,3*i)   + k(10,3);
K(3*n-2,3*j-2) = K(3*n-2,3*j-2) + k(10,4);
K(3*n-2,3*j-1) = K(3*n-2,3*j-1) + k(10,5);
K(3*n-2,3*j)   = K(3*n-2,3*j)   + k(10,6);
K(3*n-2,3*m-2) = K(3*n-2,3*m-2) + k(10,7);
K(3*n-2,3*m-1) = K(3*n-2,3*m-1) + k(10,8);
K(3*n-2,3*m)   = K(3*n-2,3*m)   + k(10,9);
K(3*n-2,3*n-2) = K(3*n-2,3*n-2) + k(10,10);
K(3*n-2,3*n-1) = K(3*n-2,3*n-1) + k(10,11);
K(3*n-2,3*n)   = K(3*n-2,3*n)   + k(10,12);
K(3*n-1,3*i-2) = K(3*n-1,3*i-2) + k(11,1);
K(3*n-1,3*i-1) = K(3*n-1,3*i-1) + k(11,2);
K(3*n-1,3*i)   = K(3*n-1,3*i)   + k(11,3);
K(3*n-1,3*j-2) = K(3*n-1,3*j-2) + k(11,4);
K(3*n-1,3*j-1) = K(3*n-1,3*j-1) + k(11,5);
K(3*n-1,3*j)   = K(3*n-1,3*j)   + k(11,6);
K(3*n-1,3*m-2) = K(3*n-1,3*m-2) + k(11,7);
K(3*n-1,3*m-1) = K(3*n-1,3*m-1) + k(11,8);
K(3*n-1,3*m)   = K(3*n-1,3*m)   + k(11,9);
K(3*n-1,3*n-2) = K(3*n-1,3*n-2) + k(11,10);
K(3*n-1,3*n-1) = K(3*n-1,3*n-1) + k(11,11);
K(3*n-1,3*n)   = K(3*n-1,3*n)   + k(11,12);
K(3*n,3*i-2)   = K(3*n,3*i-2)   + k(12,1);
K(3*n,3*i-1)   = K(3*n,3*i-1)   + k(12,2);
K(3*n,3*i)     = K(3*n,3*i)     + k(12,3);
K(3*n,3*j-2)   = K(3*n,3*j-2)   + k(12,4);
K(3*n,3*j-1)   = K(3*n,3*j-1)   + k(12,5);
K(3*n,3*j)     = K(3*n,3*j)     + k(12,6);
K(3*n,3*m-2)   = K(3*n,3*m-2)   + k(12,7);
K(3*n,3*m-1)   = K(3*n,3*m-1)   + k(12,8);
K(3*n,3*m)     = K(3*n,3*m)     + k(12,9);
K(3*n,3*n-2)   = K(3*n,3*n-2)   + k(12,10);
```

```
K(3*n,3*n-1) = K(3*n,3*n-1) + k(12,11);
K(3*n,3*n) = K(3*n,3*n) + k(12,12);
y = K;
```

```
function y = TetrahedronElementStresses
             (E,NU,x1,y1,z1,x2,y2,z2,x3,y3,z3,x4,y4,z4,u)
%TetrahedronElementStresses          This function returns the element
%                                    stress vector for a linear
%                                    tetrahedral (solid) element with
%                                    modulus of elasticity E,
%                                    Poisson's ratio NU, coordinates
%                                    of the first node (x1,y1,z1),
%                                    coordinates of the second node
%                                    (x2,y2,z2), coordinates of the
%                                    third node (x3,y3,z3),
%                                    coordinates of the fourth node
%                                    (x4,y4,z4), and element displacement
%                                    vector u.
%                                    The size of the element stress
%                                    vector is 6 x 1.
xyz = [1 x1 y1 z1 ; 1 x2 y2 z2 ; 1 x3 y3 z3 ; 1 x4 y4 z4];
V = det(xyz)/6;
mbeta1 = [1 y2 z2 ; 1 y3 z3 ; 1 y4 z4];
mbeta2 = [1 y1 z1 ; 1 y3 z3 ; 1 y4 z4];
mbeta3 = [1 y1 z1 ; 1 y2 z2 ; 1 y4 z4];
mbeta4 = [1 y1 z1 ; 1 y2 z2 ; 1 y3 z3];
mgamma1 = [1 x2 z2 ; 1 x3 z3 ; 1 x4 z4];
mgamma2 = [1 x1 z1 ; 1 x3 z3 ; 1 x4 z4];
mgamma3 = [1 x1 z1 ; 1 x2 z2 ; 1 x4 z4];
mgamma4 = [1 x1 z1 ; 1 x2 z2 ; 1 x3 z3];
mdelta1 = [1 x2 y2 ; 1 x3 y3 ; 1 x4 y4];
mdelta2 = [1 x1 y1 ; 1 x3 y3 ; 1 x4 y4];
mdelta3 = [1 x1 y1 ; 1 x2 y2 ; 1 x4 y4];
mdelta4 = [1 x1 y1 ; 1 x2 y2 ; 1 x3 y3];
beta1 = -1*det(mbeta1);
beta2 = det(mbeta2);
beta3 = -1*det(mbeta3);
beta4 = det(mbeta4);
gamma1 = det(mgamma1);
gamma2 = -1*det(mgamma2);
gamma3 = det(mgamma3);
gamma4 = -1*det(mgamma4);
delta1 = -1*det(mdelta1);
delta2 = det(mdelta2);
delta3 = -1*det(mdelta3);
delta4 = det(mdelta4);
B1 = [beta1 0 0 ; 0 gamma1 0 ; 0 0 delta1 ;
   gamma1 beta1 0 ; 0 delta1 gamma1 ; delta1 0 beta1];
B2 = [beta2 0 0 ; 0 gamma2 0 ; 0 0 delta2 ;
   gamma2 beta2 0 ; 0 delta2 gamma2 ; delta2 0 beta2];
B3 = [beta3 0 0 ; 0 gamma3 0 ; 0 0 delta3 ;
   gamma3 beta3 0 ; 0 delta3 gamma3 ; delta3 0 beta3];
```

```
B4 = [beta4 0 0 ; 0 gamma4 0 ; 0 0 delta4 ;
    gamma4 beta4 0 ; 0 delta4 gamma4 ; delta4 0 beta4];
B = [B1 B2 B3 B4]/(6*V);
D = (E/((1+NU)*(1-2*NU)))*[1-NU NU NU 0 0 0 ; NU 1-NU NU 0 0 0 ;
    NU NU 1-NU 0 0 0 ; 0 0 0 (1-2*NU)/2 0 0 ; 0 0 0 0 (1-2*NU)/2 0 ;
    0 0 0 0 0 (1-2*NU)/2];
y = D*B*u;
```

```
function y = TetrahedronElementPStresses(sigma)
%TetrahedronElementPStresses    This function returns the three
%                               principal stresses for the element
%                               given the element stress vector.
%                               The principal angles are not returned.
s1 = sigma(1) + sigma(2) + sigma(3);
s2 = sigma(1)*sigma(2) + sigma(1)*sigma(3) + sigma(2)*sigma(3) -
    sigma(4)*sigma(4) -sigma(5)*sigma(5) - sigma(6)*sigma(6);
ms3 = [sigma(1) sigma(4) sigma(6) ; sigma(4) sigma(2) sigma(5) ;
    sigma(6) sigma(5) sigma(3)];
s3 = det(ms3);
y = [s1 ; s2 ; s3];
```

Example 15.1:

Consider the thin plate subjected to a uniformly distributed load as shown in Fig. 15.2. Use five linear tetrahedral elements to solve this problem as shown in Fig. 15.3. Given $E = 210\,\text{GPa}$, $\nu = 0.3$, $t = 0.025\,\text{m}$, and $w = 3000\,\text{kN/m}^2$, determine:

1. the global stiffness matrix for the structure.
2. the displacements at nodes 3, 4, 7, and 8.
3. the reactions at nodes 1, 2, 5, and 6.
4. the stresses in each element.
5. the principal stresses for each element.

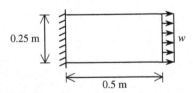

Fig. 15.2. Thin Plate for Example 15.1

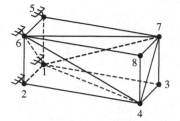

Fig. 15.3. Discretization of Thin Plate into Five Linear Tetrahedra

Solution:

Use the six steps outlined in Chap. 1 to solve this problem using the linear tetrahedral element.

Step 1 – Discretizing the Domain:

We subdivide the plate into five linear tetrahedral elements only for illustration purposes. More elements must be used in order to obtain reliable results. Thus the domain is subdivided into five elements and eight nodes as shown in Fig. 15.3. The total force due to the distributed load is divided equally between nodes 3, 4, 7, and 8 in the ratio 1 : 2 : 2 : 1. This ratio is obtained considering that nodes 4 and 7 take loads from two elements each while nodes 3 and 8 take loads from one element each. The units used in the MATLAB calculations are kN and meter. Table 15.1 shows the element connectivity for this example.

Table 15.1. Element Connectivity for Example 15.1

Element Number	Node i	Node j	Node m	Node n
1	1	2	4	6
2	1	4	3	7
3	6	5	7	1
4	6	7	8	4
5	1	6	4	7

Step 2 – Writing the Element Stiffness Matrices:

The five element stiffness matrices k_1, k_2, k_3, k_4, and k_5 are obtained by making calls to the MATLAB function *TetrahedronElementStiffness*. Each matrix has size 12×12.

```
» E=210e6

E  =

   210000000
```

```
» NU=0.3

NU =

   0.3000

» k1=TetrahedronElementStiffness
      (E,NU,0,0,0,0.025,0,0,0.025,0.5,0,0.025,0,0.25)

k1 =

   1.0e+008 *

   Columns 1 through 7

    2.3558         0          0    -2.3558     0.0505     0.1010          0
         0    0.6731          0     0.0337    -0.6731          0    -0.0337
         0         0     0.6731     0.0673          0    -0.6731          0
   -2.3558    0.0337     0.0673     2.3642    -0.0841    -0.1683    -0.0017
    0.0505   -0.6731          0    -0.0841     0.6857     0.0084     0.0337
    0.1010         0    -0.6731    -0.1683     0.0084     0.6983          0
         0   -0.0337          0    -0.0017     0.0337          0     0.0017
   -0.0505         0          0     0.0505    -0.0059    -0.0050          0
         0         0          0          0    -0.0034    -0.0017          0
         0         0    -0.0673    -0.0067          0     0.0673          0
         0         0          0          0    -0.0067    -0.0034          0
   -0.1010         0          0     0.1010    -0.0050    -0.0236          0

   Columns 8 through 12

   -0.0505         0          0          0    -0.1010
         0         0          0          0          0
         0         0    -0.0673          0          0
    0.0505         0    -0.0067          0     0.1010
   -0.0059   -0.0034          0    -0.0067    -0.0050
   -0.0050   -0.0017     0.0673    -0.0034    -0.0236
         0         0          0          0          0
    0.0059         0          0          0     0.0050
         0    0.0017          0     0.0034          0
         0         0     0.0067          0          0
         0    0.0034          0     0.0067          0
    0.0050         0          0          0     0.0236
```

» k2=TetrahedronElementStiffness
    (E,NU,0,0,0,0.025,0.5,0,0,0.5,0,0,0.5,0.25)

k2 =

    1.0e+008 *

    Columns 1 through 7

```
    0.0017         0         0         0   -0.0337         0   -0.0017
         0    0.0059         0   -0.0505         0         0    0.0505
         0         0    0.0017         0         0         0         0
         0   -0.0505         0    2.3558         0         0   -2.3558
   -0.0337         0         0         0    0.6731         0    0.0337
         0         0         0         0         0    0.6731   -0.0673
   -0.0017    0.0505         0   -2.3558    0.0337   -0.0673    2.3642
    0.0337   -0.0059    0.0034    0.0505   -0.6731         0   -0.0841
         0    0.0050   -0.0017   -0.1010         0   -0.6731    0.1683
         0         0         0         0         0    0.0673   -0.0067
         0         0   -0.0034         0         0         0         0
         0   -0.0050         0    0.1010         0         0   -0.1010
```

    Columns 8 through 12

```
    0.0337         0         0         0         0
   -0.0059    0.0050         0         0   -0.0050
    0.0034   -0.0017         0   -0.0034         0
    0.0505   -0.1010         0         0    0.1010
   -0.6731         0         0         0         0
         0   -0.6731    0.0673         0         0
   -0.0841    0.1683   -0.0067         0   -0.1010
    0.6857   -0.0084         0   -0.0067    0.0050
   -0.0084    0.6983   -0.0673    0.0034   -0.0236
         0   -0.0673    0.0067         0         0
   -0.0067    0.0034         0    0.0067         0
    0.0050   -0.0236         0         0    0.0236
```

» k3=TetrahedronElementStiffness
    (E,NU,0.025,0,0.25,0,0,0.25,0,0.5,0.25,0,0,0)

k3 =

    1.0e+008 *

    Columns 1 through 7

2.3558	0	0	-2.3558	-0.0505	0.1010	0
0	0.6731	0	-0.0337	-0.6731	0	0.0337
0	0	0.6731	0.0673	0	-0.6731	0
-2.3558	-0.0337	0.0673	2.3642	0.0841	-0.1683	-0.0017
-0.0505	-0.6731	0	0.0841	0.6857	-0.0084	-0.0337
0.1010	0	-0.6731	-0.1683	-0.0084	0.6983	0
0	0.0337	0	-0.0017	-0.0337	0	0.0017
0.0505	0	0	-0.0505	-0.0059	0.0050	0
0	0	0	0	0.0034	-0.0017	0
0	0	-0.0673	-0.0067	0	0.0673	0
0	0	0	0	-0.0067	0.0034	0
-0.1010	0	0	0.1010	0.0050	-0.0236	0

Columns 8 through 12

0.0505	0	0	0	-0.1010
0	0	0	0	0
0	0	-0.0673	0	0
-0.0505	0	-0.0067	0	0.1010
-0.0059	0.0034	0	-0.0067	0.0050
0.0050	-0.0017	0.0673	0.0034	-0.0236
0	0	0	0	0
0.0059	0	0	0	-0.0050
0	0.0017	0	-0.0034	0
0	0	0.0067	0	0
0	-0.0034	0	0.0067	0
-0.0050	0	0	0	0.0236

```
» k4=TetrahedronElementStiffness
    (E,NU,0.025,0,0.25,0,0.5,0.25,0.025,0.5,0.25,
    0.025,0.5,0)

k4 =

  1.0e+008 *

Columns 1 through 7
```

0.0017	0	0	0	0.0337	0	-0.0017
0	0.0059	0	0.0505	0	0	-0.0505
0	0	0.0017	0	0	0	0
0	0.0505	0	2.3558	0	0	-2.3558

```
   0.0337        0          0          0    0.6731          0   -0.0337
        0        0          0          0          0     0.6731   -0.0673
  -0.0017  -0.0505          0   -2.3558   -0.0337    -0.0673    2.3642
  -0.0337  -0.0059    -0.0034   -0.0505   -0.6731          0     0.0841
        0  -0.0050    -0.0017   -0.1010          0    -0.6731    0.1683
        0        0          0          0          0     0.0673   -0.0067
        0        0     0.0034          0          0          0          0
        0   0.0050          0    0.1010          0          0    -0.1010
```

Columns 8 through 12

```
  -0.0337        0          0          0          0
  -0.0059  -0.0050          0          0     0.0050
  -0.0034  -0.0017          0     0.0034          0
  -0.0505  -0.1010          0          0     0.1010
  -0.6731        0          0          0          0
        0  -0.6731     0.0673          0          0
   0.0841   0.1683    -0.0067          0    -0.1010
   0.6857   0.0084          0    -0.0067    -0.0050
   0.0084   0.6983    -0.0673    -0.0034    -0.0236
        0  -0.0673     0.0067          0          0
  -0.0067  -0.0034          0     0.0067          0
  -0.0050  -0.0236          0          0     0.0236
```

» k5=TetrahedronElementStiffness
    (E,NU,0,0,0,0.025,0,0.25,0.025,0.5,0,0,0.5,0.25)

k5 =

  1.0e+008 *

Columns 1 through 7

```
   1.1821   0.0421     0.0841   -1.1804    0.0084    -0.0841   -1.1754
   0.0421   0.3428     0.0042   -0.0084   -0.3370    -0.0008   -0.0421
   0.0841   0.0042     0.3492   -0.0841    0.0008    -0.3475   -0.0168
  -1.1804  -0.0084    -0.0841    1.1821   -0.0421     0.0841    1.1737
   0.0084  -0.3370     0.0008   -0.0421    0.3428    -0.0042   -0.0084
  -0.0841  -0.0008    -0.3475    0.0841   -0.0042     0.3492    0.0168
  -1.1754  -0.0421    -0.0168    1.1737   -0.0084     0.0168    1.1821
  -0.0421  -0.3361    -0.0008    0.0084    0.3302     0.0042    0.0421
   0.0168   0.0008    -0.3256   -0.0168    0.0042     0.3239   -0.0841
```

```
 1.1737      0.0084      0.0168     -1.1754      0.0421     -0.0168     -1.1804
-0.0084      0.3302     -0.0042      0.0421     -0.3361      0.0008      0.0084
-0.0168     -0.0042      0.3239      0.0168     -0.0008     -0.3256      0.0841
```

Columns 8 through 12

```
-0.0421      0.0168      1.1737     -0.0084     -0.0168
-0.3361      0.0008      0.0084      0.3302     -0.0042
-0.0008     -0.3256      0.0168     -0.0042      0.3239
 0.0084     -0.0168     -1.1754      0.0421      0.0168
 0.3302      0.0042      0.0421     -0.3361     -0.0008
 0.0042      0.3239     -0.0168      0.0008     -0.3256
 0.0421     -0.0841     -1.1804      0.0084      0.0841
 0.3428     -0.0042     -0.0084     -0.3370      0.0008
-0.0042      0.3492      0.0841     -0.0008     -0.3475
-0.0084      0.0841      1.1821     -0.0421     -0.0841
-0.3370     -0.0008     -0.0421      0.3428      0.0042
 0.0008     -0.3475     -0.0841      0.0042      0.3492
```

Step 3 – Assembling the Global Stiffness Matrix:

Since the structure has eight nodes, the size of the global stiffness matrix is 24×24. Therefore to obtain K we first set up a zero matrix of size 24×24 then make five calls to the MATLAB function *TetrahedronAssemble* since we have five elements in the structure. Each call to the function will assemble one element. The following are the MATLAB commands. The final result is shown only after the fifth element has been assembled.

» K=zeros(24,24);

» K=TetrahedronAssemble(K,k1,1,2,4,6);

» K=TetrahedronAssemble(K,k2,1,4,3,7);

» K=TetrahedronAssemble(K,k3,6,5,7,1);

» K=TetrahedronAssemble(K,k4,6,7,8,4);

» K=TetrahedronAssemble(K,k5,1,6,4,7)

K =

    1.0e+008 *

```
Columns 1 through 7


 3.5463     0.0421     0.0841    -2.3558     0.0505     0.1010    -0.0017
 0.0421     1.0285     0.0042     0.0337    -0.6731          0     0.0505
 0.0841     0.0042     1.0475     0.0673          0    -0.6731          0
-2.3558     0.0337     0.0673     2.3642    -0.0841    -0.1683          0
 0.0505    -0.6731          0    -0.0841     0.6857     0.0084          0
 0.1010          0    -0.6731    -0.1683     0.0084     0.6983          0
-0.0017     0.0505          0          0          0          0     2.3642
 0.0337    -0.0059     0.0034          0          0          0    -0.0841
      0     0.0050    -0.0017          0          0          0     0.1683
-1.1754    -0.1262    -0.0168    -0.0017     0.0337          0    -2.3558
-0.1262    -0.3361    -0.0008     0.0505    -0.0059    -0.0050     0.0337
 0.0168     0.0008    -0.3256          0    -0.0034    -0.0017    -0.0673
-0.0067          0     0.1010          0          0          0          0
      0    -0.0067     0.0050          0          0          0          0
 0.0673     0.0034    -0.0236          0          0          0          0
-1.1804    -0.0084    -0.2524    -0.0067          0     0.0673          0
 0.0084    -0.3370     0.0008          0    -0.0067    -0.0034          0
-0.2524    -0.0008    -0.3475     0.1010    -0.0050    -0.0236          0
 1.1737     0.0084     0.0168          0          0          0    -0.0067
-0.0084     0.3302    -0.0126          0          0          0          0
-0.0168    -0.0126     0.3239          0          0          0    -0.1010
      0          0          0          0          0          0          0
      0          0          0          0          0          0          0
      0          0          0          0          0          0          0


Columns 8 through 14


 0.0337          0    -1.1754    -0.1262     0.0168    -0.0067          0
-0.0059     0.0050    -0.1262    -0.3361     0.0008          0    -0.0067
 0.0034    -0.0017    -0.0168    -0.0008    -0.3256     0.1010     0.0050
      0          0    -0.0017     0.0505          0          0          0
      0          0     0.0337    -0.0059    -0.0034          0          0
      0          0          0    -0.0050    -0.0017          0          0
-0.0841     0.1683    -2.3558     0.0337    -0.0673          0          0
 0.6857    -0.0084     0.0505    -0.6731          0          0          0
-0.0084     0.6983    -0.1010          0    -0.6731          0          0
 0.0505    -0.1010     3.5463     0.0421    -0.0841          0          0
-0.6731          0     0.0421     1.0285    -0.0042          0          0
      0    -0.6731    -0.0841    -0.0042     1.0475          0          0
      0          0          0          0          0     2.3642     0.0841
      0          0          0          0          0     0.0841     0.6857
      0          0          0          0          0    -0.1683    -0.0084
```

0	0	1.1737	0.0084	-0.0168	-2.3558	-0.0505
0	0	-0.0084	0.3302	0.0126	-0.0337	-0.6731
0	0	0.0168	0.0126	0.3239	0.0673	0
0	-0.0673	-1.1804	-0.0084	0.2524	-0.0017	-0.0337
-0.0067	0.0034	0.0084	-0.3370	-0.0008	-0.0505	-0.0059
0.0050	-0.0236	0.2524	0.0008	-0.3475	0	0.0034
0	0	-0.0067	0	-0.1010	0	0
0	0	0	-0.0067	-0.0050	0	0
0	0	-0.0673	-0.0034	-0.0236	0	0

Columns 15 through 21

0.0673	-1.1804	0.0084	-0.2524	1.1737	-0.0084	-0.0168
0.0034	-0.0084	-0.3370	-0.0008	0.0084	0.3302	-0.0126
-0.0236	-0.2524	0.0008	-0.3475	0.0168	-0.0126	0.3239
0	-0.0067	0	0.1010	0	0	0
0	0	-0.0067	-0.0050	0	0	0
0	0.0673	-0.0034	-0.0236	0	0	0
0	0	0	0	-0.0067	0	-0.1010
0	0	0	0	0	-0.0067	0.0050
0	0	0	0	-0.0673	0.0034	-0.0236
0	1.1737	-0.0084	0.0168	-1.1804	0.0084	0.2524
0	0.0084	0.3302	0.0126	-0.0084	-0.3370	0.0008
0	-0.0168	0.0126	0.3239	0.2524	-0.0008	-0.3475
-0.1683	-2.3558	-0.0337	0.0673	-0.0017	-0.0505	0
-0.0084	-0.0505	-0.6731	0	-0.0337	-0.0059	0.0034
0.6983	0.1010	0	-0.6731	0	0.0050	-0.0017
0.1010	3.5463	-0.0421	0.0841	-1.1754	0.1262	0.0168
0	-0.0421	1.0285	-0.0042	0.1262	-0.3361	-0.0008
-0.6731	0.0841	-0.0042	1.0475	-0.0168	0.0008	-0.3256
0	-1.1754	0.1262	-0.0168	3.5463	-0.0421	-0.0841
0.0050	0.1262	-0.3361	0.0008	-0.0421	1.0285	0.0042
-0.0017	0.0168	-0.0008	-0.3256	-0.0841	0.0042	1.0475
0	-0.0017	-0.0505	0	-2.3558	-0.0337	-0.0673
0	-0.0337	-0.0059	-0.0034	-0.0505	-0.6731	0
0	0	-0.0050	-0.0017	-0.1010	0	-0.6731

Columns 22 through 24

0	0	0
0	0	0
0	0	0
0	0	0
0	0	0
0	0	0

0	0	0
0	0	0
0	0	0
-0.0067	0	-0.0673
0	-0.0067	-0.0034
-0.1010	-0.0050	-0.0236
0	0	0
0	0	0
0	0	0
-0.0017	-0.0337	0
-0.0505	-0.0059	-0.0050
0	-0.0034	-0.0017
-2.3558	-0.0505	-0.1010
-0.0337	-0.6731	0
-0.0673	0	-0.6731
2.3642	0.0841	0.1683
0.0841	0.6857	0.0084
0.1683	0.0084	0.6983

Step 4 – Applying the Boundary Conditions:

The matrix (15.10) for this structure can be written using the global stiffness matrix obtained in the previous step. The boundary conditions for this problem are given as:

$$U_{1x} = U_{1y} = U_{1z} = U_{2x} = U_{2y} = U_{2z} = 0$$
$$U_{5x} = U_{5y} = U_{5z} = U_{6x} = U_{6y} = U_{6z} = 0$$
$$F_{3x} = 0, \quad F_{3y} = 3.125, \quad F_{3z} = 0$$
$$F_{4x} = 0, \quad F_{4y} = 6.25, \quad F_{4z} = 0$$
$$F_{7x} = 0, \quad F_{7y} = 6.25, \quad F_{7z} = 0$$
$$F_{8x} = 0, \quad F_{8y} = 3.125, \quad F_{8z} = 0 \tag{15.12}$$

We next insert the above conditions into the matrix equation for this structure (not shown here) and proceed to the solution step below.

Step 5 – Solving the Equations:

Solving the resulting system of equations will be performed by partitioning (manually) and Gaussian elimination (with MATLAB). First we partition the resulting equation by extracting the submatrces in rows 7 to 12, rows 19 to 24, and columns 7 to 12, columns 19 to 24. Therefore we obtain the following equation noting that the numbers are shown to only two decimal places although MATLAB carries out the calculations using at least four decimal places.

$$10^8 \begin{bmatrix}
2.36 & -0.08 & 0.17 & -2.36 & 0.03 & -0.07 & -0.01 & 0 & -0.10 & 0 & 0 & 0 \\
-0.08 & 0.69 & -0.01 & 0.05 & -0.67 & 0 & 0 & -0.01 & 0.01 & 0 & 0 & 0 \\
0.17 & -0.01 & 0.70 & -0.10 & 0 & -0.67 & -0.07 & 0.00 & -0.02 & 0 & 0 & 0 \\
-2.36 & 0.05 & -0.10 & 3.55 & 0.04 & -0.08 & -1.18 & 0.01 & 0.25 & -0.01 & 0 & -0.07 \\
0.03 & -0.67 & 0 & 0.04 & 1.03 & -0.00 & -0.01 & -0.34 & 0.00 & 0 & -0.01 & -0.00 \\
-0.07 & 0 & -0.67 & -0.08 & -0.00 & 1.05 & 0.25 & -0.00 & -0.35 & -0.10 & -0.01 & -0.02 \\
-0.01 & 0 & -0.07 & -1.18 & -0.01 & 0.25 & 3.55 & -0.04 & -0.08 & -2.36 & -0.05 & -0.10 \\
0 & -0.01 & 0.00 & 0.01 & -0.34 & -0.00 & -0.04 & 1.03 & 0.00 & -0.03 & -0.67 & 0 \\
-0.10 & 0.01 & -0.02 & 0.25 & 0.00 & -0.35 & -0.08 & 0.00 & 1.05 & -0.07 & 0 & -0.67 \\
0 & 0 & 0 & -0.01 & 0 & -0.10 & -2.36 & -0.03 & -0.07 & 2.36 & 0.08 & 0.17 \\
0 & 0 & 0 & 0 & -0.01 & -0.01 & -0.05 & -0.67 & 0 & 0.08 & 0.69 & 0.01 \\
0 & 0 & 0 & -0.07 & -0.00 & -0.02 & -0.10 & 0 & -0.67 & 0.17 & 0.01 & 0.70
\end{bmatrix}$$

$$\begin{Bmatrix} U_{3x} \\ U_{3y} \\ U_{3z} \\ U_{4x} \\ U_{4y} \\ U_{4z} \\ U_{7x} \\ U_{7y} \\ U_{7z} \\ U_{8x} \\ U_{8y} \\ U_{8z} \end{Bmatrix} = \begin{Bmatrix} 0 \\ 3.125 \\ 0 \\ 0 \\ 6.25 \\ 0 \\ 0 \\ 6.25 \\ 0 \\ 0 \\ 3.125 \\ 0 \end{Bmatrix} \tag{15.13}$$

The solution of the above system is obtained using MATLAB as follows. Note that the backslash operator "\" is used for Gaussian elimination.

```
» k=[K(7:12,7:12) K(7:12,19:24) ; K(19:24,7:12)
    K(19:24,19:24)]

k =

    1.0e+008 *

Columns 1 through 7

    2.3642   -0.0841    0.1683   -2.3558    0.0337   -0.0673   -0.0067
   -0.0841    0.6857   -0.0084    0.0505   -0.6731         0         0
    0.1683   -0.0084    0.6983   -0.1010         0   -0.6731   -0.0673
   -2.3558    0.0505   -0.1010    3.5463    0.0421   -0.0841   -1.1804
    0.0337   -0.6731         0    0.0421    1.0285   -0.0042   -0.0084
   -0.0673         0   -0.6731   -0.0841   -0.0042    1.0475    0.2524
   -0.0067         0   -0.0673   -1.1804   -0.0084    0.2524    3.5463
         0   -0.0067    0.0034    0.0084   -0.3370   -0.0008   -0.0421
   -0.1010    0.0050   -0.0236    0.2524    0.0008   -0.3475   -0.0841
         0         0         0   -0.0067         0   -0.1010   -2.3558
```

```
    0        0        0              0     -0.0067    -0.0050    -0.0505
    0        0        0        -0.0673    -0.0034    -0.0236    -0.1010
```

Columns 8 through 12

```
         0    -0.1010          0          0          0
   -0.0067     0.0050          0          0          0
    0.0034    -0.0236          0          0          0
    0.0084     0.2524    -0.0067          0    -0.0673
   -0.3370     0.0008          0    -0.0067    -0.0034
   -0.0008    -0.3475    -0.1010    -0.0050    -0.0236
   -0.0421    -0.0841    -2.3558    -0.0505    -0.1010
    1.0285     0.0042    -0.0337    -0.6731          0
    0.0042     1.0475    -0.0673          0    -0.6731
   -0.0337    -0.0673     2.3642     0.0841     0.1683
   -0.6731          0     0.0841     0.6857     0.0084
         0    -0.6731     0.1683     0.0084     0.6983
```

» f=[0 ; 3.125 ; 0 ; 0 ; 6.25 ; 0 ; 0 ; 6.25 ; 0 ; 0 ;
    3.125 ;0]

f =

```
        0
   3.1250
        0
        0
   6.2500
        0
        0
   6.2500
        0
        0
   3.1250
        0
```

» u=k\f

u =

  1.0e-005 *

  -0.0004
   0.6082
```
```

```
    0.0090
   -0.0127
    0.6078
    0.0056
    0.0127
    0.6078
   -0.0056
    0.0004
    0.6082
   -0.0090
```

It is now clear that the horizontal displacement along the y-direction at both nodes 3 and 8 is 0.6082 m, and the horizontal displacement along the y-direction at both nodes 4 and 7 is 0.6078 m. These results are compared with the result of approximately 0.7 m obtained in previous examples and problems in Chap. 11 through 14 using other elements.

Step 6 – Post-processing:

In this step, we obtain the reactions at nodes 1, 2, 5, and 6, and the stresses in each element using MATLAB as follows. First we set up the global nodal displacement vector U, then we calculate the global nodal force vector F.

```
» U=[0;0;0;0;0;0;u(1:6);0;0;0;0;0;0;u(7:12)]

U =

   1.0e-005 *

            0
            0
            0
            0
            0
            0
      -0.0004
       0.6082
       0.0090
      -0.0127
       0.6078
       0.0056
            0
            0
            0
            0
```

```
          0
          0
     0.0127
     0.6078
    -0.0056
     0.0004
     0.6082
    -0.0090
```

» F=K*U

F =

```
   -31.3296
    -5.3492
    -9.3286
    30.7045
    -4.0258
    -3.0777
     0.0000
     3.1250
          0
     0.0000
     6.2500
     0.0000
   -30.7045
    -4.0258
     3.0777
    31.3296
    -5.3492
     9.3286
     0.0000
     6.2500
     0.0000
     0.0000
     3.1250
     0.0000
```

The force reactions along the three directions are clearly shown above. Obviously force equilibrium is satisfied for this problem. Next we set up the element nodal displacement vectors u_1, u_2, u_3, u_4, and u_5 then we calculate the element stresses *sigma1*, *sigma2*, *sigma3*, *sigma4*, and *sigma5* by making calls to the MATLAB function *TetrahedronElementStresses*.

» u1=[U(1) ; U(2) ; U(3) ; U(4) ; U(5) ; U(6) ; U(10) ;
      U(11) ; U(12) ; U(16) ; U(17) ; U(18)]

```
u1 =

   1.0e-005 *

            0
            0
            0
            0
            0
            0
      -0.0127
       0.6078
       0.0056
            0
            0
            0
```

» u2=[U(1) ; U(2) ; U(3) ; U(10) ; U(11) ; U(12) ; U(7) ;
      U(8) ;U(9) ; U(19) ; U(20) ; U(21)]

```
u2 =

   1.0e-005 *

            0
            0
            0
      -0.0127
       0.6078
       0.0056
      -0.0004
       0.6082
       0.0090
       0.0127
       0.6078
      -0.0056
```

» u3=[U(16) ; U(17) ; U(18) ; U(13) ; U(14) ; U(15) ;
      U(19) ; U(20) ; U(21) ; U(1) ; U(2) ; U(3)]

```
u3 =

   1.0e-005 *
```

```
              0
              0
              0
              0
              0
              0
         0.0127
         0.6078
        -0.0056
              0
              0
              0
```

» u4=[U(16) ; U(17) ; U(18) ; U(19) ; U(20) ; U(21) ;
        U(22) ; U(23) ; U(24) ; U(10) ; U(11) ; U(12)]

u4 =

    1.0e-005 *

              0
              0
              0
         0.0127
         0.6078
        -0.0056
         0.0004
         0.6082
        -0.0090
        -0.0127
         0.6078
         0.0056

» u5=[U(1) ; U(2) ; U(3) ; U(16) ; U(17) ; U(18) ;
        U(10) ; U(11); U(12) ; U(19) ; U(20) ; U(21)]

u5 =

    1.0e-005 *

              0
              0
              0
              0
              0

```
            0
       -0.0127
        0.6078
        0.0056
        0.0127
        0.6078
       -0.0056
```

» sigma1=TetrahedronElementStresses
   (E,NU,0,0,0,0.025,0,0,0.025,0.5,0,0.025,0,0.25,u1)

sigma1 =

   1.0e+003 *

       1.4728
       3.4365
       1.4728
      -0.0205
       0.0090
            0

» sigma2=TetrahedronElementStresses
   (E,NU,0,0,0,0.025,0.5,0,0,0.5,0,0,0.5,0.25,u2)

sigma2 =

   1.0e+003 *

       0.0064
       2.7694
       0.7102
      -0.0129
       0.0134
      -0.0704

» sigma3=TetrahedronElementStresses
   (E,NU,0.025,0,0.25,0,0,0.25,0,0.5,0.25,0,0,0,u3)

sigma3 =

   1.0e+003 *

       1.4728
       3.4365
       1.4728
```
```

```
        0.0205
       -0.0090
             0

» sigma4=TetrahedronElementStresses
        (E,NU,0.025,0,0.25,0,0.5,0.25,0.025,0.5,
          0.25,0.025,0.5,0,u4)

sigma4 =

   1.0e+003 *

        0.0064
        2.7694
        0.7102
        0.0129
       -0.0134
       -0.0704


» sigma5=TetrahedronElementStresses
        (E,NU,0,0,0,0.025,0,0.25,0.025,0.5,0,0,0.5,0.25,u5)

sigma5 =

   1.0e+003 *

        0.0096
        2.7941
        0.7945
        0.0000
        0.0000
        0.2204
```

Thus it is clear that the normal stresses σ_y along the y-direction in elements 1, 2, 3, 4, and 5 are are 3.4365 MPa (tensile), 2.7694 MPa (tensile), 3.4365 MPa (tensile), 2.7694 MPa (tensile), and 2.7941 MPa (tensile), respectively. It is clear that the stresses in the y-direction approach closely the correct value of 3 MPa (tensile). Next we calculate the principal stresses for each element by making calls to the MATLAB function *TetrahedronElementPStresses*.

```
» s1=TetrahedronElementPStresses(sigma1)

s1 =

   1.0e+009 *
```

```
        0.0000
        0.0123
        7.4534

» s2=TetrahedronElementPStresses(sigma2)

s2 =

    1.0e+006 *

        0.0035
        1.9839
       -1.2296

» s3=TetrahedronElementPStresses(sigma3)

s3 =

    1.0e+009 *

        0.0000
        0.0123
        7.4534

» s4=TetrahedronElementPStresses(sigma4)

s4 =

    1.0e+006 *

        0.0035
        1.9839
       -1.2296

» s5=TetrahedronElementPStresses(sigma5)

s5 =

    1.0e+008 *

        0.0000
        0.0221
       -1.1431
```

The principal stresses in the three directions are shown clearly above for each one of the five elements in this example.

Problems:

Problem 15.1:

Consider the thin plate problem solved in Example 15.1. Solve the problem again using six linear tetrahedral elements instead of five elements as shown in Fig. 15.4. Compare your answers for the displacements at nodes 3, 4, 7, and 8 with the answers obtained in the example. Compare also the stresses obtained for the six elements with those obtained for the five elements in the example. Compare also your answers with those obtained in the related examples and problems in Chaps. 11 to 14.

Hint: Table 15.2 shows the element connectivity for this problem.

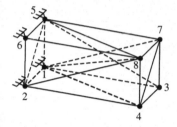

Fig. 15.4. Discretization of Thin Plate into Six Linear Tetrahedra

Table 15.2. Element Connectivity for Problem 15.1

Element Number	Node i	Node j	Node m	Node n
1	1	2	4	8
2	1	2	8	5
3	2	8	5	6
4	1	3	7	4
5	1	7	5	8
6	1	8	4	7

16 The Linear Brick (Solid) Element

16.1
Basic Equations

The linear brick (solid) element is a three-dimensional finite element with both local and global coordinates. It is characterized by linear shape functions in each of the $x, y,$ and z directions. It is also called a trilinear hexahedron. This is the third isoparametric element we deal with in this book. The linear brick element has modulus of elasticity E and Poisson's ratio ν. Each linear brick element has eight nodes with three degrees of freedom at each node as shown in Fig. 16.1. The global coordinates of the eight nodes are denoted by $(x_1, y_1, z_1), (x_2, y_2, z_2), (x_3, y_3, z_3), (x_4, y_4, z_4), (x_5, y_5, z_5), (x_6, y_6, z_6), (x_7, y_7, z_7),$ and (x_8, y_8, z_8). The order of the nodes for each element is important – they should be numbered such that the volume of the element is positive. You can actually check this by using the MATLAB function *LinearBrickElementVolume* which is written specifically for this purpose. The element is mapped to a hexahedron through the use of the natural coordinates ξ, η, and μ as shown in Fig. 16.2. In this case the element stiffness matrix is not written explicitly but calculated through symbolic integration with the aid of the MATLAB Symbolic Math Toolbox. The eight shape functions for this element are listed explicitly as follows in terms of the natural coordinates ξ, η, and μ (see [1]).

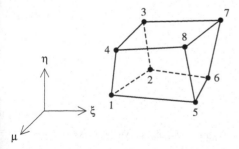

Fig. 16.1. The Linear Brick (Solid) Element

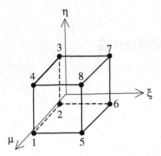

Fig. 16.2. The Linear Brick Element with Natural Coordinates

$$N_1 = \frac{1}{8}(1 - \xi)(1 - \eta)(1 + \mu)$$

$$N_2 = \frac{1}{8}(1 - \xi)(1 - \eta)(1 - \mu)$$

$$N_3 = \frac{1}{8}(1 - \xi)(1 + \eta)(1 - \mu)$$

$$N_4 = \frac{1}{8}(1 - \xi)(1 + \eta)(1 + \mu)$$

$$N_5 = \frac{1}{8}(1 + \xi)(1 - \eta)(1 + \mu)$$

$$N_6 = \frac{1}{8}(1 + \xi)(1 - \eta)(1 - \mu)$$

$$N_7 = \frac{1}{8}(1 + \xi)(1 + \eta)(1 - \mu)$$

$$N_8 = \frac{1}{8}(1 + \xi)(1 + \eta)(1 + \mu) \tag{16.1}$$

The Jacobian matrix for this element is given by

$$[J] = \begin{bmatrix} \dfrac{\partial x}{\partial \xi} & \dfrac{\partial y}{\partial \xi} & \dfrac{\partial z}{\partial \xi} \\[2mm] \dfrac{\partial x}{\partial \eta} & \dfrac{\partial y}{\partial \eta} & \dfrac{\partial z}{\partial \eta} \\[2mm] \dfrac{\partial x}{\partial \mu} & \dfrac{\partial y}{\partial \mu} & \dfrac{\partial z}{\partial \mu} \end{bmatrix} \tag{16.2}$$

where x, y, and z are given by

$$x = N_1 x_1 + N_2 x_2 + N_3 x_3 + N_4 x_4 + N_5 x_5 + N_6 x_6 + N_7 x_7 + N_8 x_8$$
$$y = N_1 y_1 + N_2 y_2 + N_3 y_3 + N_4 y_4 + N_5 y_5 + N_6 y_6 + N_7 y_7 + N_8 y_8$$
$$z = N_1 z_1 + N_2 z_2 + N_3 z_3 + N_4 z_4 + N_5 z_5 + N_6 z_6 + N_7 z_7 + N_8 z_8 \tag{16.3}$$

The $[B]$ matrix is given as follows for this element:

$$[B] = [D'][N] \tag{16.4}$$

where $[D']$ and $[N]$ are given by:

$$[D'] = \begin{bmatrix} \dfrac{\partial()}{\partial x} & 0 & 0 \\[2mm] 0 & \dfrac{\partial()}{\partial y} & 0 \\[2mm] 0 & 0 & \dfrac{\partial()}{\partial z} \\[2mm] \dfrac{\partial()}{\partial y} & \dfrac{\partial()}{\partial x} & 0 \\[2mm] 0 & \dfrac{\partial()}{\partial z} & \dfrac{\partial()}{\partial y} \\[2mm] \dfrac{\partial()}{\partial z} & 0 & \dfrac{\partial()}{\partial x} \end{bmatrix} \tag{16.5}$$

$$[N] = \begin{bmatrix} [N_0] & 0 & 0 \\ 0 & [N_0] & 0 \\ 0 & 0 & [N_0] \end{bmatrix} \tag{16.6}$$

and the submatrix $[N_0]$ is given by

$$[N_0] = \begin{bmatrix} N_1 & N_2 & N_3 & N_4 & N_5 & N_6 & N_7 & N_8 \end{bmatrix} \tag{16.7}$$

The partial derivatives in (16.5) are evaluated as follows

$$\frac{\partial f}{\partial x} = \frac{1}{|J|} \begin{vmatrix} \dfrac{\partial f}{\partial \xi} & \dfrac{\partial y}{\partial \xi} & \dfrac{\partial z}{\partial \xi} \\[2mm] \dfrac{\partial f}{\partial \eta} & \dfrac{\partial y}{\partial \eta} & \dfrac{\partial z}{\partial \eta} \\[2mm] \dfrac{\partial f}{\partial \mu} & \dfrac{\partial y}{\partial \mu} & \dfrac{\partial z}{\partial \mu} \end{vmatrix} \tag{16.8a}$$

$$\frac{\partial f}{\partial y} = \frac{1}{|J|} \begin{vmatrix} \dfrac{\partial x}{\partial \xi} & \dfrac{\partial f}{\partial \xi} & \dfrac{\partial z}{\partial \xi} \\[2mm] \dfrac{\partial x}{\partial \eta} & \dfrac{\partial f}{\partial \eta} & \dfrac{\partial z}{\partial \eta} \\[2mm] \dfrac{\partial x}{\partial \mu} & \dfrac{\partial f}{\partial \mu} & \dfrac{\partial z}{\partial \mu} \end{vmatrix} \tag{16.8b}$$

$$\frac{\partial f}{\partial z} = \frac{1}{|J|} \begin{vmatrix} \dfrac{\partial x}{\partial \xi} & \dfrac{\partial y}{\partial \xi} & \dfrac{\partial f}{\partial \xi} \\ \dfrac{\partial x}{\partial \eta} & \dfrac{\partial y}{\partial \eta} & \dfrac{\partial f}{\partial \eta} \\ \dfrac{\partial x}{\partial \mu} & \dfrac{\partial y}{\partial \mu} & \dfrac{\partial f}{\partial \mu} \end{vmatrix} \tag{16.8c}$$

where f is a dummy variable to stand for the empty parentheses () in (16.5), representing either N_1, N_2, or N_3.

For three-dimensional analysis, the matrix $[D]$ is given by

$$[D] = \frac{E}{(1+\nu)(1-2\nu)} \begin{bmatrix} 1-\nu & \nu & \nu & 0 & 0 & 0 \\ \nu & 1-\nu & \nu & 0 & 0 & 0 \\ \nu & \nu & 1-\nu & 0 & 0 & 0 \\ 0 & 0 & 0 & \dfrac{1-2\nu}{2} & 0 & 0 \\ 0 & 0 & 0 & 0 & \dfrac{1-2\nu}{2} & 0 \\ 0 & 0 & 0 & 0 & 0 & \dfrac{1-2\nu}{2} \end{bmatrix} \tag{16.9}$$

The element stiffness matrix for the linear brick element is written in terms of a triple integral as follows:

$$[k] = \int\limits_{-1}^{1} \int\limits_{-1}^{1} \int\limits_{-1}^{1} [B]^T [D][B] \, |J| \, d\xi \, d\eta d\mu \tag{16.10}$$

The partial differentiation of (16.5) and (16.8), and the double integration of (16.10) are carried out symbolically with the aid of the MATLAB Symbolic Math Toolbox. See the details of the MATLAB code for the function *LinearBrickElementStiffness* which calculates the element stiffness matrix for this element. The reader should note the calculation of this matrix will be somewhat slow due to the symbolic computations involved.

It is clear that the linear brick element has twenty-four degrees of freedom – three at each node. Consequently for a structure with n nodes, the global stiffness matrix K will be of size $3n \times 3n$ (since we have three degrees of freedom at each node). The global stiffness matrix K is assembled by making calls to the MATLAB function *LinearBrickAssemble* which is written specifically for this purpose. This process will be illustrated in detail in the example.

Once the global stiffness matrix K is obtained we have the following structure equation:

$$[K]\{U\} = \{F\} \tag{16.11}$$

where U is the global nodal displacement vector and F is the global nodal force vector. At this step the boundary conditions are applied manually to the vectors U and F. Then the matrix (16.11) is solved by partitioning and Gaussian elimination. Finally once the unknown displacements and reactions are found, the stress vector is obtained for each element as follows:

$$\{\sigma\} = [D][B]\{u\} \tag{16.12}$$

where σ is the stress vector in the element (of size 6×1) and u is the 24×1 element displacement vector. The vector σ is written for each element as $\{\sigma\} = [\sigma_x \sigma_y \sigma_z \tau_{xy} \tau_{yz} \tau_{zx}]^T$. Finally, the element volume is given by the following formula:

$$V = \int_{-1}^{1} \int_{-1}^{1} \int_{-1}^{1} |J| \, d\xi \, d\eta \, d\mu \tag{16.13}$$

16.2
MATLAB Functions Used

The five MATLAB functions used for the linear brick element are:

LinearBrickElementVolume$(x_1, y_1, z_1, x_2, y_2, z_2, x_3, y_3, z_3, x_4, y_4, z_4, x_5, y_5, z_5, x_6, y_6, z_6, x_7, y_7, z_7, x_8, y_8, z_8)$ – This function returns the element volume given the coordinates of the first node (x_1, y_1, z_1), the coordinates of the second node (x_2, y_2, z_2), the coordinates of the third node (x_3, y_3, z_3), the coordinates of the fourth node (x_4, y_4, z_4), the coordinates of the fifth node (x_5, y_5, z_5), the coordinates of the sixth node (x_6, y_6, z_6), the coordinates of the seventh node (x_7, y_7, z_7), and the coordinates of the eighth node (x_8, y_8, z_8).

LinearBrickElementStiffness$(E, NU, x_1, y_1, z_1, x_2, y_2, z_2, x_3, y_3, z_3, x_4, y_4, z_4, x_5, y_5, z_5, x_6, y_6, z_6, x_7, y_7, z_7, x_8, y_8, z_8)$ – This function calculates the element stiffness matrix for each linear brick element with modulus of elasticity E, Poisson's ratio NU, and coordinates of the first node (x_1, y_1, z_1), the coordinates of the second node (x_2, y_2, z_2), the coordinates of the third node (x_3, y_3, z_3), the coordinates of the fourth node (x_4, y_4, z_4), the coordinates of the fifth node (x_5, y_5, z_5), the coordinates of the sixth node (x_6, y_6, z_6), the coordinates of the seventh node (x_7, y_7, z_7), and the coordinates of the eighth node (x_8, y_8, z_8). It returns the 24×24 element stiffness matrix k.

LinearBrickAssemble$(K, k, i, j, m, n, p, q, r, s)$ – This function assembles the element stiffness matrix k of the linear brick element joining nodes i, j, m, n, p, q, r, and s into the global stiffness matrix K. It returns the $3n \times 3n$ global stiffness matrix K every time an element is assembled.

LinearBrickElementStresses(E, NU, x_1, y_1, z_1, x_2, y_2, z_2, x_3, y_3, z_3, x_4, y_4, z_4, x_5, y_5, z_5, x_6, y_6, z_6, x_7, y_7, z_7, x_8, y_8, z_8, u) – This function calculates the element stresses using the modulus of elasticity E, Poisson's ratio NU, the coordinates of the first node (x_1, y_1, z_1), the coordinates of the second node (x_2, y_2, z_2), the coordinates of the third node (x_3, y_3, z_3), the coordinates of the fourth node (x_4, y_4, z_4), the coordinates of the fifth node (x_5, y_5, z_5), the coordinates of the sixth node (x_6, y_6, z_6), the coordinates of the seventh node (x_7, y_7, z_7), and the coordinates of the eighth node (x_8, y_8, z_8). It returns the stress vector for the element.

LinearBrickElementPStresses(*sigma*) – This function calculates the element principal stresses using the element stress vector *sigma*. It returns a 3×1 vector in the form $[sigma1 \ \ sigma2 \ \ theta]^{\mathrm{T}}$ where *sigma1* and *sigma2* are the principal stresses for the element and *theta* is the principal angle.

The following is a listing of the MATLAB source code for each function:

```
function w =
LinearBrickElementVolume(x1,y1,z1,x2,y2,z2,x3,y3,z3,x4,y4,z4,
x5,y5,z5,x6,y6,z6,x7,y7,z7,x8,y8,z8)
% LinearBrickElementVolume      This function returns the volume
%                               of the linear brick element
%                               whose first node has coordinates
%                               (x1,y1,z1), second node has
%                               coordinates (x2,y2,z2), third node
%                               has coordinates (x3,y3,z3),
%                               fourth node has coordiantes
%                               (x4,y4,z4), fifth node has coordiantes
%                               (x5,y5,z5), sixth node has coordiantes
%                               (x6,y6,z6), seventh node has coordiantes
%                               (x7,y7,z7), and eighth node has
%                               coordiantes (x8,y8,z8).
syms s t u;
N1 = (1-s)*(1-t)*(1+u)/8;
N2 = (1-s)*(1-t)*(1-u)/8;
N3 = (1-s)*(1+t)*(1-u)/8;
N4 = (1-s)*(1+t)*(1+u)/8;
N5 = (1+s)*(1-t)*(1+u)/8;
N6 = (1+s)*(1-t)*(1-u)/8;
N7 = (1+s)*(1+t)*(1-u)/8;
N8 = (1+s)*(1+t)*(1+u)/8;
x = N1*x1 + N2*x2 + N3*x3 + N4*x4 + N5*x5 + N6*x6 + N7*x7 + N8*x8;
y = N1*y1 + N2*y2 + N3*y3 + N4*y4 + N5*y5 + N6*y6 + N7*y7 + N8*y8;
z = N1*z1 + N2*z2 + N3*z3 + N4*z4 + N5*z5 + N6*z6 + N7*z7 + N8*z8;
xs = diff(x,s);
xt = diff(x,t);
xu = diff(x,u);
ys = diff(y,s);
yt = diff(y,t);
yu = diff(y,u);
```

```
zs = diff(z,s);
zt = diff(z,t);
zu = diff(z,u);
J = xs*(yt*zu - zt*yu) - ys*(xt*zu - zt*xu) + zs*(xt*yu - yt*xu);
Jnew = simplify(J);
r = int(int(int(Jnew, u, -1, 1), t, -1, 1), s, -1, 1);
w = double(r);
```

```
function w =
LinearBrickElementStiffness(E,NU,x1,y1,z1,x2,y2,z2,x3,y3,z3,x4,y4,z4,
x5,y5,z5,x6,y6,z6,x7,y7,z7,x8,y8,z8)
% LinearBrickElementStiffness  This function returns the element
%                              stiffness matrix for a linear brick
%                              element with modulus of elasticity
%                              E, Poisson's ratio NU, coordinates of
%                              node 1 (x1,y1,z1), coordinates
%                              of node 2 (x2,y2,z2), coordinates of
%                              node 3 (x3,y3,z3), coordinates of
%                              node 4 (x4,y4,z4), coordinates of
%                              node 5 (x5,y5,z5), coordinates of
%                              node 6 (x6,y6,z6), coordinates of
%                              node 7 (x7,y7,z7), and coordinates
%                              of node 8 (x8,y8,z8).
%                              The size of the element
%                              stiffness matrix is 24 x 24.
syms s t u;
N1 = (1-s)*(1-t)*(1+u)/8;
N2 = (1-s)*(1-t)*(1-u)/8;
N3 = (1-s)*(1+t)*(1-u)/8;
N4 = (1-s)*(1+t)*(1+u)/8;
N5 = (1+s)*(1-t)*(1+u)/8;
N6 = (1+s)*(1-t)*(1-u)/8;
N7 = (1+s)*(1+t)*(1-u)/8;
N8 = (1+s)*(1+t)*(1+u)/8;
x = N1*x1 + N2*x2 + N3*x3 + N4*x4 + N5*x5 + N6*x6 + N7*x7 + N8*x8;
y = N1*y1 + N2*y2 + N3*y3 + N4*y4 + N5*y5 + N6*y6 + N7*y7 + N8*y8;
z = N1*z1 + N2*z2 + N3*z3 + N4*z4 + N5*z5 + N6*z6 + N7*z7 + N8*z8;
xs = diff(x,s);
xt = diff(x,t);
xu = diff(x,u);
ys = diff(y,s);
yt = diff(y,t);
yu = diff(y,u);
zs = diff(z,s);
zt = diff(z,t);
zu = diff(z,u);
J = xs*(yt*zu - zt*yu) - ys*(xt*zu - zt*xu) + zs*(xt*yu - yt*xu);
N1s = diff(N1,s);
N2s = diff(N2,s);
N3s = diff(N3,s);
```

```
N4s = diff(N4,s);
N5s = diff(N5,s);
N6s = diff(N6,s);
N7s = diff(N7,s);
N8s = diff(N8,s);
N1t = diff(N1,t);
N2t = diff(N2,t);
N3t = diff(N3,t);
N4t = diff(N4,t);
N5t = diff(N5,t);
N6t = diff(N6,t);
N7t = diff(N7,t);
N8t = diff(N8,t);
N1u = diff(N1,u);
N2u = diff(N2,u);
N3u = diff(N3,u);
N4u = diff(N4,u);
N5u = diff(N5,u);
N6u = diff(N6,u);
N7u = diff(N7,u);
N8u = diff(N8,u);
% The expressions below are not divided by J - they are adjusted for
later
% in the calculation of BD matrix below.
N1x = N1s*(yt*zu - zt*yu) - ys*(N1t*zu - zt*N1u) + zs*(N1t*yu - yt*N1u);
N2x = N2s*(yt*zu - zt*yu) - ys*(N2t*zu - zt*N2u) + zs*(N2t*yu - yt*N2u);
N3x = N3s*(yt*zu - zt*yu) - ys*(N3t*zu - zt*N3u) + zs*(N3t*yu - yt*N3u);
N4x = N4s*(yt*zu - zt*yu) - ys*(N4t*zu - zt*N4u) + zs*(N4t*yu - yt*N4u);
N5x = N5s*(yt*zu - zt*yu) - ys*(N5t*zu - zt*N5u) + zs*(N5t*yu - yt*N5u);
N6x = N6s*(yt*zu - zt*yu) - ys*(N6t*zu - zt*N6u) + zs*(N6t*yu - yt*N6u);
N7x = N7s*(yt*zu - zt*yu) - ys*(N7t*zu - zt*N7u) + zs*(N7t*yu - yt*N7u);
N8x = N8s*(yt*zu - zt*yu) - ys*(N8t*zu - zt*N8u) + zs*(N8t*yu - yt*N8u);
N1y = xs*(N1t*zu - zt*N1u) - N1s*(xt*zu - zt*xu) + zs*(xt*N1u - N1t*xu);
N2y = xs*(N2t*zu - zt*N2u) - N2s*(xt*zu - zt*xu) + zs*(xt*N2u - N2t*xu);
N3y = xs*(N3t*zu - zt*N3u) - N3s*(xt*zu - zt*xu) + zs*(xt*N3u - N3t*xu);
N4y = xs*(N4t*zu - zt*N4u) - N4s*(xt*zu - zt*xu) + zs*(xt*N4u - N4t*xu);
N5y = xs*(N5t*zu - zt*N5u) - N5s*(xt*zu - zt*xu) + zs*(xt*N5u - N5t*xu);
N6y = xs*(N6t*zu - zt*N6u) - N6s*(xt*zu - zt*xu) + zs*(xt*N6u - N6t*xu);
N7y = xs*(N7t*zu - zt*N7u) - N7s*(xt*zu - zt*xu) + zs*(xt*N7u - N7t*xu);
N8y = xs*(N8t*zu - zt*N8u) - N8s*(xt*zu - zt*xu) + zs*(xt*N8u - N8t*xu);
N1z = xs*(yt*N1u - N1t*yu) - ys*(xt*N1u - N1t*xu) + N1s*(xt*yu - yt*xu);
N2z = xs*(yt*N2u - N2t*yu) - ys*(xt*N2u - N2t*xu) + N2s*(xt*yu - yt*xu);
N3z = xs*(yt*N3u - N3t*yu) - ys*(xt*N3u - N3t*xu) + N3s*(xt*yu - yt*xu);
N4z = xs*(yt*N4u - N4t*yu) - ys*(xt*N4u - N4t*xu) + N4s*(xt*yu - yt*xu);
N5z = xs*(yt*N5u - N5t*yu) - ys*(xt*N5u - N5t*xu) + N5s*(xt*yu - yt*xu);
N6z = xs*(yt*N6u - N6t*yu) - ys*(xt*N6u - N6t*xu) + N6s*(xt*yu - yt*xu);
N7z = xs*(yt*N7u - N7t*yu) - ys*(xt*N7u - N7t*xu) + N7s*(xt*yu - yt*xu);
N8z = xs*(yt*N8u - N8t*yu) - ys*(xt*N8u - N8t*xu) + N8s*(xt*yu - yt*xu);
% Next, the B matrix is calculated explicitly as follows:
B = [N1x N2x N3x N4x N5x N6x N7x N8x 0 0 0 0 0 0 0 0 0 0 0 0 0 0 0 0 ;
     0 0 0 0 0 0 0 0 N1y N2y N3y N4y N5y N6y N7y N8y 0 0 0 0 0 0 0 0 ;
     0 0 0 0 0 0 0 0 0 0 0 0 0 0 0 0 N1z N2z N3z N4z N5z N6z N7z N8z ;
     N1y N2y N3y N4y N5y N6y N7y N8y N1x N2x N3x N4x N5x N6x N7x N8x 0 0 0
     0 0 0 0 0 ;
```

```
    0  0  0  0  0  0  0  N1z N2z N3z N4z N5z N6z N7z N8z N1y N2y N3y N4y N5y
N6y N7y N8y ;
   N1z N2z N3z N4z N5z N6z N7z N8z 0  0  0  0  0  0  0  0  N1x N2x N3x N4x N5x
N6x N7x N8x];
Bnew = simplify(B);
Jnew = simplify(J);
D = (E/((1+NU)*(1-2*NU)))*[1-NU NU NU 0 0 0 ; NU 1-NU NU 0 0 0 ; NU NU
1- NU 0 0 0 ;
   0  0  0  (1-2*NU)/2 0  0 ; 0 0 0 0 (1- 2*NU)/2 0 ; 0 0 0 0 0 (1- 2*NU)/2];
BD = transpose(Bnew)*D*Bnew/Jnew;
r = int(int(int(BD, u, -1, 1), t, -1, 1), s, -1, 1);
w = double(r);
```

```
function y = LinearBrickAssemble(K,k,i,j,m,n,p,q,r,s)
% LinearBrickAssemble    This function assembles the element stiffness
%                        matrix k of the linear brick (solid)
%                        element with nodes i, j, m, n, p, q, r,
%                        and s into the global stiffness matrix K.
%                        This function returns the global stiffness
%                        matrix K after the element stiffness matrix
%                        k is assembled.
K(3*i-2,3*i-2) = K(3*i-2,3*i-2) + k(1,1);
K(3*i-2,3*i-1) = K(3*i-2,3*i-1) + k(1,2);
K(3*i-2,3*i) = K(3*i-2,3*i) + k(1,3);
K(3*i-2,3*j-2) = K(3*i-2,3*j-2) + k(1,4);
K(3*i-2,3*j-1) = K(3*i-2,3*j-1) + k(1,5);
K(3*i-2,3*j) = K(3*i-2,3*j) + k(1,6);
K(3*i-2,3*m-2) = K(3*i-2,3*m-2) + k(1,7);
K(3*i-2,3*m-1) = K(3*i-2,3*m-1) + k(1,8);
K(3*i-2,3*m) = K(3*i-2,3*m) + k(1,9);
K(3*i-2,3*n-2) = K(3*i-2,3*n-2) + k(1,10);
K(3*i-2,3*n-1) = K(3*i-2,3*n-1) + k(1,11);
K(3*i-2,3*n) = K(3*i-2,3*n) + k(1,12);
K(3*i-2,3*p-2) = K(3*i-2,3*p-2) + k(1,13);
K(3*i-2,3*p-1) = K(3*i-2,3*p-1) + k(1,14);
K(3*i-2,3*p) = K(3*i-2,3*p) + k(1,15);
K(3*i-2,3*q-2) = K(3*i-2,3*q-2) + k(1,16);
K(3*i-2,3*q-1) = K(3*i-2,3*q-1) + k(1,17);
K(3*i-2,3*q) = K(3*i-2,3*q) + k(1,18);
K(3*i-2,3*r-2) = K(3*i-2,3*r-2) + k(1,19);
K(3*i-2,3*r-1) = K(3*i-2,3*r-1) + k(1,20);
K(3*i-2,3*r) = K(3*i-2,3*r) + k(1,21);
K(3*i-2,3*s-2) = K(3*i-2,3*s-2) + k(1,22);
K(3*i-2,3*s-1) = K(3*i-2,3*s-1) + k(1,23);
K(3*i-2,3*s) = K(3*i-2,3*s) + k(1,24);
K(3*i-1,3*i-2) = K(3*i-1,3*i-2) + k(2,1);
K(3*i-1,3*i-1) = K(3*i-1,3*i-1) + k(2,2);
K(3*i-1,3*i) = K(3*i-1,3*i) + k(2,3);
K(3*i-1,3*j-2) = K(3*i-1,3*j-2) + k(2,4);
K(3*i-1,3*j-1) = K(3*i-1,3*j-1) + k(2,5);
K(3*i-1,3*j) = K(3*i-1,3*j) + k(2,6);
K(3*i-1,3*m-2) = K(3*i-1,3*m-2) + k(2,7);
K(3*i-1,3*m-1) = K(3*i-1,3*m-1) + k(2,8);
K(3*i-1,3*m) = K(3*i-1,3*m) + k(2,9);
```

```
K(3*i-1,3*n-2) = K(3*i-1,3*n-2) + k(2,10);
K(3*i-1,3*n-1) = K(3*i-1,3*n-1) + k(2,11);
K(3*i-1,3*n) = K(3*i-1,3*n) + k(2,12);
K(3*i-1,3*p-2) = K(3*i-1,3*p-2) + k(2,13);
K(3*i-1,3*p-1) = K(3*i-1,3*p-1) + k(2,14);
K(3*i-1,3*p) = K(3*i-1,3*p) + k(2,15);
K(3*i-1,3*q-2) = K(3*i-1,3*q-2) + k(2,16);
K(3*i-1,3*q-1) = K(3*i-1,3*q-1) + k(2,17);
K(3*i-1,3*q) = K(3*i-1,3*q) + k(2,18);
K(3*i-1,3*r-2) = K(3*i-1,3*r-2) + k(2,19);
K(3*i-1,3*r-1) = K(3*i-1,3*r-1) + k(2,20);
K(3*i-1,3*r) = K(3*i-1,3*r) + k(2,21);
K(3*i-1,3*s-2) = K(3*i-1,3*s-2) + k(2,22);
K(3*i-1,3*s-1) = K(3*i-1,3*s-1) + k(2,23);
K(3*i-1,3*s) = K(3*i-1,3*s) + k(2,24);
K(3*i,3*i-2) = K(3*i,3*i-2) + k(3,1);
K(3*i,3*i-1) = K(3*i,3*i-1) + k(3,2);
K(3*i,3*i) = K(3*i,3*i) + k(3,3);
K(3*i,3*j-2) = K(3*i,3*j-2) + k(3,4);
K(3*i,3*j-1) = K(3*i,3*j-1) + k(3,5);
K(3*i,3*j) = K(3*i,3*j) + k(3,6);
K(3*i,3*m-2) = K(3*i,3*m-2) + k(3,7);
K(3*i,3*m-1) = K(3*i,3*m-1) + k(3,8);
K(3*i,3*m) = K(3*i,3*m) + k(3,9);
K(3*i,3*n-2) = K(3*i,3*n-2) + k(3,10);
K(3*i,3*n-1) = K(3*i,3*n-1) + k(3,11);
K(3*i,3*n) = K(3*i,3*n) + k(3,12);
K(3*i,3*p-2) = K(3*i,3*p-2) + k(3,13);
K(3*i,3*p-1) = K(3*i,3*p-1) + k(3,14);
K(3*i,3*p) = K(3*i,3*p) + k(3,15);
K(3*i,3*q-2) = K(3*i,3*q-2) + k(3,16);
K(3*i,3*q-1) = K(3*i,3*q-1) + k(3,17);
K(3*i,3*q) = K(3*i,3*q) + k(3,18);
K(3*i,3*r-2) = K(3*i,3*r-2) + k(3,19);
K(3*i,3*r-1) = K(3*i,3*r-1) + k(3,20);
K(3*i,3*r) = K(3*i,3*r) + k(3,21);
K(3*i,3*s-2) = K(3*i,3*s-2) + k(3,22);
K(3*i,3*s-1) = K(3*i,3*s-1) + k(3,23);
K(3*i,3*s) = K(3*i,3*s) + k(3,24);
K(3*j-2,3*i-2) = K(3*j-2,3*i-2) + k(4,1);
K(3*j-2,3*i-1) = K(3*j-2,3*i-1) + k(4,2);
K(3*j-2,3*i) = K(3*j-2,3*i) + k(4,3);
K(3*j-2,3*j-2) = K(3*j-2,3*j-2) + k(4,4);
K(3*j-2,3*j-1) = K(3*j-2,3*j-1) + k(4,5);
K(3*j-2,3*j) = K(3*j-2,3*j) + k(4,6);
K(3*j-2,3*m-2) = K(3*j-2,3*m-2) + k(4,7);
K(3*j-2,3*m-1) = K(3*j-2,3*m-1) + k(4,8);
K(3*j-2,3*m) = K(3*j-2,3*m) + k(4,9);
K(3*j-2,3*n-2) = K(3*j-2,3*n-2) + k(4,10);
K(3*j-2,3*n-1) = K(3*j-2,3*n-1) + k(4,11);
K(3*j-2,3*n) = K(3*j-2,3*n) + k(4,12);
K(3*j-2,3*p-2) = K(3*j-2,3*p-2) + k(4,13);
K(3*j-2,3*p-1) = K(3*j-2,3*p-1) + k(4,14);
```

```
K(3*j-2,3*p) = K(3*j-2,3*p) + k(4,15);
K(3*j-2,3*q-2) = K(3*j-2,3*q-2) + k(4,16);
K(3*j-2,3*q-1) = K(3*j-2,3*q-1) + k(4,17);
K(3*j-2,3*q) = K(3*j-2,3*q) + k(4,18);
K(3*j-2,3*r-2) = K(3*j-2,3*r-2) + k(4,19);
K(3*j-2,3*r-1) = K(3*j-2,3*r-1) + k(4,20);
K(3*j-2,3*r) = K(3*j-2,3*r) + k(4,21);
K(3*j-2,3*s-2) = K(3*j-2,3*s-2) + k(4,22);
K(3*j-2,3*s-1) = K(3*j-2,3*s-1) + k(4,23);
K(3*j-2,3*s) = K(3*j-2,3*s) + k(4,24);
K(3*j-1,3*i-2) = K(3*j-1,3*i-2) + k(5,1);
K(3*j-1,3*i-1) = K(3*j-1,3*i-1) + k(5,2);
K(3*j-1,3*i) = K(3*j-1,3*i) + k(5,3);
K(3*j-1,3*j-2) = K(3*j-1,3*j-2) + k(5,4);
K(3*j-1,3*j-1) = K(3*j-1,3*j-1) + k(5,5);
K(3*j-1,3*j) = K(3*j-1,3*j) + k(5,6);
K(3*j-1,3*m-2) = K(3*j-1,3*m-2) + k(5,7);
K(3*j-1,3*m-1) = K(3*j-1,3*m-1) + k(5,8);
K(3*j-1,3*m) = K(3*j-1,3*m) + k(5,9);
K(3*j-1,3*n-2) = K(3*j-1,3*n-2) + k(5,10);
K(3*j-1,3*n-1) = K(3*j-1,3*n-1) + k(5,11);
K(3*j-1,3*n) = K(3*j-1,3*n) + k(5,12);
K(3*j-1,3*p-2) = K(3*j-1,3*p-2) + k(5,13);
K(3*j-1,3*p-1) = K(3*j-1,3*p-1) + k(5,14);
K(3*j-1,3*p) = K(3*j-1,3*p) + k(5,15);
K(3*j-1,3*q-2) = K(3*j-1,3*q-2) + k(5,16);
K(3*j-1,3*q-1) = K(3*j-1,3*q-1) + k(5,17);
K(3*j-1,3*q) = K(3*j-1,3*q) + k(5,18);
K(3*j-1,3*r-2) = K(3*j-1,3*r-2) + k(5,19);
K(3*j-1,3*r-1) = K(3*j-1,3*r-1) + k(5,20);
K(3*j-1,3*r) = K(3*j-1,3*r) + k(5,21);
K(3*j-1,3*s-2) = K(3*j-1,3*s-2) + k(5,22);
K(3*j-1,3*s-1) = K(3*j-1,3*s-1) + k(5,23);
K(3*j-1,3*s) = K(3*j-1,3*s) + k(5,24);
K(3*j,3*i-2) = K(3*j,3*i-2) + k(6,1);
K(3*j,3*i-1) = K(3*j,3*i-1) + k(6,2);
K(3*j,3*i) = K(3*j,3*i) + k(6,3);
K(3*j,3*j-2) = K(3*j,3*j-2) + k(6,4);
K(3*j,3*j-1) = K(3*j,3*j-1) + k(6,5);
K(3*j,3*j) = K(3*j,3*j) + k(6,6);
K(3*j,3*m-2) = K(3*j,3*m-2) + k(6,7);
K(3*j,3*m-1) = K(3*j,3*m-1) + k(6,8);
K(3*j,3*m) = K(3*j,3*m) + k(6,9);
K(3*j,3*n-2) = K(3*j,3*n-2) + k(6,10);
K(3*j,3*n-1) = K(3*j,3*n-1) + k(6,11);
K(3*j,3*n) = K(3*j,3*n) + k(6,12);
K(3*j,3*p-2) = K(3*j,3*p-2) + k(6,13);
K(3*j,3*p-1) = K(3*j,3*p-1) + k(6,14);
K(3*j,3*p) = K(3*j,3*p) + k(6,15);
K(3*j,3*q-2) = K(3*j,3*q-2) + k(6,16);
K(3*j,3*q-1) = K(3*j,3*q-1) + k(6,17);
K(3*j,3*q) = K(3*j,3*q) + k(6,18);
K(3*j,3*r-2) = K(3*j,3*r-2) + k(6,19);
```

```
K(3*j,3*r-1) = K(3*j,3*r-1) + k(6,20);
K(3*j,3*r) = K(3*j,3*r) + k(6,21);
K(3*j,3*s-2) = K(3*j,3*s-2) + k(6,22);
K(3*j,3*s-1) = K(3*j,3*s-1) + k(6,23);
K(3*j,3*s) = K(3*j,3*s) + k(6,24);
K(3*m-2,3*i-2) = K(3*m-2,3*i-2) + k(7,1);
K(3*m-2,3*i-1) = K(3*m-2,3*i-1) + k(7,2);
K(3*m-2,3*i) = K(3*m-2,3*i) + k(7,3);
K(3*m-2,3*j-2) = K(3*m-2,3*j-2) + k(7,4);
K(3*m-2,3*j-1) = K(3*m-2,3*j-1) + k(7,5);
K(3*m-2,3*j) = K(3*m-2,3*j) + k(7,6);
K(3*m-2,3*m-2) = K(3*m-2,3*m-2) + k(7,7);
K(3*m-2,3*m-1) = K(3*m-2,3*m-1) + k(7,8);
K(3*m-2,3*m) = K(3*m-2,3*m) + k(7,9);
K(3*m-2,3*n-2) = K(3*m-2,3*n-2) + k(7,10);
K(3*m-2,3*n-1) = K(3*m-2,3*n-1) + k(7,11);
K(3*m-2,3*n) = K(3*m-2,3*n) + k(7,12);
K(3*m-2,3*p-2) = K(3*m-2,3*p-2) + k(7,13);
K(3*m-2,3*p-1) = K(3*m-2,3*p-1) + k(7,14);
K(3*m-2,3*p) = K(3*m-2,3*p) + k(7,15);
K(3*m-2,3*q-2) = K(3*m-2,3*q-2) + k(7,16);
K(3*m-2,3*q-1) = K(3*m-2,3*q-1) + k(7,17);
K(3*m-2,3*q) = K(3*m-2,3*q) + k(7,18);
K(3*m-2,3*r-2) = K(3*m-2,3*r-2) + k(7,19);
K(3*m-2,3*r-1) = K(3*m-2,3*r-1) + k(7,20);
K(3*m-2,3*r) = K(3*m-2,3*r) + k(7,21);
K(3*m-2,3*s-2) = K(3*m-2,3*s-2) + k(7,22);
K(3*m-2,3*s-1) = K(3*m-2,3*s-1) + k(7,23);
K(3*m-2,3*s) = K(3*m-2,3*s) + k(7,24);
K(3*m-1,3*i-2) = K(3*m-1,3*i-2) + k(8,1);
K(3*m-1,3*i-1) = K(3*m-1,3*i-1) + k(8,2);
K(3*m-1,3*i) = K(3*m-1,3*i) + k(8,3);
K(3*m-1,3*j-2) = K(3*m-1,3*j-2) + k(8,4);
K(3*m-1,3*j-1) = K(3*m-1,3*j-1) + k(8,5);
K(3*m-1,3*j) = K(3*m-1,3*j) + k(8,6);
K(3*m-1,3*m-2) = K(3*m-1,3*m-2) + k(8,7);
K(3*m-1,3*m-1) = K(3*m-1,3*m-1) + k(8,8);
K(3*m-1,3*m) = K(3*m-1,3*m) + k(8,9);
K(3*m-1,3*n-2) = K(3*m-1,3*n-2) + k(8,10);
K(3*m-1,3*n-1) = K(3*m-1,3*n-1) + k(8,11);
K(3*m-1,3*n) = K(3*m-1,3*n) + k(8,12);
K(3*m-1,3*p-2) = K(3*m-1,3*p-2) + k(8,13);
K(3*m-1,3*p-1) = K(3*m-1,3*p-1) + k(8,14);
K(3*m-1,3*p) = K(3*m-1,3*p) + k(8,15);
K(3*m-1,3*q-2) = K(3*m-1,3*q-2) + k(8,16);
K(3*m-1,3*q-1) = K(3*m-1,3*q-1) + k(8,17);
K(3*m-1,3*q) = K(3*m-1,3*q) + k(8,18);
K(3*m-1,3*r-2) = K(3*m-1,3*r-2) + k(8,19);
K(3*m-1,3*r-1) = K(3*m-1,3*r-1) + k(8,20);
K(3*m-1,3*r) = K(3*m-1,3*r) + k(8,21);
K(3*m-1,3*s-2) = K(3*m-1,3*s-2) + k(8,22);
K(3*m-1,3*s-1) = K(3*m-1,3*s-1) + k(8,23);
K(3*m-1,3*s) = K(3*m-1,3*s) + k(8,24);
```

```
K(3*m,3*i-2) = K(3*m,3*i-2) + k(9,1);
K(3*m,3*i-1) = K(3*m,3*i-1) + k(9,2);
K(3*m,3*i) = K(3*m,3*i) + k(9,3);
K(3*m,3*j-2) = K(3*m,3*j-2) + k(9,4);
K(3*m,3*j-1) = K(3*m,3*j-1) + k(9,5);
K(3*m,3*j) = K(3*m,3*j) + k(9,6);
K(3*m,3*m-2) = K(3*m,3*m-2) + k(9,7);
K(3*m,3*m-1) = K(3*m,3*m-1) + k(9,8);
K(3*m,3*m) = K(3*m,3*m) + k(9,9);
K(3*m,3*n-2) = K(3*m,3*n-2) + k(9,10);
K(3*m,3*n-1) = K(3*m,3*n-1) + k(9,11);
K(3*m,3*n) = K(3*m,3*n) + k(9,12);
K(3*m,3*p-2) = K(3*m,3*p-2) + k(9,13);
K(3*m,3*p-1) = K(3*m,3*p-1) + k(9,14);
K(3*m,3*p) = K(3*m,3*p) + k(9,15);
K(3*m,3*q-2) = K(3*m,3*q-2) + k(9,16);
K(3*m,3*q-1) = K(3*m,3*q-1) + k(9,17);
K(3*m,3*q) = K(3*m,3*q) + k(9,18);
K(3*m,3*r-2) = K(3*m,3*r-2) + k(9,19);
K(3*m,3*r-1) = K(3*m,3*r-1) + k(9,20);
K(3*m,3*r) = K(3*m,3*r) + k(9,21);
K(3*m,3*s-2) = K(3*m,3*s-2) + k(9,22);
K(3*m,3*s-1) = K(3*m,3*s-1) + k(9,23);
K(3*m,3*s) = K(3*m,3*s) + k(9,24);
K(3*n-2,3*i-2) = K(3*n-2,3*i-2) + k(10,1);
K(3*n-2,3*i-1) = K(3*n-2,3*i-1) + k(10,2);
K(3*n-2,3*i) = K(3*n-2,3*i) + k(10,3);
K(3*n-2,3*j-2) = K(3*n-2,3*j-2) + k(10,4);
K(3*n-2,3*j-1) = K(3*n-2,3*j-1) + k(10,5);
K(3*n-2,3*j) = K(3*n-2,3*j) + k(10,6);
K(3*n-2,3*m-2) = K(3*n-2,3*m-2) + k(10,7);
K(3*n-2,3*m-1) = K(3*n-2,3*m-1) + k(10,8);
K(3*n-2,3*m) = K(3*n-2,3*m) + k(10,9);
K(3*n-2,3*n-2) = K(3*n-2,3*n-2) + k(10,10);
K(3*n-2,3*n-1) = K(3*n-2,3*n-1) + k(10,11);
K(3*n-2,3*n) = K(3*n-2,3*n) + k(10,12);
K(3*n-2,3*p-2) = K(3*n-2,3*p-2) + k(10,13);
K(3*n-2,3*p-1) = K(3*n-2,3*p-1) + k(10,14);
K(3*n-2,3*p) = K(3*n-2,3*p) + k(10,15);
K(3*n-2,3*q-2) = K(3*n-2,3*q-2) + k(10,16);
K(3*n-2,3*q-1) = K(3*n-2,3*q-1) + k(10,17);
K(3*n-2,3*q) = K(3*n-2,3*q) + k(10,18);
K(3*n-2,3*r-2) = K(3*n-2,3*r-2) + k(10,19);
K(3*n-2,3*r-1) = K(3*n-2,3*r-1) + k(10,20);
K(3*n-2,3*r) = K(3*n-2,3*r) + k(10,21);
K(3*n-2,3*s-2) = K(3*n-2,3*s-2) + k(10,22);
K(3*n-2,3*s-1) = K(3*n-2,3*s-1) + k(10,23);
K(3*n-2,3*s) = K(3*n-2,3*s) + k(10,24);
K(3*n-1,3*i-2) = K(3*n-1,3*i-2) + k(11,1);
K(3*n-1,3*i-1) = K(3*n-1,3*i-1) + k(11,2);
K(3*n-1,3*i) = K(3*n-1,3*i) + k(11,3);
K(3*n-1,3*j-2) = K(3*n-1,3*j-2) + k(11,4);
K(3*n-1,3*j-1) = K(3*n-1,3*j-1) + k(11,5);
K(3*n-1,3*j) = K(3*n-1,3*j) + k(11,6);
K(3*n-1,3*m-2) = K(3*n-1,3*m-2) + k(11,7);
```

```
K(3*n-1,3*m-1) = K(3*n-1,3*m-1) + k(11,8);
K(3*n-1,3*m) = K(3*n-1,3*m) + k(11,9);
K(3*n-1,3*n-2) = K(3*n-1,3*n-2) + k(11,10);
K(3*n-1,3*n-1) = K(3*n-1,3*n-1) + k(11,11);
K(3*n-1,3*n) = K(3*n-1,3*n) + k(11,12);
K(3*n-1,3*p-2) = K(3*n-1,3*p-2) + k(11,13);
K(3*n-1,3*p-1) = K(3*n-1,3*p-1) + k(11,14);
K(3*n-1,3*p) = K(3*n-1,3*p) + k(11,15);
K(3*n-1,3*q-2) = K(3*n-1,3*q-2) + k(11,16);
K(3*n-1,3*q-1) = K(3*n-1,3*q-1) + k(11,17);
K(3*n-1,3*q) = K(3*n-1,3*q) + k(11,18);
K(3*n-1,3*r-2) = K(3*n-1,3*r-2) + k(11,19);
K(3*n-1,3*r-1) = K(3*n-1,3*r-1) + k(11,20);
K(3*n-1,3*r) = K(3*n-1,3*r) + k(11,21);
K(3*n-1,3*s-2) = K(3*n-1,3*s-2) + k(11,22);
K(3*n-1,3*s-1) = K(3*n-1,3*s-1) + k(11,23);
K(3*n-1,3*s) = K(3*n-1,3*s) + k(11,24);
K(3*n,3*i-2) = K(3*n,3*i-2) + k(12,1);
K(3*n,3*i-1) = K(3*n,3*i-1) + k(12,2);
K(3*n,3*i) = K(3*n,3*i) + k(12,3);
K(3*n,3*j-2) = K(3*n,3*j-2) + k(12,4);
K(3*n,3*j-1) = K(3*n,3*j-1) + k(12,5);
K(3*n,3*j) = K(3*n,3*j) + k(12,6);
K(3*n,3*m-2) = K(3*n,3*m-2) + k(12,7);
K(3*n,3*m-1) = K(3*n,3*m-1) + k(12,8);
K(3*n,3*m) = K(3*n,3*m) + k(12,9);
K(3*n,3*n-2) = K(3*n,3*n-2) + k(12,10);
K(3*n,3*n-1) = K(3*n,3*n-1) + k(12,11);
K(3*n,3*n) = K(3*n,3*n) + k(12,12);
K(3*n,3*p-2) = K(3*n,3*p-2) + k(12,13);
K(3*n,3*p-1) = K(3*n,3*p-1) + k(12,14);
K(3*n,3*p) = K(3*n,3*p) + k(12,15);
K(3*n,3*q-2) = K(3*n,3*q-2) + k(12,16);
K(3*n,3*q-1) = K(3*n,3*q-1) + k(12,17);
K(3*n,3*q) = K(3*n,3*q) + k(12,18);
K(3*n,3*r-2) = K(3*n,3*r-2) + k(12,19);
K(3*n,3*r-1) = K(3*n,3*r-1) + k(12,20);
K(3*n,3*r) = K(3*n,3*r) + k(12,21);
K(3*n,3*s-2) = K(3*n,3*s-2) + k(12,22);
K(3*n,3*s-1) = K(3*n,3*s-1) + k(12,23);
K(3*n,3*s) = K(3*n,3*s) + k(12,24);
K(3*p-2,3*i-2) = K(3*p-2,3*i-2) + k(13,1);
K(3*p-2,3*i-1) = K(3*p-2,3*i-1) + k(13,2);
K(3*p-2,3*i) = K(3*p-2,3*i) + k(13,3);
K(3*p-2,3*j-2) = K(3*p-2,3*j-2) + k(13,4);
K(3*p-2,3*j-1) = K(3*p-2,3*j-1) + k(13,5);
K(3*p-2,3*j) = K(3*p-2,3*j) + k(13,6);
K(3*p-2,3*m-2) = K(3*p-2,3*m-2) + k(13,7);
K(3*p-2,3*m-1) = K(3*p-2,3*m-1) + k(13,8);
K(3*p-2,3*m) = K(3*p-2,3*m) + k(13,9);
K(3*p-2,3*n-2) = K(3*p-2,3*n-2) + k(13,10);
K(3*p-2,3*n-1) = K(3*p-2,3*n-1) + k(13,11);
K(3*p-2,3*n) = K(3*p-2,3*n) + k(13,12);
K(3*p-2,3*p-2) = K(3*p-2,3*p-2) + k(13,13);
```

```
K(3*p-2,3*p-1) = K(3*p-2,3*p-1) + k(13,14);
K(3*p-2,3*p) = K(3*p-2,3*p) + k(13,15);
K(3*p-2,3*q-2) = K(3*p-2,3*q-2) + k(13,16);
K(3*p-2,3*q-1) = K(3*p-2,3*q-1) + k(13,17);
K(3*p-2,3*q) = K(3*p-2,3*q) + k(13,18);
K(3*p-2,3*r-2) = K(3*p-2,3*r-2) + k(13,19);
K(3*p-2,3*r-1) = K(3*p-2,3*r-1) + k(13,20);
K(3*p-2,3*r) = K(3*p-2,3*r) + k(13,21);
K(3*p-2,3*s-2) = K(3*p-2,3*s-2) + k(13,22);
K(3*p-2,3*s-1) = K(3*p-2,3*s-1) + k(13,23);
K(3*p-2,3*s) = K(3*p-2,3*s) + k(13,24);
K(3*p-1,3*i-2) = K(3*p-1,3*i-2) + k(14,1);
K(3*p-1,3*i-1) = K(3*p-1,3*i-1) + k(14,2);
K(3*p-1,3*i) = K(3*p-1,3*i) + k(14,3);
K(3*p-1,3*j-2) = K(3*p-1,3*j-2) + k(14,4);
K(3*p-1,3*j-1) = K(3*p-1,3*j-1) + k(14,5);
K(3*p-1,3*j) = K(3*p-1,3*j) + k(14,6);
K(3*p-1,3*m-2) = K(3*p-1,3*m-2) + k(14,7);
K(3*p-1,3*m-1) = K(3*p-1,3*m-1) + k(14,8);
K(3*p-1,3*m) = K(3*p-1,3*m) + k(14,9);
K(3*p-1,3*n-2) = K(3*p-1,3*n-2) + k(14,10);
K(3*p-1,3*n-1) = K(3*p-1,3*n-1) + k(14,11);
K(3*p-1,3*n) = K(3*p-1,3*n) + k(14,12);
K(3*p-1,3*p-2) = K(3*p-1,3*p-2) + k(14,13);
K(3*p-1,3*p-1) = K(3*p-1,3*p-1) + k(14,14);
K(3*p-1,3*p) = K(3*p-1,3*p) + k(14,15);
K(3*p-1,3*q-2) = K(3*p-1,3*q-2) + k(14,16);
K(3*p-1,3*q-1) = K(3*p-1,3*q-1) + k(14,17);
K(3*p-1,3*q) = K(3*p-1,3*q) + k(14,18);
K(3*p-1,3*r-2) = K(3*p-1,3*r-2) + k(14,19);
K(3*p-1,3*r-1) = K(3*p-1,3*r-1) + k(14,20);
K(3*p-1,3*r) = K(3*p-1,3*r) + k(14,21);
K(3*p-1,3*s-2) = K(3*p-1,3*s-2) + k(14,22);
K(3*p-1,3*s-1) = K(3*p-1,3*s-1) + k(14,23);
K(3*p-1,3*s) = K(3*p-1,3*s) + k(14,24);
K(3*p,3*i-2) = K(3*p,3*i-2) + k(15,1);
K(3*p,3*i-1) = K(3*p,3*i-1) + k(15,2);
K(3*p,3*i) = K(3*p,3*i) + k(15,3);
K(3*p,3*j-2) = K(3*p,3*j-2) + k(15,4);
K(3*p,3*j-1) = K(3*p,3*j-1) + k(15,5);
K(3*p,3*j) = K(3*p,3*j) + k(15,6);
K(3*p,3*m-2) = K(3*p,3*m-2) + k(15,7);
K(3*p,3*m-1) = K(3*p,3*m-1) + k(15,8);
K(3*p,3*m) = K(3*p,3*m) + k(15,9);
K(3*p,3*n-2) = K(3*p,3*n-2) + k(15,10);
K(3*p,3*n-1) = K(3*p,3*n-1) + k(15,11);
K(3*p,3*n) = K(3*p,3*n) + k(15,12);
K(3*p,3*p-2) = K(3*p,3*p-2) + k(15,13);
K(3*p,3*p-1) = K(3*p,3*p-1) + k(15,14);
K(3*p,3*p) = K(3*p,3*p) + k(15,15);
K(3*p,3*q-2) = K(3*p,3*q-2) + k(15,16);
K(3*p,3*q-1) = K(3*p,3*q-1) + k(15,17);
K(3*p,3*q) = K(3*p,3*q) + k(15,18);
K(3*p,3*r-2) = K(3*p,3*r-2) + k(15,19);
```

```
K(3*p,3*r-1) = K(3*p,3*r-1) + k(15,20);
K(3*p,3*r) = K(3*p,3*r) + k(15,21);
K(3*p,3*s-2) = K(3*p,3*s-2) + k(15,22);
K(3*p,3*s-1) = K(3*p,3*s-1) + k(15,23);
K(3*p,3*s) = K(3*p,3*s) + k(15,24);
K(3*q-2,3*i-2) = K(3*q-2,3*i-2) + k(16,1);
K(3*q-2,3*i-1) = K(3*q-2,3*i-1) + k(16,2);
K(3*q-2,3*i) = K(3*q-2,3*i) + k(16,3);
K(3*q-2,3*j-2) = K(3*q-2,3*j-2) + k(16,4);
K(3*q-2,3*j-1) = K(3*q-2,3*j-1) + k(16,5);
K(3*q-2,3*j) = K(3*q-2,3*j) + k(16,6);
K(3*q-2,3*m-2) = K(3*q-2,3*m-2) + k(16,7);
K(3*q-2,3*m-1) = K(3*q-2,3*m-1) + k(16,8);
K(3*q-2,3*m) = K(3*q-2,3*m) + k(16,9);
K(3*q-2,3*n-2) = K(3*q-2,3*n-2) + k(16,10);
K(3*q-2,3*n-1) = K(3*q-2,3*n-1) + k(16,11);
K(3*q-2,3*n) = K(3*q-2,3*n) + k(16,12);
K(3*q-2,3*p-2) = K(3*q-2,3*p-2) + k(16,13);
K(3*q-2,3*p-1) = K(3*q-2,3*p-1) + k(16,14);
K(3*q-2,3*p) = K(3*q-2,3*p) + k(16,15);
K(3*q-2,3*q-2) = K(3*q-2,3*q-2) + k(16,16);
K(3*q-2,3*q-1) = K(3*q-2,3*q-1) + k(16,17);
K(3*q-2,3*q) = K(3*q-2,3*q) + k(16,18);
K(3*q-2,3*r-2) = K(3*q-2,3*r-2) + k(16,19);
K(3*q-2,3*r-1) = K(3*q-2,3*r-1) + k(16,20);
K(3*q-2,3*r) = K(3*q-2,3*r) + k(16,21);
K(3*q-2,3*s-2) = K(3*q-2,3*s-2) + k(16,22);
K(3*q-2,3*s-1) = K(3*q-2,3*s-1) + k(16,23);
K(3*q-2,3*s) = K(3*q-2,3*s) + k(16,24);
K(3*q-1,3*i-2) = K(3*q-1,3*i-2) + k(17,1);
K(3*q-1,3*i-1) = K(3*q-1,3*i-1) + k(17,2);
K(3*q-1,3*i) = K(3*q-1,3*i) + k(17,3);
K(3*q-1,3*j-2) = K(3*q-1,3*j-2) + k(17,4);
K(3*q-1,3*j-1) = K(3*q-1,3*j-1) + k(17,5);
K(3*q-1,3*j) = K(3*q-1,3*j) + k(17,6);
K(3*q-1,3*m-2) = K(3*q-1,3*m-2) + k(17,7);
K(3*q-1,3*m-1) = K(3*q-1,3*m-1) + k(17,8);
K(3*q-1,3*m) = K(3*q-1,3*m) + k(17,9);
K(3*q-1,3*n-2) = K(3*q-1,3*n-2) + k(17,10);
K(3*q-1,3*n-1) = K(3*q-1,3*n-1) + k(17,11);
K(3*q-1,3*n) = K(3*q-1,3*n) + k(17,12);
K(3*q-1,3*p-2) = K(3*q-1,3*p-2) + k(17,13);
K(3*q-1,3*p-1) = K(3*q-1,3*p-1) + k(17,14);
K(3*q-1,3*p) = K(3*q-1,3*p) + k(17,15);
K(3*q-1,3*q-2) = K(3*q-1,3*q-2) + k(17,16);
K(3*q-1,3*q-1) = K(3*q-1,3*q-1) + k(17,17);
K(3*q-1,3*q) = K(3*q-1,3*q) + k(17,18);
K(3*q-1,3*r-2) = K(3*q-1,3*r-2) + k(17,19);
K(3*q-1,3*r-1) = K(3*q-1,3*r-1) + k(17,20);
K(3*q-1,3*r) = K(3*q-1,3*r) + k(17,21);
K(3*q-1,3*s-2) = K(3*q-1,3*s-2) + k(17,22);
K(3*q-1,3*s-1) = K(3*q-1,3*s-1) + k(17,23);
```

```
K(3*q-1,3*s) = K(3*q-1,3*s) + k(17,24);
K(3*q,3*i-2) = K(3*q,3*i-2) + k(18,1);
K(3*q,3*i-1) = K(3*q,3*i-1) + k(18,2);
K(3*q,3*i) = K(3*q,3*i) + k(18,3);
K(3*q,3*j-2) = K(3*q,3*j-2) + k(18,4);
K(3*q,3*j-1) = K(3*q,3*j-1) + k(18,5);
K(3*q,3*j) = K(3*q,3*j) + k(18,6);
K(3*q,3*m-2) = K(3*q,3*m-2) + k(18,7);
K(3*q,3*m-1) = K(3*q,3*m-1) + k(18,8);
K(3*q,3*m) = K(3*q,3*m) + k(18,9);
K(3*q,3*n-2) = K(3*q,3*n-2) + k(18,10);
K(3*q,3*n-1) = K(3*q,3*n-1) + k(18,11);
K(3*q,3*n) = K(3*q,3*n) + k(18,12);
K(3*q,3*p-2) = K(3*q,3*p-2) + k(18,13);
K(3*q,3*p-1) = K(3*q,3*p-1) + k(18,14);
K(3*q,3*p) = K(3*q,3*p) + k(18,15);
K(3*q,3*q-2) = K(3*q,3*q-2) + k(18,16);
K(3*q,3*q-1) = K(3*q,3*q-1) + k(18,17);
K(3*q,3*q) = K(3*q,3*q) + k(18,18);
K(3*q,3*r-2) = K(3*q,3*r-2) + k(18,19);
K(3*q,3*r-1) = K(3*q,3*r-1) + k(18,20);
K(3*q,3*r) = K(3*q,3*r) + k(18,21);
K(3*q,3*s-2) = K(3*q,3*s-2) + k(18,22);
K(3*q,3*s-1) = K(3*q,3*s-1) + k(18,23);
K(3*q,3*s) = K(3*q,3*s) + k(18,24);
K(3*r-2,3*i-2) = K(3*r-2,3*i-2) + k(19,1);
K(3*r-2,3*i-1) = K(3*r-2,3*i-1) + k(19,2);
K(3*r-2,3*i) = K(3*r-2,3*i) + k(19,3);
K(3*r-2,3*j-2) = K(3*r-2,3*j-2) + k(19,4);
K(3*r-2,3*j-1) = K(3*r-2,3*j-1) + k(19,5);
K(3*r-2,3*j) = K(3*r-2,3*j) + k(19,6);
K(3*r-2,3*m-2) = K(3*r-2,3*m-2) + k(19,7);
K(3*r-2,3*m-1) = K(3*r-2,3*m-1) + k(19,8);
K(3*r-2,3*m) = K(3*r-2,3*m) + k(19,9);
K(3*r-2,3*n-2) = K(3*r-2,3*n-2) + k(19,10);
K(3*r-2,3*n-1) = K(3*r-2,3*n-1) + k(19,11);
K(3*r-2,3*n) = K(3*r-2,3*n) + k(19,12);
K(3*r-2,3*p-2) = K(3*r-2,3*p-2) + k(19,13);
K(3*r-2,3*p-1) = K(3*r-2,3*p-1) + k(19,14);
K(3*r-2,3*p) = K(3*r-2,3*p) + k(19,15);
K(3*r-2,3*q-2) = K(3*r-2,3*q-2) + k(19,16);
K(3*r-2,3*q-1) = K(3*r-2,3*q-1) + k(19,17);
K(3*r-2,3*q) = K(3*r-2,3*q) + k(19,18);
K(3*r-2,3*r-2) = K(3*r-2,3*r-2) + k(19,19);
K(3*r-2,3*r-1) = K(3*r-2,3*r-1) + k(19,20);
K(3*r-2,3*r) = K(3*r-2,3*r) + k(19,21);
K(3*r-2,3*s-2) = K(3*r-2,3*s-2) + k(19,22);
K(3*r-2,3*s-1) = K(3*r-2,3*s-1) + k(19,23);
K(3*r-2,3*s) = K(3*r-2,3*s) + k(19,24);
K(3*r-1,3*i-2) = K(3*r-1,3*i-2) + k(20,1);
K(3*r-1,3*i-1) = K(3*r-1,3*i-1) + k(20,2);
K(3*r-1,3*i) = K(3*r-1,3*i) + k(20,3);
K(3*r-1,3*j-2) = K(3*r-1,3*j-2) + k(20,4);
K(3*r-1,3*j-1) = K(3*r-1,3*j-1) + k(20,5);
```

```
K(3*r-1,3*j) = K(3*r-1,3*j) + k(20,6);
K(3*r-1,3*m-2) = K(3*r-1,3*m-2) + k(20,7);
K(3*r-1,3*m-1) = K(3*r-1,3*m-1) + k(20,8);
K(3*r-1,3*m) = K(3*r-1,3*m) + k(20,9);
K(3*r-1,3*n-2) = K(3*r-1,3*n-2) + k(20,10);
K(3*r-1,3*n-1) = K(3*r-1,3*n-1) + k(20,11);
K(3*r-1,3*n) = K(3*r-1,3*n) + k(20,12);
K(3*r-1,3*p-2) = K(3*r-1,3*p-2) + k(20,13);
K(3*r-1,3*p-1) = K(3*r-1,3*p-1) + k(20,14);
K(3*r-1,3*p) = K(3*r-1,3*p) + k(20,15);
K(3*r-1,3*q-2) = K(3*r-1,3*q-2) + k(20,16);
K(3*r-1,3*q-1) = K(3*r-1,3*q-1) + k(20,17);
K(3*r-1,3*q) = K(3*r-1,3*q) + k(20,18);
K(3*r-1,3*r-2) = K(3*r-1,3*r-2) + k(20,19);
K(3*r-1,3*r-1) = K(3*r-1,3*r-1) + k(20,20);
K(3*r-1,3*r) = K(3*r-1,3*r) + k(20,21);
K(3*r-1,3*s-2) = K(3*r-1,3*s-2) + k(20,22);
K(3*r-1,3*s-1) = K(3*r-1,3*s-1) + k(20,23);
K(3*r-1,3*s) = K(3*r-1,3*s) + k(20,24);
K(3*r,3*i-2) = K(3*r,3*i-2) + k(21,1);
K(3*r,3*i-1) = K(3*r,3*i-1) + k(21,2);
K(3*r,3*i) = K(3*r,3*i) + k(21,3);
K(3*r,3*j-2) = K(3*r,3*j-2) + k(21,4);
K(3*r,3*j-1) = K(3*r,3*j-1) + k(21,5);
K(3*r,3*j) = K(3*r,3*j) + k(21,6);
K(3*r,3*m-2) = K(3*r,3*m-2) + k(21,7);
K(3*r,3*m-1) = K(3*r,3*m-1) + k(21,8);
K(3*r,3*m) = K(3*r,3*m) + k(21,9);
K(3*r,3*n-2) = K(3*r,3*n-2) + k(21,10);
K(3*r,3*n-1) = K(3*r,3*n-1) + k(21,11);
K(3*r,3*n) = K(3*r,3*n) + k(21,12);
K(3*r,3*p-2) = K(3*r,3*p-2) + k(21,13);
K(3*r,3*p-1) = K(3*r,3*p-1) + k(21,14);
K(3*r,3*p) = K(3*r,3*p) + k(21,15);
K(3*r,3*q-2) = K(3*r,3*q-2) + k(21,16);
K(3*r,3*q-1) = K(3*r,3*q-1) + k(21,17);
K(3*r,3*q) = K(3*r,3*q) + k(21,18);
K(3*r,3*r-2) = K(3*r,3*r-2) + k(21,19);
K(3*r,3*r-1) = K(3*r,3*r-1) + k(21,20);
K(3*r,3*r) = K(3*r,3*r) + k(21,21);
K(3*r,3*s-2) = K(3*r,3*s-2) + k(21,22);
K(3*r,3*s-1) = K(3*r,3*s-1) + k(21,23);
K(3*r,3*s) = K(3*r,3*s) + k(21,24);
K(3*s-2,3*i-2) = K(3*s-2,3*i-2) + k(22,1);
K(3*s-2,3*i-1) = K(3*s-2,3*i-1) + k(22,2);
K(3*s-2,3*i) = K(3*s-2,3*i) + k(22,3);
K(3*s-2,3*j-2) = K(3*s-2,3*j-2) + k(22,4);
K(3*s-2,3*j-1) = K(3*s-2,3*j-1) + k(22,5);
K(3*s-2,3*j) = K(3*s-2,3*j) + k(22,6);
K(3*s-2,3*m-2) = K(3*s-2,3*m-2) + k(22,7);
K(3*s-2,3*m-1) = K(3*s-2,3*m-1) + k(22,8);
K(3*s-2,3*m) = K(3*s-2,3*m) + k(22,9);
K(3*s-2,3*n-2) = K(3*s-2,3*n-2) + k(22,10);
K(3*s-2,3*n-1) = K(3*s-2,3*n-1) + k(22,11);
```

```
K(3*s-2,3*n) = K(3*s-2,3*n) + k(22,12);
K(3*s-2,3*p-2) = K(3*s-2,3*p-2) + k(22,13);
K(3*s-2,3*p-1) = K(3*s-2,3*p-1) + k(22,14);
K(3*s-2,3*p) = K(3*s-2,3*p) + k(22,15);
K(3*s-2,3*q-2) = K(3*s-2,3*q-2) + k(22,16);
K(3*s-2,3*q-1) = K(3*s-2,3*q-1) + k(22,17);
K(3*s-2,3*q) = K(3*s-2,3*q) + k(22,18);
K(3*s-2,3*r-2) = K(3*s-2,3*r-2) + k(22,19);
K(3*s-2,3*r-1) = K(3*s-2,3*r-1) + k(22,20);
K(3*s-2,3*r) = K(3*s-2,3*r) + k(22,21);
K(3*s-2,3*s-2) = K(3*s-2,3*s-2) + k(22,22);
K(3*s-2,3*s-1) = K(3*s-2,3*s-1) + k(22,23);
K(3*s-2,3*s) = K(3*s-2,3*s) + k(22,24);
K(3*s-1,3*i-2) = K(3*s-1,3*i-2) + k(23,1);
K(3*s-1,3*i-1) = K(3*s-1,3*i-1) + k(23,2);
K(3*s-1,3*i) = K(3*s-1,3*i) + k(23,3);
K(3*s-1,3*j-2) = K(3*s-1,3*j-2) + k(23,4);
K(3*s-1,3*j-1) = K(3*s-1,3*j-1) + k(23,5);
K(3*s-1,3*j) = K(3*s-1,3*j) + k(23,6);
K(3*s-1,3*m-2) = K(3*s-1,3*m-2) + k(23,7);
K(3*s-1,3*m-1) = K(3*s-1,3*m-1) + k(23,8);
K(3*s-1,3*m) = K(3*s-1,3*m) + k(23,9);
K(3*s-1,3*n-2) = K(3*s-1,3*n-2) + k(23,10);
K(3*s-1,3*n-1) = K(3*s-1,3*n-1) + k(23,11);
K(3*s-1,3*n) = K(3*s-1,3*n) + k(23,12);
K(3*s-1,3*p-2) = K(3*s-1,3*p-2) + k(23,13);
K(3*s-1,3*p-1) = K(3*s-1,3*p-1) + k(23,14);
K(3*s-1,3*p) = K(3*s-1,3*p) + k(23,15);
K(3*s-1,3*q-2) = K(3*s-1,3*q-2) + k(23,16);
K(3*s-1,3*q-1) = K(3*s-1,3*q-1) + k(23,17);
K(3*s-1,3*q) = K(3*s-1,3*q) + k(23,18);
K(3*s-1,3*r-2) = K(3*s-1,3*r-2) + k(23,19);
K(3*s-1,3*r-1) = K(3*s-1,3*r-1) + k(23,20);
K(3*s-1,3*r) = K(3*s-1,3*r) + k(23,21);
K(3*s-1,3*s-2) = K(3*s-1,3*s-2) + k(23,22);
K(3*s-1,3*s-1) = K(3*s-1,3*s-1) + k(23,23);
K(3*s-1,3*s) = K(3*s-1,3*s) + k(23,24);
K(3*s,3*i-2) = K(3*s,3*i-2) + k(24,1);
K(3*s,3*i-1) = K(3*s,3*i-1) + k(24,2);
K(3*s,3*i) = K(3*s,3*i) + k(24,3);
K(3*s,3*j-2) = K(3*s,3*j-2) + k(24,4);
K(3*s,3*j-1) = K(3*s,3*j-1) + k(24,5);
K(3*s,3*j) = K(3*s,3*j) + k(24,6);
K(3*s,3*m-2) = K(3*s,3*m-2) + k(24,7);
K(3*s,3*m-1) = K(3*s,3*m-1) + k(24,8);
K(3*s,3*m) = K(3*s,3*m) + k(24,9);
K(3*s,3*n-2) = K(3*s,3*n-2) + k(24,10);
K(3*s,3*n-1) = K(3*s,3*n-1) + k(24,11);
K(3*s,3*n) = K(3*s,3*n) + k(24,12);
K(3*s,3*p-2) = K(3*s,3*p-2) + k(24,13);
K(3*s,3*p-1) = K(3*s,3*p-1) + k(24,14);
K(3*s,3*p) = K(3*s,3*p) + k(24,15);
K(3*s,3*q-2) = K(3*s,3*q-2) + k(24,16);
```

```
K(3*s,3*q-1) = K(3*s,3*q-1) + k(24,17);
K(3*s,3*q) = K(3*s,3*q) + k(24,18);
K(3*s,3*r-2) = K(3*s,3*r-2) + k(24,19);
K(3*s,3*r-1) = K(3*s,3*r-1) + k(24,20);
K(3*s,3*r) = K(3*s,3*r) + k(24,21);
K(3*s,3*s-2) = K(3*s,3*s-2) + k(24,22);
K(3*s,3*s-1) = K(3*s,3*s-1) + k(24,23);
K(3*s,3*s) = K(3*s,3*s) + k(24,24);
y = K;
```

```
function w = LinearBrickElementStresses
              (E,NU,x1,y1,z1,x2,y2,z2,x3,y3,z3,x4,y4,z4,x5,y5,z5,x6,y6,
              z6,x7,y7,z7,x8,y8,z8,u)
%LinearBrickElementStresses This function returns the element
%                           stress vector for a linear brick
%                           (solid) element with modulus
%                           of elasticity E, Poisson's ratio
%                           NU, coordinates of
%                           node 1 (x1,y1,z1), coordinates
%                           of node 2 (x2,y2,z2), coordinates of
%                           node 3 (x3,y3,z3), coordinates of
%                           node 4 (x4,y4,z4), coordinates of
%                           node 5 (x5,y5,z5), coordinates
%                           of node 6 (x6,y6,z6), coordinates of
%                           node 7 (x7,y7,z7), and coordinates of
%                           node 8 (x8,y8,z8).
syms s t u;
N1 = (1-s)*(1-t)*(1+u)/8;
N2 = (1-s)*(1-t)*(1-u)/8;
N3 = (1-s)*(1+t)*(1-u)/8;
N4 = (1-s)*(1+t)*(1+u)/8;
N5 = (1+s)*(1-t)*(1+u)/8;
N6 = (1+s)*(1-t)*(1-u)/8;
N7 = (1+s)*(1+t)*(1-u)/8;
N8 = (1+s)*(1+t)*(1+u)/8;
x = N1*x1 + N2*x2 + N3*x3 + N4*x4 + N5*x5 + N6*x6 + N7*x7 + N8*x8;
y = N1*y1 + N2*y2 + N3*y3 + N4*y4 + N5*y5 + N6*y6 + N7*y7 + N8*y8;
z = N1*z1 + N2*z2 + N3*z3 + N4*z4 + N5*z5 + N6*z6 + N7*z7 + N8*z8;
xs = diff(x,s);
xt = diff(x,t);
xu = diff(x,u);
ys = diff(y,s);
yt = diff(y,t);
yu = diff(y,u);
zs = diff(z,s);
zt = diff(z,t);
zu = diff(z,u);
J = xs*(yt*zu - zt*yu) - ys*(xt*zu - zt*xu) + zs*(xt*yu - yt*xu);
N1s = diff(N1,s);
N2s = diff(N2,s);
N3s = diff(N3,s);
N4s = diff(N4,s);
N5s = diff(N5,s);
N6s = diff(N6,s);
```

```
N7s = diff(N7,s);
N8s = diff(N8,s);
N1t = diff(N1,t);
N2t = diff(N2,t);
N3t = diff(N3,t);
N4t = diff(N4,t);
N5t = diff(N5,t);
N6t = diff(N6,t);
N7t = diff(N7,t);
N8t = diff(N8,t);
N1u = diff(N1,u);
N2u = diff(N2,u);
N3u = diff(N3,u);
N4u = diff(N4,u);
N5u = diff(N5,u);
N6u = diff(N6,u);
N7u = diff(N7,u);
N8u = diff(N8,u);
% The expressions below are not divided by J - they are adjusted for
later
N1x = N1s*(yt*zu - zt*yu) - ys*(N1t*zu - zt*N1u) + zs*(N1t*yu - yt*N1u);
N2x = N2s*(yt*zu - zt*yu) - ys*(N2t*zu - zt*N2u) + zs*(N2t*yu - yt*N2u);
N3x = N3s*(yt*zu - zt*yu) - ys*(N3t*zu - zt*N3u) + zs*(N3t*yu - yt*N3u);
N4x = N4s*(yt*zu - zt*yu) - ys*(N4t*zu - zt*N4u) + zs*(N4t*yu - yt*N4u);
N5x = N5s*(yt*zu - zt*yu) - ys*(N5t*zu - zt*N5u) + zs*(N5t*yu - yt*N5u);
N6x = N6s*(yt*zu - zt*yu) - ys*(N6t*zu - zt*N6u) + zs*(N6t*yu - yt*N6u);
N7x = N7s*(yt*zu - zt*yu) - ys*(N7t*zu - zt*N7u) + zs*(N7t*yu - yt*N7u);
N8x = N8s*(yt*zu - zt*yu) - ys*(N8t*zu - zt*N8u) + zs*(N8t*yu - yt*N8u);
N1y = xs*(N1t*zu - zt*N1u) - N1s*(xt*zu - zt*xu) + zs*(xt*N1u - N1t*xu);
N2y = xs*(N2t*zu - zt*N2u) - N2s*(xt*zu - zt*xu) + zs*(xt*N2u - N2t*xu);
N3y = xs*(N3t*zu - zt*N3u) - N3s*(xt*zu - zt*xu) + zs*(xt*N3u - N3t*xu);
N4y = xs*(N4t*zu - zt*N4u) - N4s*(xt*zu - zt*xu) + zs*(xt*N4u - N4t*xu);
N5y = xs*(N5t*zu - zt*N5u) - N5s*(xt*zu - zt*xu) + zs*(xt*N5u - N5t*xu);
N6y = xs*(N6t*zu - zt*N6u) - N6s*(xt*zu - zt*xu) + zs*(xt*N6u - N6t*xu);
N7y = xs*(N7t*zu - zt*N7u) - N7s*(xt*zu - zt*xu) + zs*(xt*N7u - N7t*xu);
N8y = xs*(N8t*zu - zt*N8u) - N8s*(xt*zu - zt*xu) + zs*(xt*N8u - N8t*xu);
N1z = xs*(yt*N1u - N1t*yu) - ys*(xt*N1u - N1t*xu) + N1s*(xt*yu - yt*xu);
N2z = xs*(yt*N2u - N2t*yu) - ys*(xt*N2u - N2t*xu) + N2s*(xt*yu - yt*xu);
N3z = xs*(yt*N3u - N3t*yu) - ys*(xt*N3u - N3t*xu) + N3s*(xt*yu - yt*xu);
N4z = xs*(yt*N4u - N4t*yu) - ys*(xt*N4u - N4t*xu) + N4s*(xt*yu - yt*xu);
N5z = xs*(yt*N5u - N5t*yu) - ys*(xt*N5u - N5t*xu) + N5s*(xt*yu - yt*xu);
N6z = xs*(yt*N6u - N6t*yu) - ys*(xt*N6u - N6t*xu) + N6s*(xt*yu - yt*xu);
N7z = xs*(yt*N7u - N7t*yu) - ys*(xt*N7u - N7t*xu) + N7s*(xt*yu - yt*xu);
N8z = xs*(yt*N8u - N8t*yu) - ys*(xt*N8u - N8t*xu) + N8s*(xt*yu - yt*xu);
% Next, the B matrix is calculated explicitly as follows:
B = [N1x N2x N3x N4x N5x N6x N7x N8x 0 0 0 0 0 0 0 0 0 0 0 0 0 0 0 0 ;
     0 0 0 0 0 0 0 0 N1y N2y N3y N4y N5y N6y N7y N8y 0 0 0 0 0 0 0 0 ;
     0 0 0 0 0 0 0 0 0 0 0 0 0 0 0 0 N1z N2z N3z N4z N5z N6z N7z N8z ;
     N1y N2y N3y N4y N5y N6y N7y N8y N1x N2x N3x N4x N5x N6x N7x N8x 0 0
     0 0 0 0 0 0 ;
     0 0 0 0 0 0 0 0 N1z N2z N3z N4z N5z N6z N7z N8z N1y N2y N3y N4y N5y
     N6y N7y N8y ;
     N1z N2z N3z N4z N5z N6z N7z N8z 0 0 0 0 0 0 0 0 N1x N2x N3x N4x N5x
     N6x N7x N8x];
```

```
Bnew = simplify(B);
Jnew = simplify(J);
D = (E/((1+NU)*(1-2*NU)))*[1-NU NU NU 0 0 0 ; NU 1-NU NU 0 0 0 ;
    1- NU 0 0 0 ;
    0 0 0 (1-2*NU)/2 0 0 ; 0 0 0 0 (1-2*NU)/2 0 ; 0 0 0 0 0
    (1-2*NU)/2];
Bfinal = Bnew/Jnew;
w = D*Bfinal*u
%
% We also calculate the stresses at the centroid of the element
%
wcent = subs(w, {s,t,u}, {0,0,0});
w = double(wcent);
```

```
function y = LinearBrickElementPStresses(sigma)
%LinearBrickElementPStresses    This function returns the three
%                               principal stresses for the element
%                               given the element stress vector.
%                               The principal angles are not returned.
s1 = sigma(1) + sigma(2) + sigma(3);
s2 = sigma(1)*sigma(2) + sigma(1)*sigma(3) + sigma(2)*sigma(3)
-sigma(4)*sigma(4) -sigma(5)*sigma(5) -sigma(6)*sigma(6);
ms3 = [sigma(1) sigma(4) sigma(6) ; sigma(4) sigma(2) sigma(5) ;
sigma(6) sigma(5) sigma(3)];
s3 = det(ms3);
y = [s1 ; s2 ; s3];
```

Example 16.1:

Consider the thin plate subjected to a uniformly distributed load as shown in Fig. 16.3. Use one linear brick element to solve this problem as shown in Fig. 16.4. Given $E = 210\,\text{GPa}$, $\nu = 0.3$, $t = 0.025\,\text{m}$, and $w = 3000\,\text{kN/m}^2$, determine:

1. the global stiffness matrix for the structure.
2. the displacements at nodes 5, 6, 7, and 8.

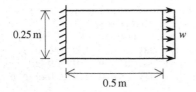

Fig. 16.3. Thin Plate for Example 16.1

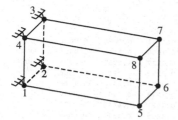

Fig. 16.4. Discretization of Thin Plate into One Linear Brick

Solution:

Use the six steps outlined in Chap. 1 to solve this problem using the linear brick element.

Step 1 – Discretizing the Domain:

We subdivide the plate into one linear brick element only for illustration purposes. More elements must be used in order to obtain reliable results. Thus the domain is subdivided into one element and eight nodes as shown in Fig. 16.4. The total force due to the distributed load is divided equally between nodes 5, 6, 7, and 8. The units used in the MATLAB calculations are kN and meter. Table 16.1 shows the element connectivity for this example.

Table 16.1. Element Connectivity for Example 16.1

Element Number	Node i	Node j	Node m	Node n	Node p	Node q	Node r	Node s
1	1	2	3	4	5	6	7	8

Step 2 – Writing the Element Stiffness Matrices:

The element stiffness matrix k_1 is obtained by making a call to the MATLAB function *LinearBrickElementStiffness*. This matrix has size 24 × 24.

```
» k1 = LinearBrickElementStiffness
        (210e6,0.3,0,0,0.025,0,0,0,0,0.25,0,0,0.25,0.0,25,
        0.5,0,0.025,0.5,0,0,0.5,0.25,0,0.5,0.25,0.025)

k1 =

    1.0e+008  *
```

Columns 1 through 10

```
 0.4571  -0.4445  -0.2256   0.2218   0.2227  -0.2252  -0.1143   0.1080   0.0042   0.0021
-0.4445   0.4571   0.2218  -0.2256  -0.2252   0.2227   0.1080  -0.1143   0.0021   0.0042
-0.2256   0.2218   0.4571  -0.4445  -0.1143   0.1080   0.2227  -0.2252   0.0004   0.0008
 0.2218  -0.2256  -0.4445   0.4571   0.1080  -0.1143  -0.2252   0.2227   0.0008   0.0004
 0.2227  -0.2252  -0.1143   0.1080   0.4571  -0.4445  -0.2256   0.2218  -0.0008  -0.0004
-0.2252   0.2227   0.1080  -0.1143  -0.4445   0.4571   0.2218  -0.2256  -0.0004  -0.0008
-0.1143   0.1080   0.2227  -0.2252  -0.2256   0.2218   0.4571  -0.4445  -0.0021  -0.0042
 0.1080  -0.1143  -0.2252   0.2227   0.2218  -0.2256  -0.4445   0.4571  -0.0042  -0.0021
 0.0042   0.0021   0.0004   0.0008  -0.0008  -0.0004  -0.0021  -0.0042   0.4655  -0.4403
 0.0021   0.0042   0.0008   0.0004  -0.0004  -0.0008  -0.0042  -0.0021  -0.4403   0.4655
-0.0004  -0.0008  -0.0042  -0.0021   0.0021   0.0042   0.0008   0.0004  -0.2319   0.2092
-0.0008  -0.0004  -0.0021  -0.0042   0.0042   0.0021   0.0004   0.0008   0.2092  -0.2319
 0.0008   0.0004   0.0021   0.0042  -0.0042  -0.0021  -0.0004  -0.0008   0.2311  -0.2210
 0.0004   0.0008   0.0042   0.0021  -0.0021  -0.0042  -0.0008  -0.0004  -0.2210   0.2311
-0.0021  -0.0042  -0.0008  -0.0004   0.0004   0.0008   0.0042   0.0021  -0.1164   0.1038
-0.0042  -0.0021  -0.0004  -0.0008   0.0008   0.0004   0.0021   0.0042   0.1038  -0.1164
-0.0421  -0.0084  -0.0042  -0.0210   0.0084   0.0421   0.0210   0.0042  -0.0841  -0.0168
 0.0084   0.0421   0.0210   0.0042  -0.0421  -0.0084  -0.0042  -0.0210   0.0168   0.0841
 0.0042   0.0210   0.0421   0.0084  -0.0210  -0.0042  -0.0084  -0.0421   0.0841   0.0168
-0.0210  -0.0042  -0.0084  -0.0421   0.0042   0.0210   0.0421   0.0084  -0.0168  -0.0841
-0.0084  -0.0421  -0.0210  -0.0042   0.0421   0.0084   0.0042   0.0210  -0.0421  -0.0084
 0.0421   0.0084   0.0042   0.0210  -0.0084  -0.0421  -0.0210  -0.0042   0.0084   0.0421
 0.0210   0.0042   0.0084   0.0421  -0.0042  -0.0210  -0.0421  -0.0084   0.0421   0.0084
-0.0042  -0.0210  -0.0421  -0.0084   0.0210   0.0042   0.0084   0.0421  -0.0084  -0.0421
```

Columns 11 through 20

```
-0.0004  -0.0008   0.0008   0.0004  -0.0021  -0.0042  -0.0421   0.0084   0.0042  -0.0210
-0.0008  -0.0004   0.0004   0.0008  -0.0042  -0.0021  -0.0084   0.0421   0.0210  -0.0042
-0.0042  -0.0021   0.0021   0.0042  -0.0008  -0.0004  -0.0042   0.0210   0.0421  -0.0084
-0.0021  -0.0042   0.0042   0.0021  -0.0004  -0.0008  -0.0210   0.0042   0.0084  -0.0421
 0.0021   0.0042  -0.0042  -0.0021   0.0004   0.0008   0.0084  -0.0421  -0.0210   0.0042
 0.0042   0.0021  -0.0021  -0.0042   0.0008   0.0004   0.0421  -0.0084  -0.0042   0.0210
 0.0008   0.0004  -0.0004  -0.0008   0.0042   0.0021   0.0210  -0.0042  -0.0084   0.0421
 0.0004   0.0008  -0.0008  -0.0004   0.0021   0.0042   0.0042  -0.0210  -0.0421   0.0084
-0.2319   0.2092   0.2311  -0.2210  -0.1164   0.1038  -0.0841   0.0168   0.0841  -0.0168
 0.2092  -0.2319  -0.2210   0.2311   0.1038  -0.1164  -0.0168   0.0841   0.0168  -0.0841
 0.4655  -0.4403  -0.1164   0.1038   0.2311  -0.2210   0.0841  -0.0168  -0.0841   0.0168
-0.4403   0.4655   0.1038  -0.1164  -0.2210   0.2311   0.0168  -0.0841  -0.0168   0.0841
-0.1164   0.1038   0.4655  -0.4403  -0.2319   0.2092  -0.0421   0.0084   0.0421  -0.0084
 0.1038  -0.1164  -0.4403   0.4655   0.2092  -0.2319  -0.0084   0.0421   0.0084  -0.0421
 0.2311  -0.2210  -0.2319   0.2092   0.4655  -0.4403   0.0421  -0.0084  -0.0421   0.0084
-0.2210   0.2311   0.2092  -0.2319  -0.4403   0.4655   0.0084  -0.0421  -0.0084   0.0421
 0.0841   0.0168  -0.0421  -0.0084   0.0421   0.0084   1.5761  -1.5677  -0.7872   0.7813
-0.0168  -0.0841   0.0084   0.0421  -0.0084  -0.0421  -1.5677   1.5761   0.7813  -0.7872
-0.0841  -0.0168   0.0421   0.0084  -0.0421  -0.0084  -0.7872   0.7813   1.5761  -1.5677
 0.0168   0.0841  -0.0084  -0.0421   0.0084   0.0421   0.7813  -0.7872  -1.5677   1.5761
 0.0421   0.0084  -0.0841  -0.0168   0.0841   0.0168   0.7864  -0.7847  -0.3940   0.3898
-0.0084  -0.0421   0.0168   0.0841  -0.0168  -0.0841  -0.7847   0.7864   0.3898  -0.3940
-0.0421  -0.0084   0.0841   0.0168  -0.0841  -0.0168  -0.3940   0.3898   0.7864  -0.7847
 0.0084   0.0421  -0.0168  -0.0841   0.0168   0.0841   0.3898  -0.3940  -0.7847   0.7864
```

Columns 21 through 24

```
-0.0084   0.0421   0.0210  -0.0042
-0.0421   0.0084   0.0042  -0.0210
-0.0210   0.0042   0.0084  -0.0421
```

```
-0.0042     0.0210     0.0421    -0.0084
 0.0421    -0.0084    -0.0042     0.0210
 0.0084    -0.0421    -0.0210     0.0042
 0.0042    -0.0210    -0.0421     0.0084
 0.0210    -0.0042    -0.0084     0.0421
-0.0421     0.0084     0.0421    -0.0084
-0.0084     0.0421     0.0084    -0.0421
 0.0421    -0.0084    -0.0421     0.0084
 0.0084    -0.0421    -0.0084     0.0421
-0.0841     0.0168     0.0841    -0.0168
-0.0168     0.0841     0.0168    -0.0841
 0.0841    -0.0168    -0.0841     0.0168
 0.0168    -0.0841    -0.0168     0.0841
 0.7864    -0.7847    -0.3940     0.3898
-0.7847     0.7864     0.3898    -0.3940
-0.3940     0.3898     0.7864    -0.7847
 0.3898    -0.3940    -0.7847     0.7864
 1.5761    -1.5677    -0.7872     0.7813
-1.5677     1.5761     0.7813    -0.7872
-0.7872     0.7813     1.5761    -1.5677
 0.7813    -0.7872    -1.5677     1.5761
```

Step 3 – Assembling the Global Stiffness Matrix:

Since the structure has eight nodes, the size of the global stiffness matrix is 24×24. Therefore to obtain K we first set up a zero matrix of size 24×24 then make one call to the MATLAB function *LinearBrickAssemble* since we have only one element in the structure. In this case, it is clear that the element stiffness matrix is the same as the global stiffness matrix. The following are the MATLAB commands.

```
» K = zeros(24,24);
```

```
» K = LinearBrickAssemble(K,k1,1,2,3,4,5,6,7,8)
```

```
K =

   1.0e+008 *

Columns 1 through 10

  0.4571  -0.4445  -0.2256   0.2218   0.2227  -0.2252  -0.1143   0.1080   0.0042   0.0021
 -0.4445   0.4571   0.2218  -0.2256  -0.2252   0.2227   0.1080  -0.1143   0.0021   0.0042
 -0.2256   0.2218   0.4571  -0.4445  -0.1143   0.1080   0.2227  -0.2252   0.0004   0.0008
  0.2218  -0.2256  -0.4445   0.4571   0.1080  -0.1143  -0.2252   0.2227   0.0008   0.0004
```

```
 0.2227  -0.2252  -0.1143   0.1080   0.4571  -0.4445  -0.2256   0.2218  -0.0008  -0.0004
-0.2252   0.2227   0.1080  -0.1143  -0.4445   0.4571   0.2218  -0.2256  -0.0004  -0.0008
-0.1143   0.1080   0.2227  -0.2252  -0.2256   0.2218   0.4571  -0.4445  -0.0021  -0.0042
 0.1080  -0.1143  -0.2252   0.2227   0.2218  -0.2256  -0.4445   0.4571  -0.0042  -0.0021
 0.0042   0.0021   0.0004   0.0008  -0.0008  -0.0004  -0.0021  -0.0042   0.4655  -0.4403
 0.0021   0.0042   0.0008   0.0004  -0.0004  -0.0008  -0.0042  -0.0021  -0.4403   0.4655
-0.0004  -0.0008  -0.0042  -0.0021   0.0021   0.0042   0.0008   0.0004  -0.2319   0.2092
-0.0008  -0.0004  -0.0021  -0.0042   0.0042   0.0021   0.0004   0.0008   0.2092  -0.2319
 0.0008   0.0004   0.0021   0.0042  -0.0042  -0.0021  -0.0004  -0.0008   0.2311  -0.2210
 0.0004   0.0008   0.0042   0.0021  -0.0021  -0.0042  -0.0008  -0.0004  -0.2210   0.2311
-0.0021  -0.0042  -0.0008  -0.0004   0.0004   0.0008   0.0042   0.0021  -0.1164   0.1038
-0.0042  -0.0021  -0.0004  -0.0008   0.0008   0.0004   0.0021   0.0042   0.1038  -0.1164
-0.0421  -0.0084  -0.0042  -0.0210   0.0084   0.0421   0.0210   0.0042  -0.0841  -0.0168
 0.0084   0.0421   0.0210   0.0042  -0.0421  -0.0084  -0.0042  -0.0210   0.0168   0.0841
 0.0042   0.0210   0.0421   0.0084  -0.0210  -0.0042  -0.0084  -0.0421   0.0841   0.0168
-0.0210  -0.0042  -0.0084  -0.0421   0.0042   0.0210   0.0421   0.0084  -0.0168  -0.0841
-0.0084  -0.0421  -0.0210  -0.0042   0.0421   0.0084   0.0042   0.0210  -0.0421  -0.0084
 0.0421   0.0084   0.0042   0.0210  -0.0084  -0.0421  -0.0210  -0.0042   0.0084   0.0421
 0.0210   0.0042   0.0084   0.0421  -0.0042  -0.0210  -0.0421  -0.0084   0.0421   0.0084
-0.0042  -0.0210  -0.0421  -0.0084   0.0210   0.0042   0.0084   0.0421  -0.0084  -0.0421
```

Columns 11 through 20

```
-0.0004  -0.0008   0.0008   0.0004  -0.0021  -0.0042  -0.0421   0.0084   0.0042  -0.0210
-0.0008  -0.0004   0.0004   0.0008  -0.0042  -0.0021  -0.0084   0.0421   0.0210  -0.0042
-0.0042  -0.0021   0.0021   0.0042  -0.0008  -0.0004  -0.0042   0.0210   0.0421  -0.0084
-0.0021  -0.0042   0.0042   0.0021  -0.0004  -0.0008  -0.0210   0.0042   0.0084  -0.0421
 0.0021   0.0042  -0.0042  -0.0021   0.0004   0.0008   0.0084  -0.0421  -0.0210   0.0042
 0.0042   0.0021  -0.0021  -0.0042   0.0008   0.0004   0.0421  -0.0084  -0.0042   0.0210
 0.0008   0.0004  -0.0004  -0.0008   0.0042   0.0021   0.0210  -0.0042  -0.0084   0.0421
 0.0004   0.0008  -0.0008  -0.0004   0.0021   0.0042   0.0042  -0.0210  -0.0421   0.0084
-0.2319   0.2092   0.2311  -0.2210  -0.1164   0.1038  -0.0841   0.0168   0.0841  -0.0168
 0.2092  -0.2319  -0.2210   0.2311   0.1038  -0.1164  -0.0168   0.0841   0.0168  -0.0841
 0.4655  -0.4403  -0.1164   0.1038   0.2311  -0.2210   0.0841  -0.0168  -0.0841   0.0168
-0.4403   0.4655   0.1038  -0.1164  -0.2210   0.2311   0.0168  -0.0841  -0.0168   0.0841
-0.1164   0.1038   0.4655  -0.4403  -0.2319   0.2092  -0.0421   0.0084   0.0421  -0.0084
 0.1038  -0.1164  -0.4403   0.4655   0.2092  -0.2319  -0.0084   0.0421   0.0084  -0.0421
 0.2311  -0.2210  -0.2319   0.2092   0.4655  -0.4403   0.0421  -0.0084  -0.0421   0.0084
-0.2210   0.2311   0.2092  -0.2319  -0.4403   0.4655   0.0084  -0.0421  -0.0084   0.0421
 0.0841   0.0168  -0.0421  -0.0084   0.0421   0.0084   1.5761  -1.5677  -0.7872   0.7813
-0.0168  -0.0841   0.0084   0.0421  -0.0084  -0.0421  -1.5677   1.5761   0.7813  -0.7872
-0.0841  -0.0168   0.0421   0.0084  -0.0421  -0.0084  -0.7872   0.7813   1.5761  -1.5677
 0.0168   0.0841  -0.0084  -0.0421   0.0084   0.0421   0.7813  -0.7872  -1.5677   1.5761
 0.0421   0.0084  -0.0841  -0.0168   0.0841   0.0168   0.7864  -0.7847  -0.3940   0.3898
-0.0084  -0.0421   0.0168   0.0841  -0.0168  -0.0841  -0.7847   0.7864   0.3898  -0.3940
-0.0421  -0.0084   0.0841   0.0168  -0.0841  -0.0168  -0.3940   0.3898   0.7864  -0.7847
 0.0084   0.0421  -0.0168  -0.0841   0.0168   0.0841   0.3898  -0.3940  -0.7847   0.7864
```

Columns 21 through 24

```
-0.0084   0.0421   0.0210  -0.0042
-0.0421   0.0084   0.0042  -0.0210
-0.0210   0.0042   0.0084  -0.0421
-0.0042   0.0210   0.0421  -0.0084
 0.0421  -0.0084  -0.0042   0.0210
 0.0084  -0.0421  -0.0210   0.0042
 0.0042  -0.0210  -0.0421   0.0084
 0.0210  -0.0042  -0.0084   0.0421
```

```
-0.0421     0.0084     0.0421    -0.0084
-0.0084     0.0421     0.0084    -0.0421
 0.0421    -0.0084    -0.0421     0.0084
 0.0084    -0.0421    -0.0084     0.0421
-0.0841     0.0168     0.0841    -0.0168
-0.0168     0.0841     0.0168    -0.0841
 0.0841    -0.0168    -0.0841     0.0168
 0.0168    -0.0841    -0.0168     0.0841
 0.7864    -0.7847    -0.3940     0.3898
-0.7847     0.7864     0.3898    -0.3940
-0.3940     0.3898     0.7864    -0.7847
 0.3898    -0.3940    -0.7847     0.7864
 1.5761    -1.5677    -0.7872     0.7813
-1.5677     1.5761     0.7813    -0.7872
-0.7872     0.7813     1.5761    -1.5677
 0.7813    -0.7872    -1.5677     1.5761
```

Step 4 – Applying the Boundary Conditions:

The matrix (16.11) for this structure can be written using the global stiffness matrix obtained in the previous step. The boundary conditions for this problem are given as:

$$U_{1x} = U_{1y} = U_{1z} = U_{2x} = U_{2y} = U_{2z} = 0$$
$$U_{3x} = U_{3y} = U_{3z} = U_{4x} = U_{4y} = U_{4z} = 0$$
$$F_{5x} = 4.6875 \,, \; F_{5y} = 0 \,, \; F_{5z} = 0$$
$$F_{6x} = 4.6875 \,, \; F_{6y} = 0 \,, \; F_{6z} = 0$$
$$F_{7x} = 4.6875 \,, \; F_{7y} = 0 \,, \; F_{7z} = 0$$
$$F_{8x} = 4.6875 \,, \; F_{8y} = 0 \,, \; F_{8z} = 0 \tag{16.14}$$

We next insert the above conditions into the matrix equation for this structure (not shown here) and proceed to the solution step below.

Step 5 – Solving the Equations:

Solving the resulting system of equations will be performed by partitioning (manually) and Gaussian elimination (with MATLAB). First we partition the resulting equation by extracting the submatrix in rows 13 to 24, and columns 13 to 24. Therefore we obtain the following equation noting that the numbers are shown to only two decimal places although MATLAB carries out the calculations using at least four decimal places.

$$
10^8 \begin{bmatrix}
0.47 & -0.44 & -0.23 & 0.21 & -0.04 & 0.01 & 0.04 & -0.01 & -0.08 & 0.02 & 0.08 & 0-0.02 \\
-0.44 & 0.47 & 0.21 & -0.23 & -0.01 & 0.04 & 0.01 & -0.04 & -0.02 & 0.08 & 0.02 & -0.08 \\
-0.23 & 0.21 & 0.47 & -0.44 & 0.04 & -0.01 & -0.04 & 0.01 & 0.08 & -0.02 & -0.08 & 0.02 \\
0.21 & -0.23 & -0.44 & 0.47 & 0.01 & -0.04 & -0.01 & 0.04 & 0.02 & -0.08 & -0.02 & 0.08 \\
-0.04 & -0.01 & 0.04 & 0.01 & 1.58 & -1.57 & -0.79 & 0.78 & 0.79 & -0.78 & -0.39 & 0.39 \\
0.01 & 0.04 & -0.01 & -0.04 & -1.57 & 1.58 & 0.78 & -0.79 & -0.78 & 0.79 & 0.39 & -0.39 \\
0.04 & 0.01 & -0.04 & -0.01 & -0.79 & 0.78 & 1.58 & -1.57 & -0.39 & 0.39 & 0.79 & -0.78 \\
-0.01 & -0.04 & 0.01 & 0.04 & 0.78 & -0.79 & -1.57 & 1.58 & 0.39 & -0.39 & -0.78 & 0.79 \\
-0.08 & -0.02 & 0.08 & 0.02 & 0.79 & -0.78 & -0.39 & 0.39 & 1.58 & -1.57 & -0.79 & 0.78 \\
0.02 & 0.08 & -0.02 & -0.08 & -0.78 & 0.79 & 0.39 & -0.39 & -1.57 & 1.58 & 0.78 & -0.79 \\
0.08 & 0.02 & -0.08 & -0.02 & -0.39 & 0.39 & 0.79 & -0.78 & -0.79 & 0.78 & 1.58 & -1.57 \\
-0.02 & -0.08 & 0.02 & 0.08 & 0.39 & -0.39 & -0.78 & 0.79 & 0.78 & -0.79 & -1.57 & 1.58
\end{bmatrix}
$$

$$
\begin{Bmatrix}
U_{5x} \\
U_{5y} \\
U_{5z} \\
U_{6x} \\
U_{6y} \\
U_{6z} \\
U_{7x} \\
U_{7y} \\
U_{7z} \\
U_{8x} \\
U_{8y} \\
U_{8z}
\end{Bmatrix}
=
\begin{Bmatrix}
4.6875 \\
0 \\
0 \\
4.6875 \\
0 \\
0 \\
4.6875 \\
0 \\
0 \\
4.6875 \\
0 \\
0
\end{Bmatrix}
\tag{16.15}
$$

The solution of the above system is obtained using MATLAB as follows. Note that the backslash operator "\" is used for Gaussian elimination.

```
» k = K(13:24,13:24)

k =

    1.0e+008 *

    Columns 1 through 10
```

0.4655	-0.4403	-0.2319	0.2092	-0.0421	0.0084	0.0421	-0.0084	-0.0841	0.0168
-0.4403	0.4655	0.2092	-0.2319	-0.0084	0.0421	0.0084	-0.0421	-0.0168	0.0841
-0.2319	0.2092	0.4655	-0.4403	0.0421	-0.0084	-0.0421	0.0084	0.0841	-0.0168
0.2092	-0.2319	-0.4403	0.4655	0.0084	-0.0421	-0.0084	0.0421	0.0168	-0.0841
-0.0421	-0.0084	0.0421	0.0084	1.5761	-1.5677	-0.7872	0.7813	0.7864	-0.7847
0.0084	0.0421	-0.0084	-0.0421	-1.5677	1.5761	0.7813	-0.7872	-0.7847	0.7864
0.0421	0.0084	-0.0421	-0.0084	-0.7872	0.7813	1.5761	-1.5677	-0.3940	0.3898
-0.0084	-0.0421	0.0084	0.0421	0.7813	-0.7872	-1.5677	1.5761	0.3898	-0.3940
-0.0841	-0.0168	0.0841	0.0168	0.7864	-0.7847	-0.3940	0.3898	1.5761	-1.5677
0.0168	0.0841	-0.0168	-0.0841	-0.7847	0.7864	0.3898	-0.3940	-1.5677	1.5761
0.0841	0.0168	-0.0841	-0.0168	-0.3940	0.3898	0.7864	-0.7847	-0.7872	0.7813
-0.0168	-0.0841	0.0168	0.0841	0.3898	-0.3940	-0.7847	0.7864	0.7813	-0.7872

Columns 11 through 12

```
      0.0841 -0.0168
      0.0168 -0.0841
     -0.0841  0.0168
     -0.0168  0.0841
     -0.3940  0.3898
      0.3898 -0.3940
      0.7864 -0.7847
     -0.7847  0.7864
     -0.7872  0.7813
      0.7813 -0.7872
      1.5761 -1.5677
     -1.5677  1.5761
```

```
» f = [4.6875 ; 0 ; 0 ; 4.6875 ; 0 ; 0 ; 4.6875 ; 0 ; 0 ;
       4.6875 ; 0 ; 0]
```

f =

```
    4.6875
         0
         0
    4.6875
         0
         0
    4.6875
         0
         0
    4.6875
         0
         0
```

```
» u = k\f
```

Warning: Matrix is close to singular or badly scaled.
         Results may be inaccurate.
         RCOND = 7.687107e-018.

u =

   1.0e+008 *

     0.0000
     0.0000

0.0000
0.0000
2.0972
2.0972
2.0972
2.0972
2.0972
2.0972
2.0972
2.0972
2.0972

It is clear that the stiffness matrix is singular or badly scaled. This is due to using one element only in this example. This is also due to the element having a bad aspect ratio. Therefore, the results obtained above for the displacements are not reliable, even totally wrong. It is anticipated that using more elements will result in reliable answers to the displacements. This will be done by the reader in Problem 16.1.

Problems:

Problem 16.1:

Consider the thin plate problem solved in Example 16.1. Solve the problem again using two linear brick elements instead of one element as shown in Fig. 16.5. Compare your answers for the displacements at nodes 9, 10, 11, and 12 with the answers obtained in the example. Compare also your answers with those obtained in the related examples and problems in Chaps. 11 to 15.

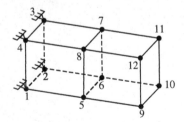

Fig. 16.5. Discretization of Thin Plate into Two Linear Bricks

Hint: Table 16.2 shows the element connectivity for this problem.

Table 16.2. Element connectivity for problem 16.2

Element Number	Node i	Node j	Node m	Node n	Node p	Node q	Node r	Node s
1	1	2	3	4	5	6	7	8
2	5	6	7	8	9	10	11	12

17 Other Elements

In this chapter we review the applications of finite elements in other areas like fluid flow, heat transfer, electro-magnetics, etc. This is intended to be a concluding chapter to the book. The first sixteen chapters were mainly concerned with finite elements in structural analysis and mechanics. In this chapter, I show how the same methodology in terms of consistency and simplicity can be used to formulate and code other types of elements in MATLAB. First, we review the applications of finite elements in other areas, then we provide a sample example with its MATLAB code, namely the fluid flow one-dimensional element. The reader may then be able to code other types of elements (more complicated or in other areas) following the same format and procedure. The six steps used in this book can still be used to write the MATLAB code for any other type of element.

17.1
Applications of Finite Elements in Other Areas

Finite elements have been used in this book in the sole area of linear elastic structural analysis and mechanics. However, finite elements are also used in other diverse areas like fluid flow, heat transfer, mass transport, electro-magnetics, geotechnical engineering, structural dynamics, plasticity and visco-plasticity, etc. In this section, we review the use of finite elements in these areas.

Finite elements are used to model membrane, plate and shell problems in structural mechanics. We have shown in this book different types of membrane elements (plane stress and plane strain elements) but have not shown the plate and shell elements. Different types of triangular, rectangular, and quadrilateral plate bending elements can be formulated using both the Kirchoff and Reissner plate theories. Shell elements of different types can also be used. For example, flat shell elements can be developed by superimposing the stiffness of the membrane element and plate bending element.

Other types of structural elements can still be developed. Axisymmetric elements may be formulated to be used in different types of problems like pressure vessels. Finite elements may also be used in fracture mechanics and damage mechanics to simulate cracks, micro-cracks, voids and other types of defects. The areas of structural dynamics and earthquake engineering are also important areas for finite element analysis. Finally, in structural mechanics, nonlinear finite elements may be used in problems in plasticity, visco-plasticity and stability of structures.

In geotechnical engineering, finite elements are used extensively in solving geotechnical problems. They are used in modeling foundations, slopes, soil-structure interaction, consolidation, seepage and flow nets, excavation, earth pressures, slope stability and other complex boundary value problems in geotechnical engineering.

Finite elements are also used in mechanical design especially in machines and components. In addition, they are used in heat transfer and mass transport to solve problems in conduction, radiation, etc. They can be used to solve both linear and nonlinear problems, and both steady-state and transient processes. All types of thermal boundary conditions may be modeled including isotropic and orthotropic properties, heat loads, etc.

The area of electrical engineering is also important for finite elements. Different types of elements may be used to in electromagnetic analysis. They may be used in studying antennas, radar, microwaves engineering, high-speed and high-frequency circuits, wireless communications, electro-optical engineering, remote sensing, bio-electromagnetics and geo-electromagnetics. The finite element method provides a very powerful technique for solving problems in circuits and circuit analysis.

A very important area for the use of finite elements is the area of water resources and the global atmosphere. Finite elements can be used in groundwater hydrology, turbulent flow problems, fluid mechanics, hydrodynamic flows, flow separation patterns, incompressible Navier-Stokes equations in three dimensions, oceanic general circulation modeling, wave propagation, typhoon surge analysis, mesh generation of groundwater flow, non-steady seepage, coastal aquifers, soil-moisture flow, contaminant transport, and sediment transport.

Finally, the finite element method is a recognized method used in the numerical analysis branch of mathematics and mathematical physics. In addition to finite differences, finite elements are used extensively in solving partial differential equations of all types.

In the next two sections, we provide as an example the full formulation and MATLAB code for the one-dimensional fluid flow element. The MATLAB formulation and code for other elements may be written in a similar fashion.

17.2
Basic Equations of the Fluid Flow 1D Element

The fluid flow 1D element is a one-dimensional finite element where the local and global coordinates coincide. It is characterized by linear shape functions and is identical to the spring and linear bar elements. The fluid flow 1D element has permeability coefficient K_{xx} (in x-direction), cross-sectional area A, and length L. Each fluid flow 1D element has two nodes as shown in Fig. 17.1. In this case the element stiffness matrix is given by (see [1]).

$$k = \begin{bmatrix} \dfrac{K_{xx}A}{L} & -\dfrac{K_{xx}A}{L} \\[2ex] -\dfrac{K_{xx}A}{L} & \dfrac{K_{xx}A}{L} \end{bmatrix} \tag{17.1}$$

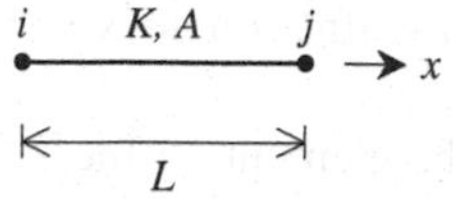

Fig. 17.1. The Fluid Flow 1D Element

Obviously the element stiffness matrix for the fluid flow 1D element is similar to that of the spring element with the stiffness replaced by $K_{xx}A/L$. It is clear that the fluid flow 1D element has only two degrees of freedom – one at each node. Consequently for a system with n nodes, the global stiffness matrix K will be of size $n \times n$ (since we have one degree of freedom at each node). The global stiffness matrix K is assembled by making calls to the MATLAB function *FluidFlow1DAssemble* which is written specifically for this purpose.

Once the global stiffness matrix K is obtained we have the following system equation:

$$[K]\{P\} = \{F\} \tag{17.2}$$

where P is the global nodal potential (fluid head) vector and F is the global nodal volumetric flow rate vector. At this step the boundary conditions are applied manually to the vectors P and F. Then the matrix (17.2) is solved by partitioning and Gaussian elimination. Finally once the unknown potentials are found, the element velocities are obtained for each element as follows:

$$v_x = -K_{xx} \begin{bmatrix} -\dfrac{1}{L} & \dfrac{1}{L} \end{bmatrix} \{p\} \tag{17.3}$$

where p is the 2×1 element potential (fluid head) vector and v_x is the element velocity. The element volumetric flow rate Q_f is obtained by multiplying the element velocity by the cross-sectional area A as follows:

$$Q_f = v_x A \tag{17.4}$$

17.3
MATLAB Functions Used in the Fluid Flow 1D Element

The four MATLAB functions used for the fluid flow 1D element are:

FluidFlow1DElementStiffness(K_{xx}, A, L) – This function calculates the element stiffness matrix for each fluid flow 1D element with permeability coefficient K_{xx}, cross-sectional area A, and length L. It returns the 2×2 element stiffness matrix k.

FluidFlow1DAssemble(K, k, i, j) – This functions assembles the element stiffness matrix k of the fluid flow 1D joining nodes i (at the left end) and j (at the right end) into the global stiffness matrix K. It returns the $n \times n$ global stiffness matrix K every time an element is assembled.

FluidFlow1DElementVelocities(K_{xx}, L, p) – This function calculates the element velocity vector using the element permeability coefficient K_{xx}, the element length L, and the element potential (fluid head) vector p. It returns the element velocity v_x as a scalar.

FluidFlow1DElementVFR(K_{xx}, L, p, A) – This function calculates the element volumetric flow rate Q_f using the element permeability coefficient K_{xx}, the element length L, the element potential (fluid head) vector p and the cross-sectional area A. It returns the element volumetric flow rate as a scalar.

The following is a listing of the MATLAB source code for each function:

```
function y = FluidFlow1DElementStiffness(Kxx,A,L)
%FluidFlow1DElementStiffness        This function returns the element
%                                   stiffness matrix for a fluid flow
%                                   1D element with coefficient of
%                                   permeability Kxx, cross-sectional
%                                   area A, and length L. The size of
%                                   the element stiffness matrix is
%                                   2 x 2.
y = [Kxx*A/L -Kxx*A/L ; -Kxx*A/L Kxx*A/L];
```

```
function y = FluidFlow1DAssemble(K,k,i,j)
%FluidFlow1DAssemble        This function assembles the element
%                           stiffness matrix k of the fluid
%                           flow 1D element with nodes i and j
%                           into the global stiffness matrix K.
%                           This function returns the global stiffness
%                           matrix K after the element stiffness matrix
%                           k is assembled.
K(i,i) = K(i,i) + k(1,1);
K(i,j) = K(i,j) + k(1,2);
```

```
K(j,i) = K(j,i) + k(2,1);
K(j,j) = K(j,j) + k(2,2);
y = K;
```

```
function y = FluidFlow1DElementVelocities(Kxx,L,p)
%FluidFlow1DElementVelocities        This function returns the element
%                                    velocity given the element
%                                    permeability coefficient Kxx, the
%                                    element length L, and the element
%                                    nodal potential (fluid head)
%                                    vector p.
y = -Kxx * [-1/L 1/L] * p;
```

```
function y = FluidFlow1DElementVFR(Kxx,L,p,A)
%FluidFlow1DElementVFR        This function returns the
%                            element volumetric flow rate given the
%                            element permeability coefficient Kxx, the
%                            element length L, the element
%                            nodal potential (fluid head) vector
%                            p, and the element cross-sectional
%                            area A.
y = -Kxx * [-1/L 1/L] * p *A;
```

In concluding this book, we emphasize that the six-step strategy followed in this book can be used to formulate and code MATLAB functions for any other type of finite element that is formulated.

References

Finite Element Analysis Books

1. Logan, D., A First Course in the Finite Element Method, Second Edition, PWS Publishing Company, ITP, 1993.
2. Logan, D., A First Course in the Finite Element Method Using Algor, Second Edition, Brooks/Cole Publishing Company, ITP, 2001.
3. Cook, R., Malkus, D. and Plesha, M., Concepts and Applications of Finite Element Analysis, Third Edition, John Wiley & Sons, 1989.
4. Cook, R., Finite Element Modeling for Stress Analysis, John Wiley & Sons, 1995.
5. Zienkiewicz, O. and Taylor, R., The Finite Element Method: Volumes I, II, and III, Fifth Edition, Butterworth-Heinemann, 2000.
6. Reddy, J., An Introduction to the Finite Element Method, Second Edition, McGraw-Hill, 1993.
7. Bathe, K-J., Finite Element Procedures, Prentice Hall, 1996.
8. Chandrupatla, T. and Belegundu, A., Introduction to Finite Elements in Engineering, Second Edition, Prentice Hall, 1997.
9. Lewis, P. and Ward, J., The Finite Element Method: Principles and Applications, Addison-Wesley, 1991.
10. Henwood, D. and Bonet, J., Finite Elements: A Gentle Introduction, Macmillan Press, 1996.
11. Fagan, M., Finite Element Analysis: Theory and Practice, Longman, 1992.
12. Knight, C., The Finite Element Method in Mechanical Design, PWS Publishing Company, ITP, 1993.
13. Burnett, D., Finite Element Analysis: From Concepts to Applications, Addison-Wesley, 1988.
14. Bickford, W., A First Course in the Finite Element Method, Second Edition, Irwin Publishing, 1994.
15. Pintur, D., Finite Element Beginnings: Mathcad Electronic Book, MathSoft, 1994.
16. Buchanan, G., Schaum's Outlines: Finite Element Analysis, Second Edition, McGraw-Hill, 1995.
17. Kwon, Y. and Bang, H., The Finite Element Method Using MATLAB, Second Edition, CRC Press, 2000.
18. Sennett, R., Matrix Analysis of Structures, Prentice Hall, 1994.

MATLAB Books

19. Etter, D., Engineering Problem Solving with MATLAB, Second Edition, Prentice Hall, 1996.
20. Part-Enander, E. and Sjoberg, A., The MATLAB 5 Handbook, Addison-Wesley, 1999.
21. Biran, A. and Breiner, M., MATLAB 5 for Engineers, Addison-Wesley, 1999.
22. Higham, D. and Higham, N., MATLAB Guide, SIAM, 2000.
23. Hanselman, D. and Littlefield, B., Mastering MATLAB 6, Prentice Hall, 2000.

24. Palm, W., Introduction to MATLAB 6 for Engineers, McGraw-Hill, 2000.
25. Chen, K., Giblin, P. and Irving, A., Mathematical Explorations with MATLAB, Cambridge University Press, 1999.
26. Polking, J. and Arnold, D., Ordinary Differential Equations Using MATLAB, Second Edition, Prentice Hall, 1999.
27. Fausett, L., Applied Numerical Analysis Using MATLAB, Prentice Hall, 1999.

MATLAB Tutorials on the Internet

28. Cavers, I., An Introductory Guide to MATLAB, Second Edition, 1998,
    http: // www.cs.ubc.ca / spider/cavers/MatlabGuide/guide.html
29. Gockenback, M., A Practical Introduction to MATLAB: Updated for MATLAB 5,
    http://www.math.mtu.edu/~msgocken/intro/intro.html
30. Huber, T., MATLAB Tutorials: Envision-It! Workshop 1997,
    http://physics.gac.edu/~huber/envision/ tutor2/
31. Maneval, J., Helpful Information for Using MATLAB,
    http://www.facstaff.bucknell.edu/maneval/ help211/helpmain.html
32. Recktenwald, G., MATLAB Hypertext Reference, 1996,
    http://www.me.pdx.edu/~gerry/ MATLAB/
33. University of Texas, MATLAB Online Reference Documentation,
    http://www.utexas.edu/ cc/math/Matlab/Manual/ReferenceTOC.html
34. Southern Illinois University at Carbondale, MATLAB Tutorials,
    http://www.math.siu.edu/matlab/ tutorials.html
35. MATLAB Educational Sites,
    http://www.eece.maine.edu/mm/matweb.html

Answers to Problems

Answers to all the problems are given below. For each problem two results are given:
the displacements and the reactions.

Problem 2.1:

```
u =

      0.0222

F =

    -4.4444
    10.0000
    -5.5556
```

Problem 2.2:

```
u =

      0.1471
      0.2206
      0.3676

F =

   -25.0000
     0.0000
     0.0000
    25.0000
```

Problem 3.1:

u =

   1.0e-003 *

      0.0143
      0.1000
      0.1429

F =

      -5.0000
     -10.0000
      -0.0000
      15.0000

Problem 3.2:

u =

   1.0e-004 *

      -0.4582
      -0.3554
      -0.2819
      -0.2248
      -0.1780
      -0.1385
      -0.1042
      -0.0739
      -0.0469
      -0.0224

It is clear from the results above that the displacement at the free end is −0.04582 mm which is very close to that obtained in Example 3.2. However, the result obtained in this problem is more accurate because we have used more elements in the discretization.

Problem 3.3:

u =

   2.4975e-005

F =

    -24.9750
     25.0000
     -0.0250

Problem 4.1:

u =

    1.0e-004 *

    -0.4211
    -0.2782
    -0.1353
    -0.0677

Thus it is clear that the displacement at the free end is -0.4211×10^{-4} m or -0.04211 mm which is very close to that obtained in Example 3.2 which was -0.04517 mm and that obtained in the solution of Problem 3.2 which was -0.04582 mm.

Problem 4.2:

u =

     0.0075
     0.0079
     0.0081

F =

    -15.0000
      0.0000
     10.0000
      5.0000

Problem 5.1:

u =

    1.0e-003 *

     0.2083
    -0.0333
     0.0106
    -0.0333

```
      0.1766
      0.0107
      0.0212
     -0.0516
```

F =

```
     -8.8889
     -9.3333
     20.0000
      0.0000
      0.0000
           0
      0.0000
           0
           0
     -0.0000
    -11.1111
      9.3333
```

Problem 5.2:

u =

```
      0.0000
     -0.0000
      0.0034
```

F =

```
     -2.6489
     -1.9866
     -5.3645
           0
     -1.9866
      1.9866
     -0.0000
     -0.0000
     10.0000
```

Problem 6.1:

u =

```
     1.0e-003 *
```

```
         0.2351
         0.0003
        -0.3671
```

F =

```
         0.0000
        16.6471
         9.9883
        -6.4151
       -10.6919
        -0.0000
        -0.0000
       -16.6862
        10.0117
        -8.5849
        10.7311
         0.0000
        15.0000
        -0.0000
       -20.0000
```

Problem 7.1:

u =

```
    1.0e-004  *

         0.2273
        -0.4545
         0.2273
```

F =

```
        -1.5584
              0
        -3.2143
       -15.0000
         4.7727
              0
```

Problem 7.2:

u =

```
         0.0006
        -0.0012
         0.0020
```

F =

        3.9946
        3.9946
       -8.4783
        7.5000
        7.7242
      -15.0000
       -3.2405
       15.0000

Problem 7.3:

u =

       -0.0016
       -0.0002
        0.0009

F =

        1.5225
        2.4913
      -10.0000
        0.0000
        0.6920
             0
        7.7855

Problem 8.1:

u =

        0.1865
        0.0000
       -0.0298
        0.1865
        0.0149

F =

      -20.0000
       -4.6875
       46.2501

```
    0.0000
         0
   15.0000
   20.0000
    4.6875
   -0.0000
```

Problem 8.2:

u =

```
    0.0013
   -0.0009
   -0.0005
    0.0012
    0.0008
    0.0003
```

F =

```
    2.0283
    8.4058
    8.0296
   20.0000
  -12.5000
  -10.4170
   -0.0000
  -12.5000
   10.4170

  -22.0283
   16.5942
   15.1226
```

Problem 8.3:

u =

```
    0.0001
   -0.0056
    0.0021
```

F =

```
     -0.0000
    -10.0000
     -0.0000
    -13.0903
      0.1822
     -0.7290
     13.0903
      9.8178
```

Problem 9.1:

u =

```
     -0.0048
      0.0000
     -0.0018
```

F =

```
    -10.0000
      0.0000
           0
      5.0000
    -13.8905
     20.0000
      5.0000
     13.8905
     20.0000
```

Problem 10.1:

u =

```
     -0.0004
      0.0000
     -0.0006
      0.0000
     -0.0004
      0.0000
     -0.0021
      0.0000
     -0.0006
```

```
      0.0000
     -0.0004
      0.0002
     -0.0021
      0.0000
      0.0006
      0.0000
     -0.0004
      0.0002
     -0.0004
      0.0000
      0.0006
      0.0000
     -0.0004
      0.0000
```

F =

```
      1.1599
      2.5054
      1.0091
      2.6719
      0.3008
     -3.2737
      6.3324
      5.7484
      1.0091
      2.6719
      0.3008
    -17.6937
      6.3481
     -5.7484
     -1.0091
     -2.6719
      0.3019
    -17.7439
      1.1596
     -2.5054
     -1.0091
     -2.6719
      0.3019
     -3.2733
           0
      0.0000
```

```
                0
           0.0000
           0.0000
           0.0000
                0
           0.0000
                0
           0.0000
           0.0000
           0.0000
         -15.0000
           0.0000
           0.0000
           0.0000
           0.0000
           0.0000
                0
                0
           0.0000
           0.0000
           0.0000
           0.0000
```

Problem 11.1:

```
u =

    1.0e-005 *

       0.6928
       0.0714
       0.6928
      -0.0714
       0.3271
       0.0000




F =

      -9.3750
      -3.7540
       9.3750
       0.0000
```

```
   9.3750
   0.0000
  -9.3750
   3.7540
        0
   0.0000
```

Problem 11.2:

u =

   1.0e-003 *

```
  -0.0200
  -0.0225
  -0.0291
  -0.0581
  -0.0305
  -0.0854
  -0.0008
  -0.0173
  -0.0072
  -0.0585
  -0.0077
  -0.0867
   0.0001
  -0.0176
   0.0010
  -0.0639
   0.0064
  -0.0960
   0.0207
  -0.0199
   0.0346
  -0.0635
   0.0356
  -0.1167
```

F =

```
  18.8054
   1.0788
        0
```

```
     0.0000
     0.0000
     0.0000
     0.0000
     0.0000
     1.3366
     9.2538
     0.0000
     0.0000
     0.0000
     0.0000
     0.0000
     0.0000
     0.9105
     0.2247
          0
     0.0000
     0.0000
     0.0000
     0.0000
     0.0000
   -21.0525
     9.4427
     0.0000
          0
          0
     0.0000
     0.0000
   -20.0000
```

Problem 11.3:

u =

```
     0.0002
     0.0044
     0.0002
     0.0044
     0.0002
     0.0044
     0.0002
     0.0044
```

F =

```
        0
  17.5000
        0
  17.5000
   0.0000
   0.0000
   0.0000
   0.0000
 -17.5000
 -17.5000
```

Problem 12.1:

u =

```
  1.0e-005 *

    0.3500
    0.0590
    0.7006
    0.0415
    0.1653
    0.0172
    0.5286
    0.0288
    0.3454
    0.0000
    0.7080
    0.0000
    0.1653
   -0.0172
    0.5286
   -0.0288
    0.3500
   -0.0590
    0.7006
   -0.0415
```

F =

```
   -3.4469
   -1.5335
```

```
              0
         0.0000
         3.1250
         0.0000
              0
         0.0000
         0.0000
         0.0000
       -11.8562
         0.0000
         0.0000
         0.0000
        12.5000
         0.0000
         0.0000
         0.0000
         0.0000
              0
        -3.4469
         1.5335
         0.0000
         0.0000
         3.1250
              0
```

Problem 13.1:

```
u =

    1.0e-005 *

        0.1768
        0.0552
        0.3500
        0.0548
        0.5284
        0.0536
        0.7071
        0.0535
        0.1648
        0.0000
        0.3496
        0.0000
        0.5287
```

```
        0.0000
        0.7071
        0.0000
        0.1768
       -0.0552
        0.3500
       -0.0548
        0.5284
       -0.0536
        0.7071
       -0.0535
```

F =

```
       -4.9836
       -1.2580
        0.0000
        0.0000
        0.0000
        0.0000
        0.0000
        0.0000
        4.6875
        0.0000
       -8.7829
        0.0000
        0.0000
        0.0000
            0
        0.0000
        0.0000
        0.0000
        9.3750
        0.0000
       -4.9836
        1.2580
        0.0000
            0
        0.0000
        0.0000
        0.0000
        0.0000
        4.6875
        0.0000
```

Problem 13.2:

u =

    1.0e-003 *

      -0.0299
      -0.0284
      -0.0402
      -0.0753
      -0.0386
      -0.1102
       0.0015
      -0.0203
      -0.0068
      -0.0800
      -0.0123
      -0.1088
      -0.0021
      -0.0185
      -0.0023
      -0.0824
       0.0047
      -0.1224
       0.0307
      -0.0260
       0.0489
      -0.0758
       0.0565
      -0.1589

F =

      17.6570
       3.4450
       0.0000
       0.0000
       0.0000
       0.0000
       0.0000
       0.0000
       7.4806
       7.0314
       0.0000
       0.0000

```
      0.0000
      0.0000
      0.0000
      0.0000
     -7.9321
      6.7416
      0.0000
      0.0000
      0.0000
      0.0000
      0.0000
      0.0000
    -17.2054
      2.7819
         0
         0
      0.0000
      0.0000
      0.0000
    -20.0000
```

Problem 13.3:

u =

```
     -0.0003
      0.0029
     -0.0003
      0.0029
     -0.0003
      0.0029
     -0.0003
      0.0029
     -0.0003
      0.0029
     -0.0003
      0.0029
     -0.0003
      0.0029
```

F =

```
      0.0000
      8.7500
      0.0000
```

```
        17.5000
         0.0000
         8.7500
         0.0000
         0.0000
         0.0000
         0.0000
              0
         0.0000
       -11.6553
       -11.6894
       -11.6553
```

Problem 14.1:

u =

```
        -0.0002
         0.0029
        -0.0002
         0.0029
        -0.0002
         0.0029
        -0.0002
         0.0029
        -0.0002
         0.0029
        -0.0002
         0.0029
        -0.0002
         0.0029
        -0.0002
         0.0029
```

F =

```
             0
        5.8333
        0.0000
       23.3333
        0.0000
        5.8333
        0.0000
        0.0000
```

```
         0
         0
    0.0000
    0.0000
    0.0000
    0.0000
    0.0000
    0.0000
  -11.6419
  -11.7162
  -11.6419
```

Problem 15.1:

u =

```
    1.0e-005 *

        0.0185
        0.6710
        0.1485
        0.0091
        0.6699
        0.1489
        0.0183
        0.5809
        0.0319
        0.0074
        0.5795
        0.0317
```

F =

```
    -51.1925
     -3.0565
     -4.4842
     51.2090
     -6.3185
     -3.0368
      0.0000
      3.1250
      0.0000
      0.0000
      6.2500
```

```
    0.0000
  -29.2553
   -6.3185
    4.6493
   29.2388
   -3.0565
    2.8717
    0.0000
    6.2500
         0
    0.0000
    3.1250
         0
```

Problem 16.1:

u =

   1.0e+008 *

```
    0.0000
    0.0000
    0.0000
    0.0000
    0.0000
    0.0000
    0.0000
    0.0000
    0.0000
    0.0000
    0.0000
    0.0000
    0.0000
    0.0000
    0.0000
    0.0000
    1.5729
    1.5729
    1.5729
    1.5729
    1.5729
    1.5729
    1.5729
    1.5729
```

Contents of the Accompanying CD-ROM

The accompanying CD-ROM includes two folders as follows:

M-Files: This folder includes the 84 MATLAB functions written specifically to be used with this book. In order to use them they should be copied to the working directory in your MATLAB folder on the hard disk or you can set the MATLAB path to the correct folder that includes these files.

Solutions Manual: This folder includes the Solutions Manual for the book. Two versions of the Solutions Manual are provided. The file "Solutions to Problems" includes detailed solutions to all the problems in the book and is about 360 pages in length. The other file "Solutions to Problems2" is a reduced version of the solutions in about 65 pages only. The two Solutions Manuals are provided in both DOC and RTF formats.

Index